Understanding and Using Technology

Understanding and Using Technology

Ronald D. Todd *New York University*

David L. McCrory *West Virginia University*

Karen I. Todd *Montclair State College*

Paul W. DeVore *Consulting Editor*

Davis Publications, Inc. *Worcester, Massachusetts*

DEDICATION

To our parents and to our children, Roni, Kelly, Scott, Stephanie and Gordon.

Davis Publications, Inc.
Worcester, Massachusetts U.S.A.

Printed in the United States of America
Library of Congress Catalog Card Number:
ISBN: 0-87192-165-0

10 9 8 7 6 5 4 3

Contents

Preface

We have all become more aware of the impact of technical means on the nature of our society and the quality of our lives. The impacts have become increasingly acute during the twentieth century. In earlier times humankind had to know about and understand the natural environment in order to survive. Today, it is necessary to know about and understand the created technological environment and the relationship of technical means to human beings, society and the environment. This new form of knowing and understanding requires a new form of literacy, a technological literacy.

In the past decisions made by individuals and groups about the creation and use of tools, machines, techniques or technical systems generally had little impact beyond the immediate group or local environment. The power of the technical means was limited. Today, however, decisions made by individuals or groups about technological systems can impact on people throughout the world, often in adverse, destructive and unplanned ways. Thus, the need for technologically literate citizens, in a free democratic society, is more critical than ever before.

Those who use this textbook will be living in the twenty-first century. The kind of life we will live in the future will depend on the decisions each of us makes today and tomorrow about the nature and characteristics of our technological systems. These decisions will depend on how well each of us understands the behavior of our technological systems and how they relate to our existence and purpose as humans.

Creating a society in which technological systems better serve human purposes will require a new level of knowledge and understanding on the part of citizens. It is for this reason that Davis Publications initiated, with considerable insight and foresight, a series of publications designed to serve this critical educational need in society; the need for technologically literate citizens.

This textbook is one of three. The primary purpose of each of these publications is to provide an organized study of technology progressing from the creation of technology (*People Create Technology*), through industrialization (*Exploring Technology*), to contemporary technical systems, including automation and cybernetics (*Understanding and Using Technology*). These three texts are part of a series designed to support a program for the study of technology as part of basic education, kindergarten through college.

The goal of this textbook and each publication in the series is to contribute to our knowledge and understanding in managing our technical means so we can create a more humane future and enter the twenty-first century by choice, not chance.

Morgantown, West Virginia
May 1985

Paul W. Devore
Consulting Editor

Chapter 1 Introduction

simulation

Before we venture into the study of technology, we would like you to participate in two simulations. A simulation is a form of "serious game" in which you have the opportunity to see how a part of the real or imagined world works. The first simulation involves an imagined world.

Through a freak cosmic event, a wandering planet, quite like our Earth but uninhabited by humans, is expected to pass through our solar system two years from now. After swinging around our sun, this spacefaring planet will spin away toward the center of our galaxy, to return again in forty years.

You have been chosen as a member of a giant hitchhiking adventure. There will be time for several space craft to leave earth and rendezvous with the traveling planet. You decide to join the group of 101 space pilgrims.

One of our unmanned space vehicles landed on the planet two months ago. It discovered that the planet has an atmosphere much like Earth but apparently has no human inhabitants. There are absolutely no signs of civilization.

The goal of the space hikers will be to settle and develop this new world. It is possible that later generations will return to Earth when the traveling planet comes near again. Our present space technology makes it easy for us to transport people. The short time available and small space vehicles allow us to send only a limited payload. Materials and devices must be about the amount that can be packed into twenty large moving vans.

The new planet has no evidence of technology. The settlers

must be self-supporting until they can use resources found on the planet. The scanning telescopes and mobile science lab on the unmanned research vehicle show that the planet is much like the earth was when the first Stone-Age humans appeared. In fact, you will get to develop a new world, just as Earth's prehistoric people did. You have the advantage of the knowledge learned throughout all the years since cave-people.

Now you must decide what you will need for a life support system on the new planet. Pay close attention to what you will need during those first few months and years. What will you take?

We have no answers to the simulation above. What you decide to take may or may not be good choices. Think about it. Perhaps this book will help you make better choices.

The second simulation is more directly related to the world in which we now live. This case involves a world in which we might live in the future.

The President of the United States has just announced that energy sources on Earth may be depleted before alternative sources are developed. Congress, in an instance of clear "future thinking," has passed new laws. One law requires that we all participate in a kind of "lifeboat drill."

For one week in the summer and one week in the winter, all electricity, gas, and other energy sources will be shut off for all homes. Hospitals, essential businesses and industries, and key aspects of government will be exempted from the exercise.

During those two weeks you must be prepared to live with no help from all the technological devices that we are so accustomed to having. Our only sources of energy for these weeks are our own muscles and the muscles of our animals.

Your first task is to identify which devices in your home will not be usable during the drill. How will you get along during those times when the lights are out? What other devices will you be without during those two weeks? What will you use to replace them? Could you manage to get along if the lifeboat drill runs longer than the predicted time?

For this simulation there are some answers. We won't provide them just yet, however. There are complex answers for the relatively simple problem in Simulation Number Two. It will take a whole book to consider them. This book is about technology—its parts, its activities, the ways to change technology and, perhaps most important, the impact of technology on our lives.

impact

Because of technology, many of us have a standard of living that is higher than ever before in history. Because of technology, we are faced with dire consequences not even dreamed of in previous times. Technology has been described by some as a guardian saint that will provide for and protect us; by others as a demon of destruction that will consume humanity. Obviously, both opinions cannot be right. The reality of technology probably falls somewhere between the two extremes.

Our challenge is to figure out what technology is and where it is headed. We want to know if technology can be predicted or controlled. We need to gain some control over the impact of technology so we won't be forced to live with whatever comes along.

Seeing technology in its relation to science and the arts will help us understand and control these impacts. Science provides us the means to understand "what is", or how nature and the physical world operate. The arts help us explore what "might be." The arts search for what is beautiful and good. Technology draws upon and supports science and the arts as we discover what "can be."

Most people do not know very much about technology. Many are uncertain about the meaning of "technology." This book is intended to help you learn about technology by (a) studying its parts, (b) seeing those parts put together, (c) examining ways in which technology changes and can be controlled, and (d) looking at some of the effects of technology on you and your world. We define technology, at least for now, as **the use of our knowledge, tools and skills to solve practical problems and to extend human capabilities.**

technology

Resources provide the basic ingredients of all technological events. They are required if we are to solve practical problems. For our needs in this book, resources are defined as land, labor, capital (money), knowledge and time.

This raises one problem you will face in this book. As with many other books you have read, you must learn a new vocabulary, the vocabulary of technology. You may already know some terms that are important to technology. These may need only to be expanded in meaning. For example, you

Technology expands our world and presents us with new challenges and problems to be solved.

probably have used the term "systems" on different occasions. You may have used it to talk about the systems of your body, or the electrical system of an automobile, or the stereo system in your home or the traffic system in your town.

system

As you progress through Unit 1, you will see how the six **elements** of technology work together as a system. The term "system" is commonly used in technology. **A system is a combination of parts that work together to accomplish a goal.** An automatic washing machine is a system of parts (water piping, controls, timer, valves, motor, drain) that work together to convert dirty clothes to clean ones. The washing machine can be viewed as part of a larger system. The washer must receive water from the water system, energy from the

The six related elements of technology.

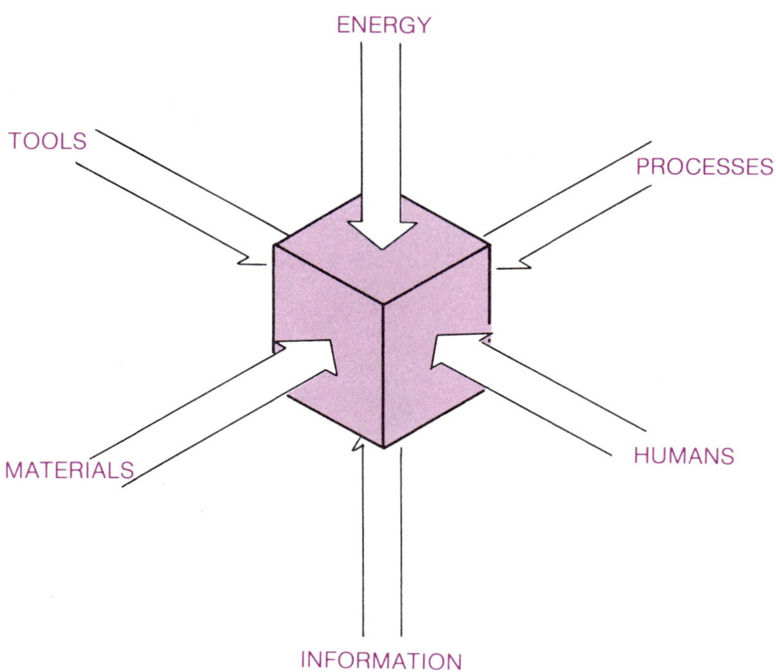

electricity system, and soap from the humans who control other larger systems.

activities

In Unit II, we will look at some of the major activities and functioning systems of technology. Four activities are considered: constructing, transporting, communicating and producing. Through Unit II, we try to show how these are parts of an even larger system. In Unit III, chapters on exploring, developing and controlling show how the systems can become self-supporting. They can help each other grow stronger and more complete.

Unit IV deals with the impact of technology. We will look at the ways our society is changing because of technology. Then we present a brief glimpse of some possible futures. This is followed by thoughts about technology's consequences. We will consider how to study and control the effects of technology in the future.

The final chapter explores decisions and decision-making. We hope that you can begin to decide individually and in groups how to control some aspects of technology that influence your life.

Your Study of Technology

To learn about something new, you can look at its parts. If you want to know how a bicycle works, you could take it apart to see how it operates. Then, if you want to fix it or create a new bicycle, you would know more about how to do it. In other words, you can look at all the parts and try to see how they go together. You would want to know how each part makes another part work.

We can do the same thing to understand technology. In the pages that follow we will take technology apart, identify its pieces and see how they fit and work together. If we are successful, you will be able to look at any example of technology and understand it because you will know what parts are there.

Key Concepts and Terms

activities	impact	system
elements	simulation	technology

UNIT I The Elements of Technology

Introduction

Before we begin to study the elements of technology, let us consider an example. Do you know that when you listen to a record player you are involved with technology? Think about it for a minute. You are using things that were mass produced by other machines. When you use a record player you are simply listening to a sophisticated vibration machine. By taking it apart and examining it, you could find out how that record player works. Maybe you have already tried that before.

The record player is a **machine** and only a part of our example of technology. What else do we need to complete our instance of technology? The diagram below shows the different parts that we need to play a record.

The machine (the record player) is one important part. A second important part is someone who operates the machine and listens to it. That makes you, the **human,** a part of this example. Without the human there would be no technology. In this case, there would be no one to listen to the record nor would there be a record.

A third part we now require is the record. The record is the **material** on which the sound is stored. The sound is placed on the surface of the record in the form of very small, irregular grooves. We now have three parts to our example of technology (machine, human and material).

The fourth part is more difficult to identify. The sound that is stored on the record is similar to information stored on a computer tape or in a book. Any example of technology, in-

Playing a record (an instance of technology) requires the six basic elements of technology. The player is the *tool* (machine), the record and player are made of *materials*, playing the record is the *process*, electricity is the *energy* required, the grooves of the record hold the *information*, and the *humans* are the listeners.

cluding our listening to a record player, will have some form of **information.** In some examples of technology, the information is not obvious. But information is one of the major parts of all examples of technology.

The fifth part that we need is something to make the record player run. This element can be a battery if the player is portable, or can be electricity from a wall outlet. In both cases we have added **energy** as an element of our example. When your parents or grandparents were young they listened to record players that were wound up like a mechanical toy. The energy used by an old-fashioned record player was mechanical energy.

One last part needs to be mentioned. We have the machine plugged into a source of energy. The sound is stored as information in the form of grooves in the record. The humans are ready to listen. As simple as it may sound, we must turn the player *on* so that it can actually begin to play the record. In this example of technology, the playing of the record represents the sixth part, called the **process.** The process is the method by which a change occurs.

Each of the chapters in this unit focuses on one part or element of technology. As you study each part, remember that all the elements work together in a system. Thus, through technology we try to solve problems and obtain what we want.

Key Concepts and Terms

activities	impact	system
elements	simulation	technology

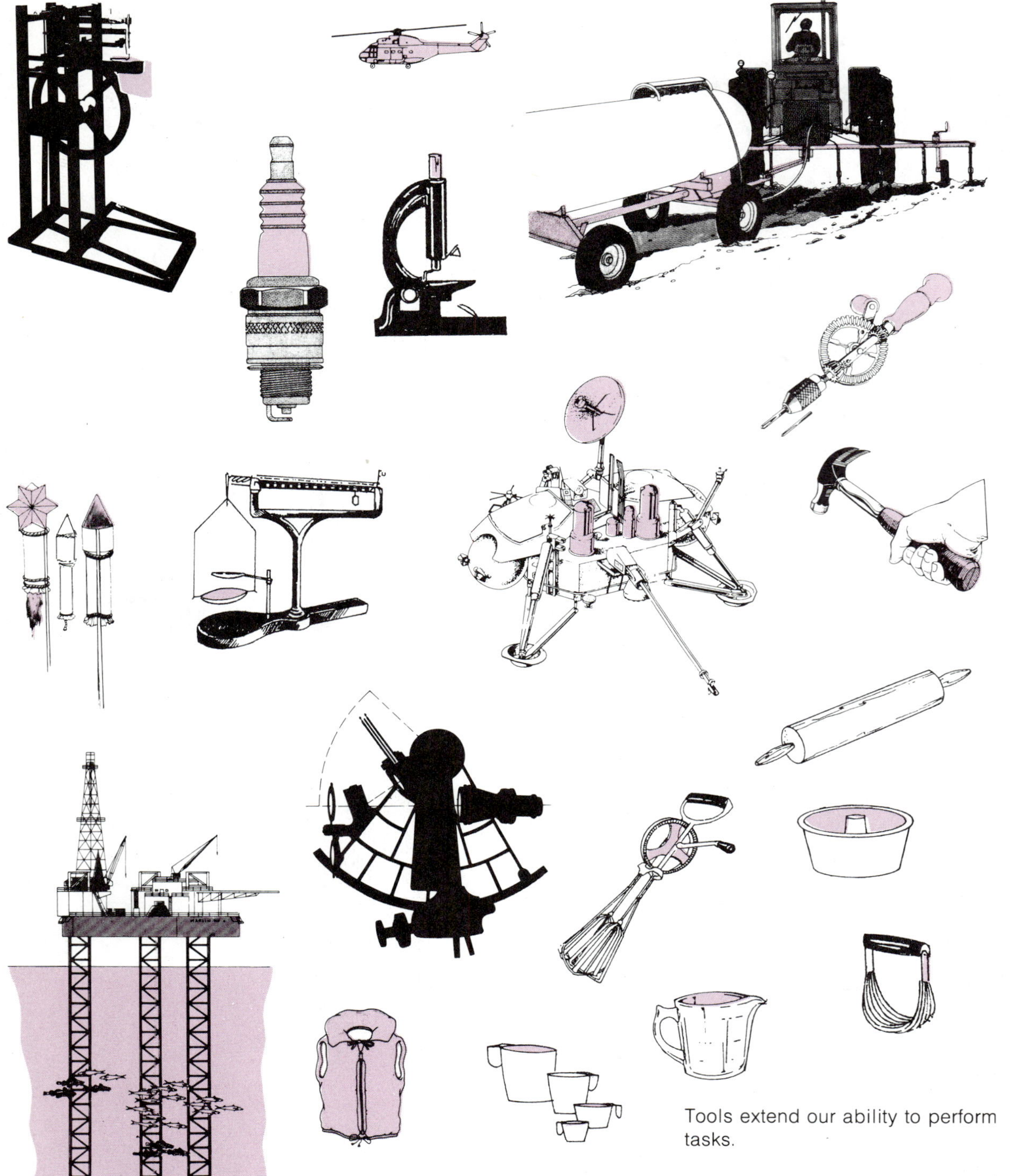

Tools extend our ability to perform tasks.

Chapter 2 Tools

From the beginning, people have used tools to improve their lives. Archeologists who study prehistoric remains look for evidence of tools, to know if humans once lived there.

Some scientists believe that it was the use of tools that helped our ape-like ancestors evolve into human beings. Some other animals also use simple tools, but humans use tools to solve much more complex problems.

Tools Extend our Abilities

Why do people use tools? Tools are devices, instruments and
devices machines that help people do more. **Devices** such as a rake or a flyswatter increase our ability to reach. Eyeglasses, electron scanning microscopes, telescopes and television sets extend our ability to see. Hearing aids such as those which can be attached to your ear, "bugging" devices used by spies and radioscopes turned toward outer space are tools used to increase our hearing capability.

Tools also are used to extend our abilities to smell and
instruments taste. **Instruments** such as smoke detectors and heat sensing devices give us early warning of home fires. Some devices extend our ability to touch. Thickness gauges and other measuring devices help us determine size, weight and volume much more accurately than we can with our unaided senses. There are tools to detect movement and tools to protect our

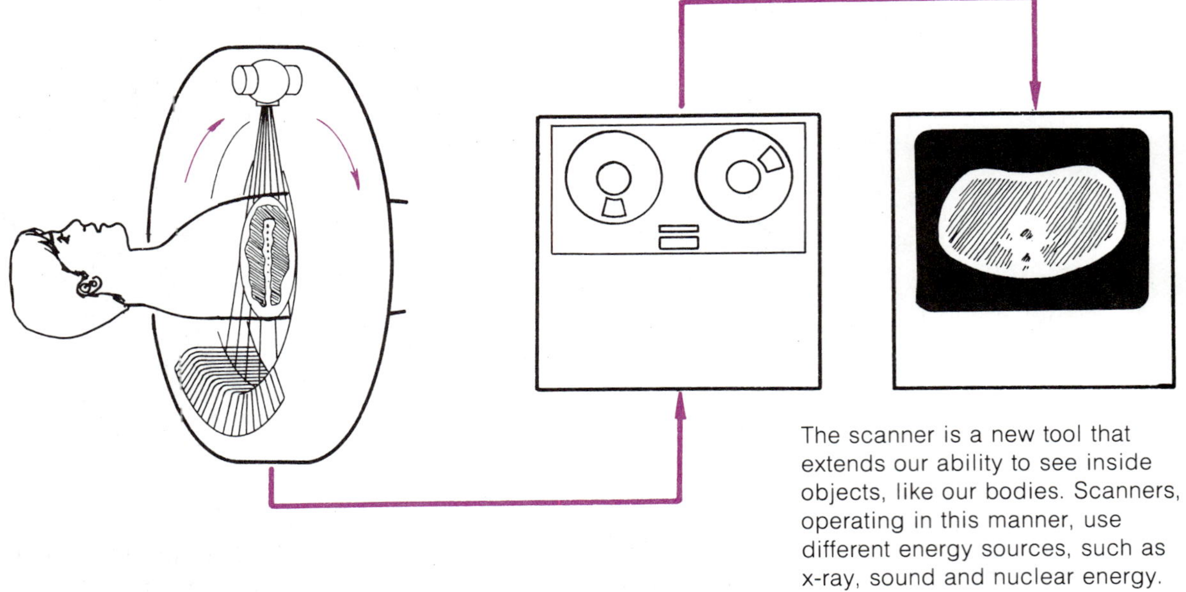

The scanner is a new tool that extends our ability to see inside objects, like our bodies. Scanners, operating in this manner, use different energy sources, such as x-ray, sound and nuclear energy.

bodies from many kinds of discomfort and danger. We use tools to live more enjoyable lives.

One recently-developed tool, the computer, has as much impact on people as did the wheel and the steam engine. Computers and electronic calculators are machines designed to extend our memories and our ability to think logically. Recent developments in microprocessors (inside computers) have led to smaller and smaller electronic devices. Tiny bionic parts for humans can now be made and attached to the body. Attaching the parts requires intricate operations, using specialized tools.

As you think through your daily activities at home and at school, you probably can remember many tools and machines that you have used. The use of modern tools makes your life very different from what it might have been if you had lived in earlier times.

tools
machines

The term **tools** includes all kinds of devices, instruments and machines. **Machines** are tools that work using moving parts. Machines usually are more complicated than are **devices. Instruments** are designed to be precise and exact. All of these words may be used to indicate the same thing—tools which extend our abilities.

The computer allows us to look at problems in different ways. Because the computer has a much better memory, large amounts of information can be stored for a long time and processed fast. Computers are better at tasks like solving complex mathematical problems.

How Tools Work

People often use tools without knowing how they work. Today many problems are solved through the use of tools and machines. If you ask people why a machine works, they may not know. Even the people who design the machine may not know. They only know that it *does* work! This has been true for centuries.

People used to make tools and later tried to figure out how they worked. Only recently has a scientific principle been discovered *first* and then a tool fashioned using the idea. The laser is an example of this. The laser was developed by scientists studying light. They realized the laser's practical uses after the operation of the laser was understood. Only then were laser tools developed.

Often, a tool was developed before the underlying function was understood. Early humans, for example, used stones to crush nuts, to stun or kill animals or enemies and to build shelters. This was long before they understood the principle of kinetic energy, trajectories or gravity. Through trial and error and, later, through scientific experimentation, people began to understand the "whys" of tools. This knowledge allows us to be much more precise as we develop new technology.

The Basic Principles

Most tools are designed to increase our ability to do work. Tools may be very simple or very complex. Even complicated machines can be broken down into simple parts. All mechanical tools function using a few basic principles. Many other tools use these same principles.

lever

The principle of the **lever** is a key element to understanding tools. Over two thousand years ago the ancient Greeks identified the importance of the lever. They showed how it could be used to make other basic devices. It is interesting to note, as the Greeks did, that all mechanical tools are built on the concept of the lever. That means that all machines spring from the same ancestor. All machines are related.

load
fulcrum
force

The lever, an amazingly simple device, involves three related ideas: load, fulcrum and force. The **load** is the weight to be lifted. The **fulcrum** is the pivot. The **force** is the effort that is required to move the load. A simple lever is like a seesaw.

You can think of the fulcrum (pivot) as the equals sign (=) in an equation. The force multiplied by the distance from the fulcrum equals the load multiplied by its distance from the fulcrum. For example, a load of 20 pounds that is 1 foot from the fulcrum could be moved by a force of 5 pounds placed 4 feet from the fulcrum ($20 \times 1 = 5 \times 4$). The same load could be moved by a force of 20 pounds placed 1 foot from the fulcrum ($20 \times 1 = 20 \times 1$), or by a force of 2 pounds placed 10 feet from the fulcrum ($20 \times 1 = 2 \times 10$).

wheel and axle
wedge

There are three ways a lever can be used: as a **lever,** as a **wheel and axle,** and as a **wedge.** Other devices are developed by combining these three.

mechanical advantage

We can extend the force of our muscles by using levers and pulleys in various combinations. (A pulley is a wheel and axle, another form of lever.) Look around you to see examples of machines that use levers or pulleys for **mechanical advantage.**

The lever, one of the first machines, provides a mechanical advantage for moving heavy weights. A small force applied to the long arm results in a large force on the short arm of the lever.

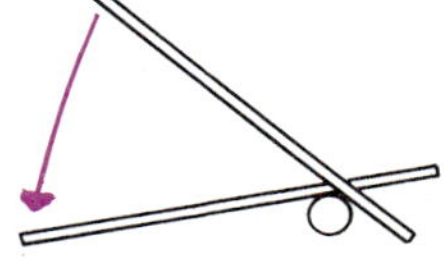

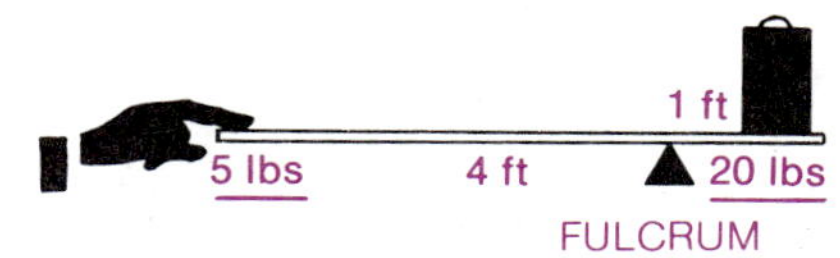

By moving a lever around a fulcrum in different ways you are able to form a wheel and a wedge.

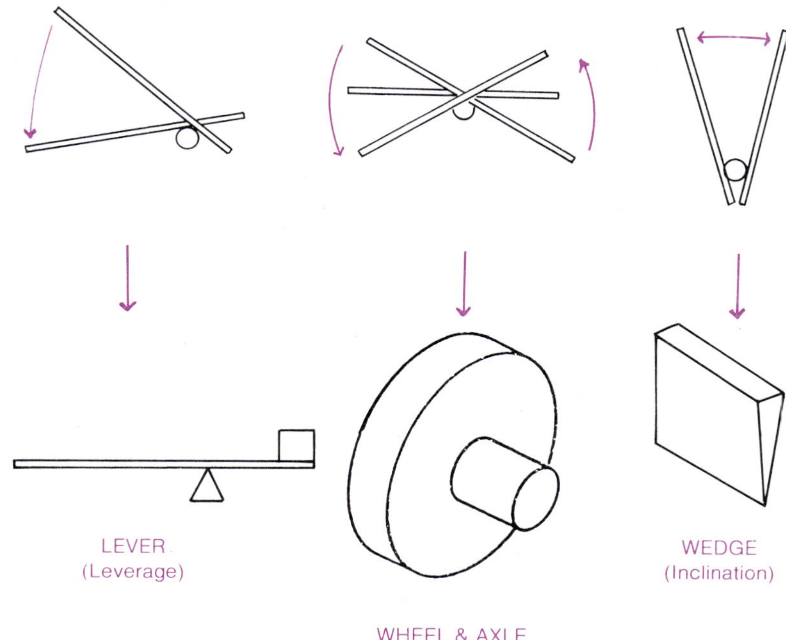

The ancient Greeks applied the lever principle to develop the crank, pulley, inclined plane and screw, wheel and wedge.

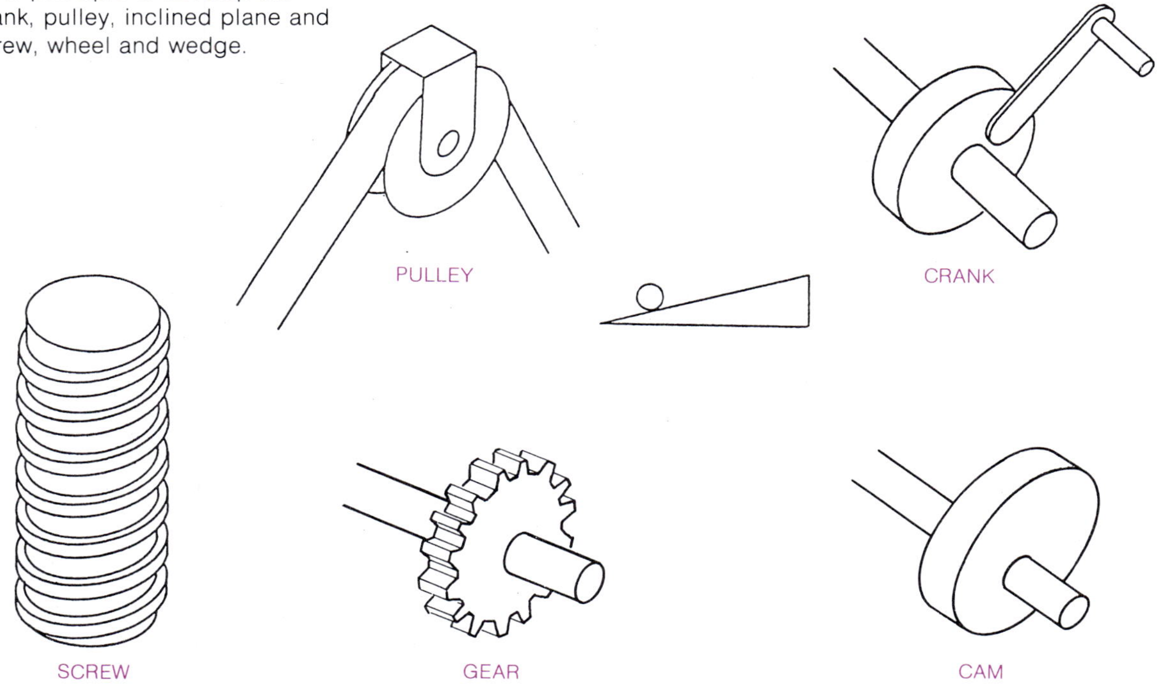

PULLEY

CRANK

SCREW

GEAR

CAM

You may find simple tools such as the "claw" at the back of a carpenter's hammer, for pulling nails. Try to pull a nail from a board with your hands; then try it with a hammer. Which is easier? Try using your fingers to pry open a bottle cap. Now use a bottle opener.

To benefit from the mechanical advantage of the lever, devices are constructed for specific purposes. They are combined and used in many machines and other tools.

As you have learned, most of the early tools were designed to increase human muscle power. Humans served as the ma-

The lever takes different forms in the machines used to lighten our work. Pictured here are two applications that use the lever for lifting water.

This ancient machine was used to drive large posts (piles) as the foundation for buildings and other structures. Many modern pile drivers operate in a similar manner. An electric motor or a gasoline engine replaces the energy of humans or horses.

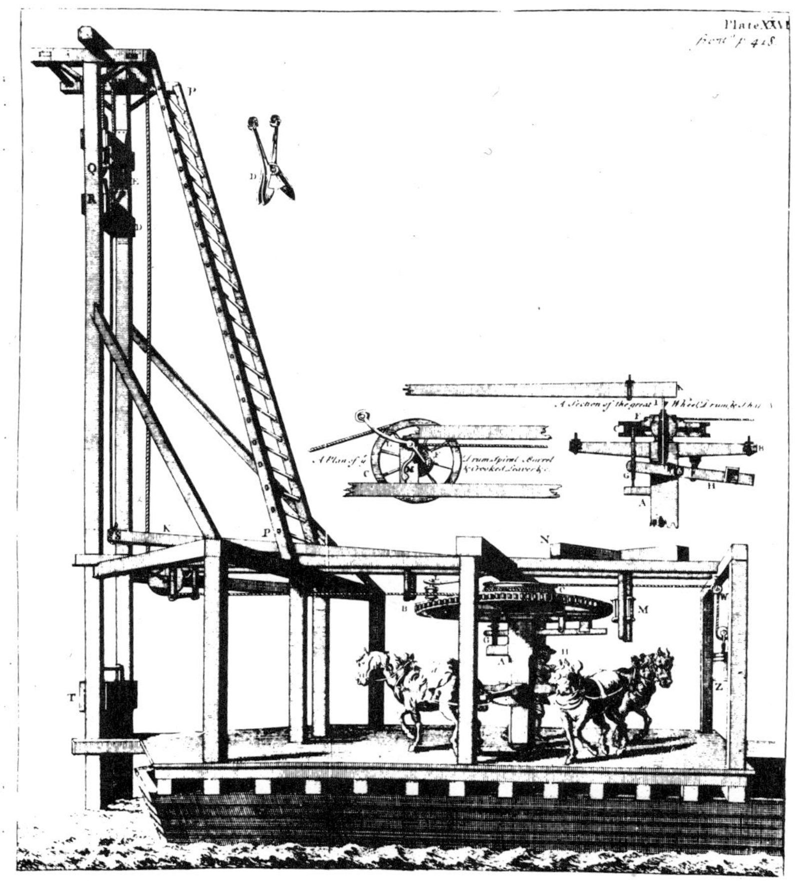

jor source of energy. Later, people learned to attach animals to tools. Then the people controlled the animals and guided them to do work. The energy to pull or push the tools came from animals.

Tools were improved so they were more efficient. Improvements in production and use of energy were even more important. Now that there are steam engines and electrical energy, people no longer have to depend on themselves or on animals to provide power. Simple tools have evolved into complicated machines. Even these complicated machines operate using the principle of the lever. Each of these machines includes three basic components. These are the lever, the wheel and axle, and the wedge.

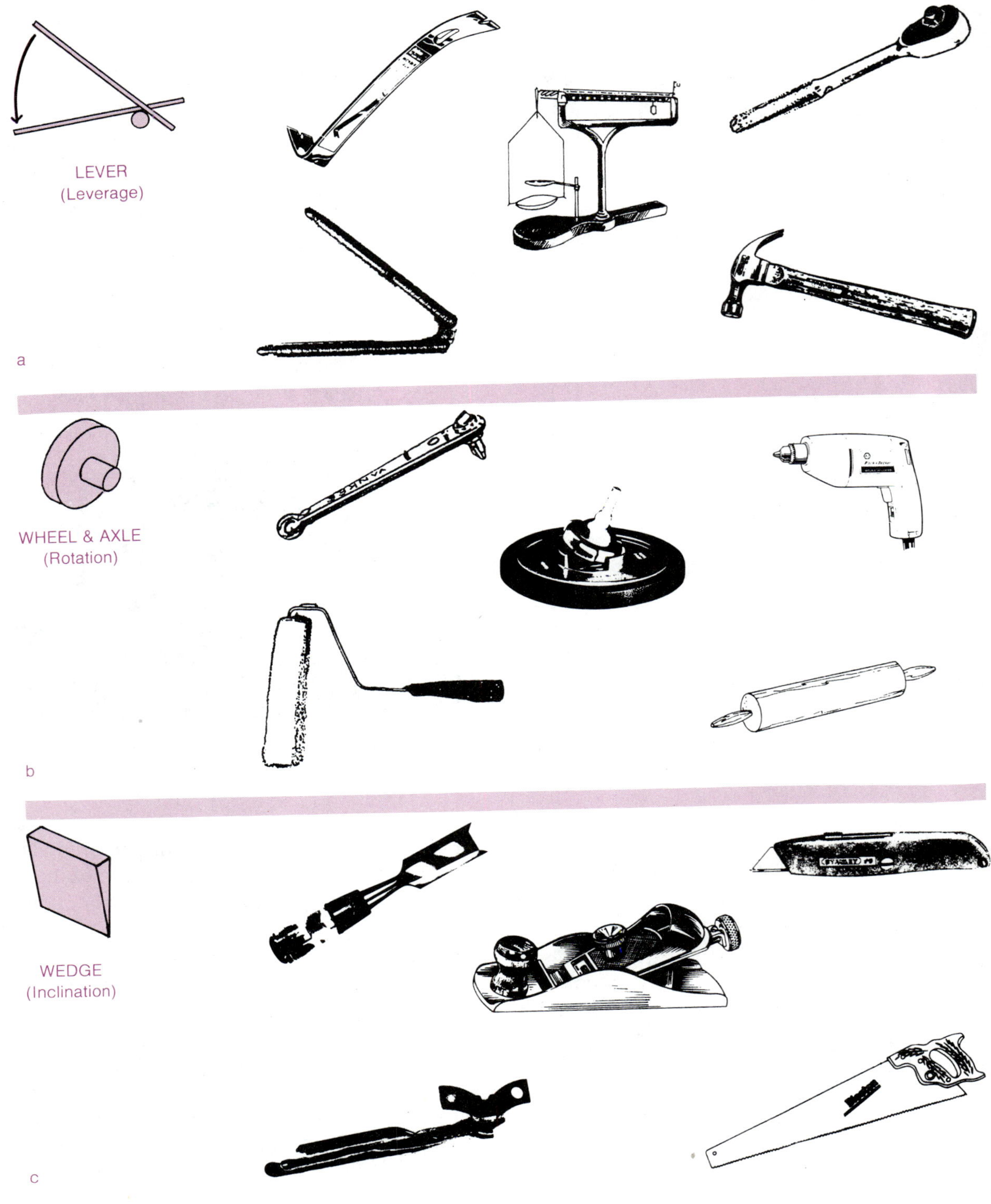
LEVER
(Leverage)
a
WHEEL & AXLE
(Rotation)
b
WEDGE
(Inclination)
c

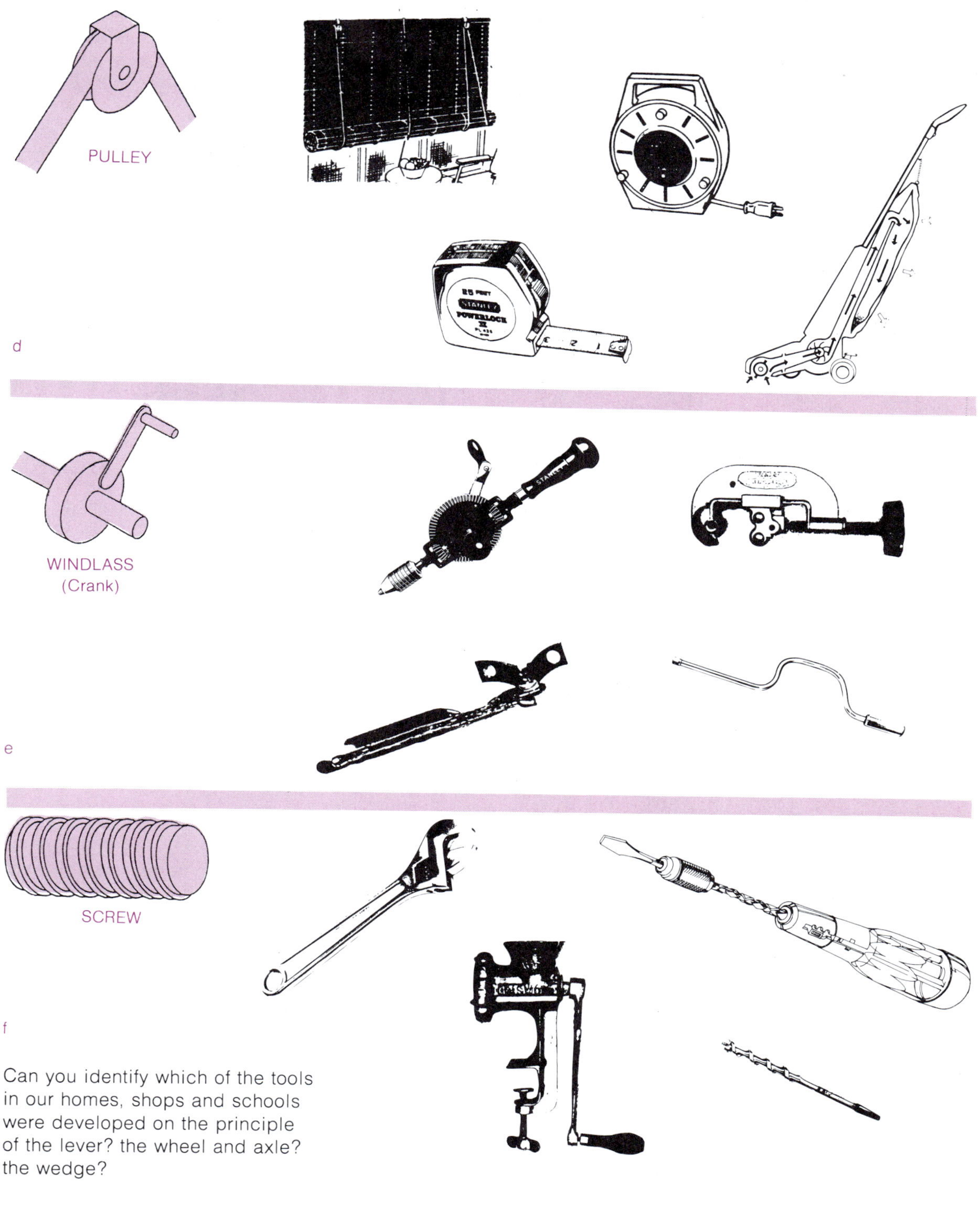

Can you identify which of the tools in our homes, shops and schools were developed on the principle of the lever? the wheel and axle? the wedge?

Before electrical energy, large machines, like these "Corliss Twins," provided the energy for industrial growth by harnessing steam. This exhibition took place in Philadelphia in 1876.

The Parts of a Machine

If you look around, you will see that we have come a long way from the simple machines such as the lever, wheel and wedge. It may surprise you to know that all machines are similar, in fact almost alike. All machines, from electric can openers to jet aircraft, have similar parts. When we know about these similarities, we can understand better how machines operate and how to use them.

To help us understand how machines work, let us look at a machine that may be familiar, the bicycle. Some parts of a bicycle hold up, or *support,* the rider and the bike. Some parts are used to cover the chain, sprocket and tires. These parts make up the **support and cover system** of the bike. The cover parts of the system provide protection for the user and the

All machines must have a means to hold themselves up, to protect their parts, to move energy to where it is to be used and to provide for their direction and control. In other words, all machines must have systems for *support and cover, energy transmission*, and *guidance and control*.

The support and cover system of this bike must hold up the weight of the rider and the bike. It must provide protection for important parts and the rider.

machine. The support parts of the system provide strength and give the machine much of its appearance.

Any machine, including a bike, must also have a system to make it go. These parts are called the **energy transmission system.** The energy transmission system transmits energy from one place to another so the machine can do its work. This system does not actually *provide* the power. On a bicycle, the rider does that.

The bicycle's energy transmission system uses the energy provided by the legs of the rider. The energy is transmitted from the pedals to the rear wheel and on to the ground, so the bike will go forward. Note that the rear wheel provides both support and power. Often one part does several jobs, as you will see in other examples.

If a bike were designed with only the support and cover and the energy transmission systems, it would probably hold you up, and you could make it go forward by pedalling. But you would have some serious problems with no way to steer or to stop it. Brakes and steering devices are examples of a **guidance and control system.** In the bicycle, some parts provide guidance and control as well as support, cover and energy transmission.

The bike has gone through countless changes. The first bike had no pedals or brakes. What do you think it was like to ride one of these? The guidance and control system of this

The energy transmission system of the bike must move the energy from its source to the point where it will be used.

The guidance and control system of the bike must provide means of guiding or directing the bike. It must control the bike's speed, as well as starting and stopping it.

early bike was very primitive. Shown here is part of a very unusual modern bike. You can be certain that this bike requires a very efficient energy transmission system, a very light support and cover system, and a very dependable guidance and control system. Why? Because it is a bike that flies!

We have seen that there are three major parts of the bicycle and of any machine. These are:

Support and cover to provide protection, strength and shape.

Energy transmission to move the energy to the proper places and enable the machine to do its work.

Guidance and control to provide a means for directing and operating of the machine.

Types of Machines

All machines are used to change or convert energy, information and materials. If you wanted, your bike could be used for all three purposes.

A bike is most often used to change the rider's energy into energy for moving something (the rider) from one place to another. Because the bike is used to change the form of energy, it could be called an **energy conversion machine.** The bike changes your muscle energy into motion.

energy conversion

You could also use your bike for another very different

Bicycles like other tools and machines, go through many changes and improvements. This unique bike is part of the cockpit of the Gossamer Albatross. It converts energy to fly!

purpose. Let us imagine that you are on a long-distance bike ride. You have an accident and break a leg, making it impossible to ride any farther. You are alone on a wilderness bike trail but there is a small town some distance away.

The headlight on your bike is operated by a small generator driven by the back wheel. In this situation your bike could be used to send a message for help. By turning the bike upside down and cranking the pedals by hand, you can generate enough energy to operate your headlight. You can cover and uncover the light with your hand in a pattern of dot-dot-dot; dash-dash-dash; dot-dot-dot. The message (... ——— ...) is an international signal for help. The signal stands for SOS in Morse code. SOS is said to mean "Save Our Ship." Perhaps in this case it is more appropriate to have SOS stand for "Save Our Skin." In sending that message your bike has been used to change your muscles' energy into the energy of light.

information conversion

You have used the light to send information (the call for help). The bike in this case was used as an **information conversion machine.**

materials conversion machine

Machines are also used to change materials. An ingenious person combined a bike with an old-fashioned ice cream freezer to make ice cream. The combination of the bike and the freezer then became a new machine. This machine allows the operator to turn the freezer more easily than by hand. The bike uses muscular energy to help convert the cream mixture into ice cream. The bike is now operating as a **materials conversion machine.**

All machines serve one of the three purposes: (1) energy conversion, (2) information conversion or (3) materials conversion. The specific purpose of a machine is determined by what you or another user intend to do with it.

This combination of a two-way television with the telephone is an example of an information conversion machine. It helps us send messages (words and images) from one point to another.

The old sawmill uses water power to convert materials (cutting timber into lumber). The weights are part of the guidance and control system of the machine. Can you figure out how the system works?

The power generated by pedaling a bicycle can provide energy for the conversion of materials—in this case making ice cream.

Some machines are developed to convert energy to a different form. Others convert energy to change materials or information. The machine may convert energy (a) by changing its direction, or (b) by converting it to a different form.

Functions of Machines

Usually a machine is designed for only one major purpose. An example of an **energy conversion machine** is shown below. The solar collector is designed to catch the radiant energy from the sun to heat the water in the storage units. Be-

Energy conversion machines may take many different forms. This one uses energy from the sun to heat water.

How solar collectors use the sun's energy to heat water.

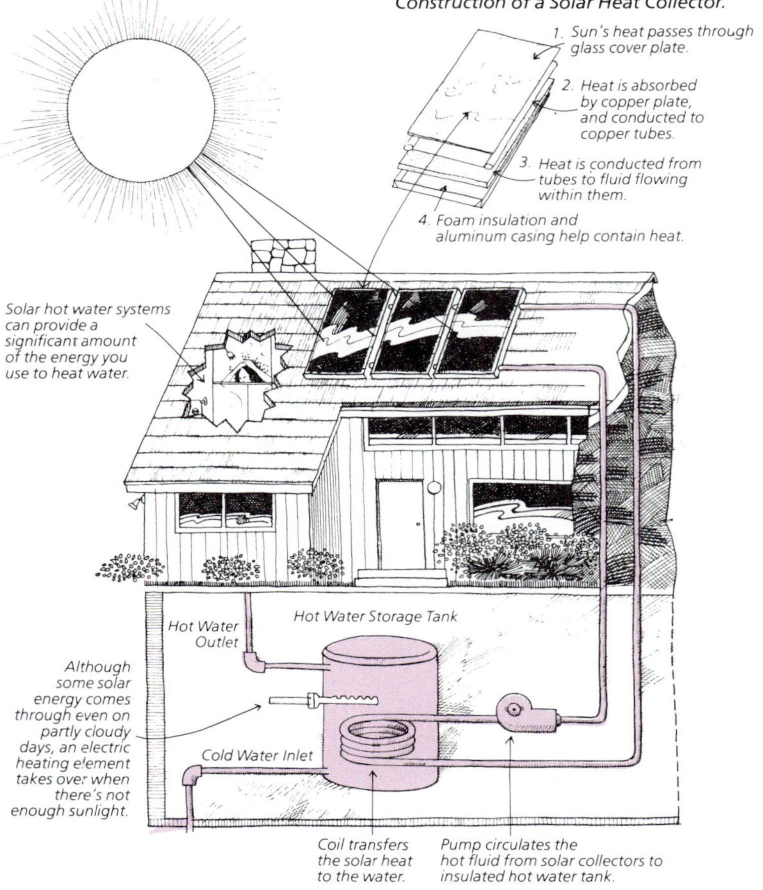

Solar water heating systems are most economical when built into a new home, but can be added to existing homes as well.

cause water heats and cools more slowly than does air, the energy from the sun is stored and released slowly. After sundown, the warm water will release its energy in the form of heat.

This solar collector is an example of a *direct* energy conversion machine. The energy goes directly from its original form to the form you want (in this case, from sunlight to heat). Direct conversion minimizes **entropy,** or the loss of usable energy. This will be discussed later in this book.

Information conversion machines are designed to tell us something. Energy is provided by you when you wind up the spring of a mechanical clock. The energy stored in the spring is converted to small movements in the parts of the clock. These move the clock's hands. The energy your hands put out (winding the spring) is converted by the clock to provide information, showing what time it is.

The "tick-tock, tick-tock" that you hear in a mechanical clock is the sound made as small amounts of energy are allowed to escape from the spring. The device that controls the slow escape of that energy is called, of all things, an "escapement."

The clock, as a machine, "talks" to us through visual and audible **symbols.** If we wind up the spring for the alarm and set it to go off at 6:00 a.m., it will tell us the time in the form of sound. The bell will ring. We hear the sound and interpret it as "Wake up, sleepyhead, it is now six o'clock."

Have you ever thought about machines being able to talk to people? Some do. Some machines also talk to other ma-

The clock was an early machine that converted energy into information. This clock used the *verge and follet* to change the mechanical energy of the suspended weight into regular movements, each one second in times.

chines. We can use a timer to turn on the coffee pot to begin heating. This is a simple example of one machine "talking" to another machine.

In some cities you can dial a number to hear a recorded message telling the time. This is another example of machines "talking." The tape, recorded by the phone company, is "told" to turn on when your phone call activates it.

When signals are sent from one machine to another, it is sometimes called machine-to-machine communication. Computers are the most sophisticated information conversion machines and can talk to people.

materials

A third type of machine is used for **materials conversion.**

An Example of a Machine

Each day thousands of plastic bottles and jugs are used to hold milk, juice, antifreeze and more. Do you know what kind of machine is used to make plastic bottles? Can you guess about how it works?

Earlier we pointed out that any whole technological event requires a machine (tool), material, energy, processes, information and humans. All are needed for technology to function. In this chapter we focus on the machine. The plastic jug machine uses electricity (energy) to heat plastic pellets (material). It then blows (processes) the warm plastic into a mold to form a bottle. The mold provides the size and shape (information) that the plastic will have when it cools. The machine operator (human) controls the machine's heat, speed and temperature. Thus, even the machines that look complicated can be understood when you identify the elements used in their basic operation.

This bottle-making machine converts plastic pellets into useful containers for milk and other liquids.

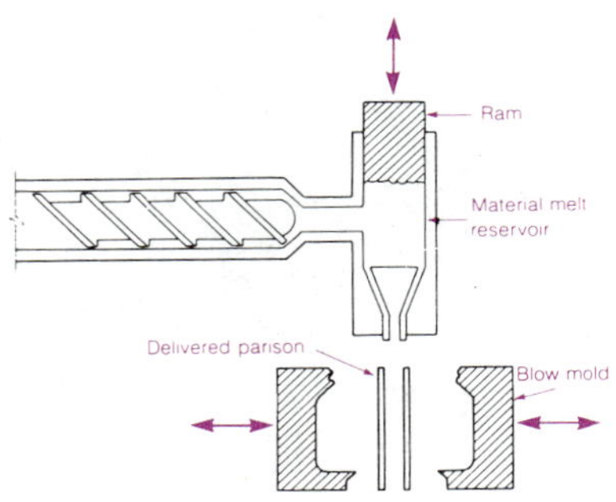

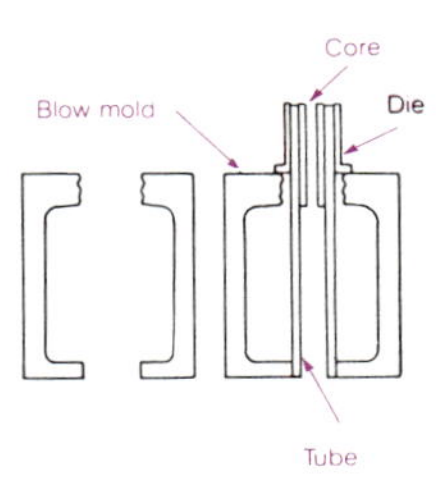

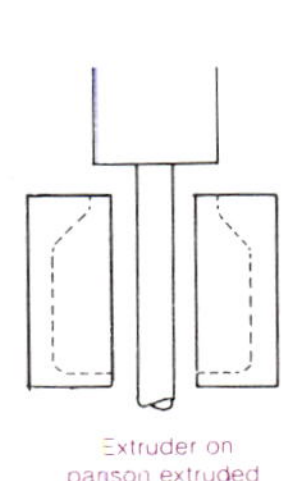

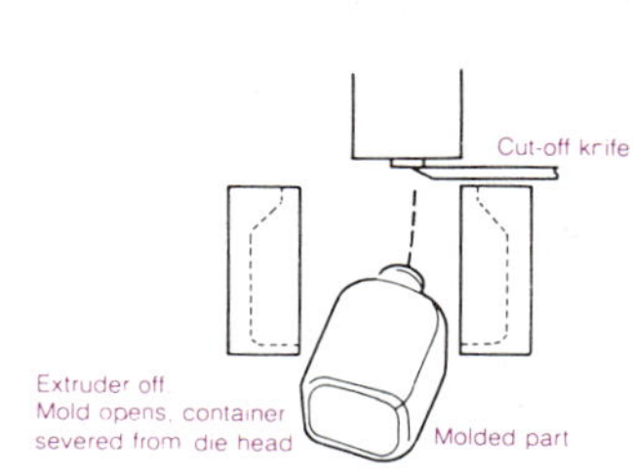

We could call the bottle-making machine a **material conversion machine.** Its purpose is to convert raw plastic pellets into a form that can be used as a container for milk.

A pencil sharpener is a machine used to change or convert materials. It makes sharp pencils out of dull or broken ones. We place the material (dull pencil) into the machine (the sharpener). We then turn the crank to supply energy.

How a Machine Uses Energy

There is another important set of ideas that may help you understand unfamiliar machines. Although machines serve us in many ways, there are only four major ways that machines use energy. As we have seen, all machines have an energy transmission system. This sends energy from one place to another. Machines can (1) transmit energy directly, with **no change in the power,** (2) **change the direction of energy,** (3) **change the form of energy,** and (4) **change the intensity of energy.**

direct use, no change
change of direction
change in form
change in intensity

A pencil sharpener is a simple machine that converts or changes materials. Like other machines, it changes energy in several ways. It transmits or moves energy from one point to another. It changes the form and direction of the energy. It can also change the intensity of the energy.

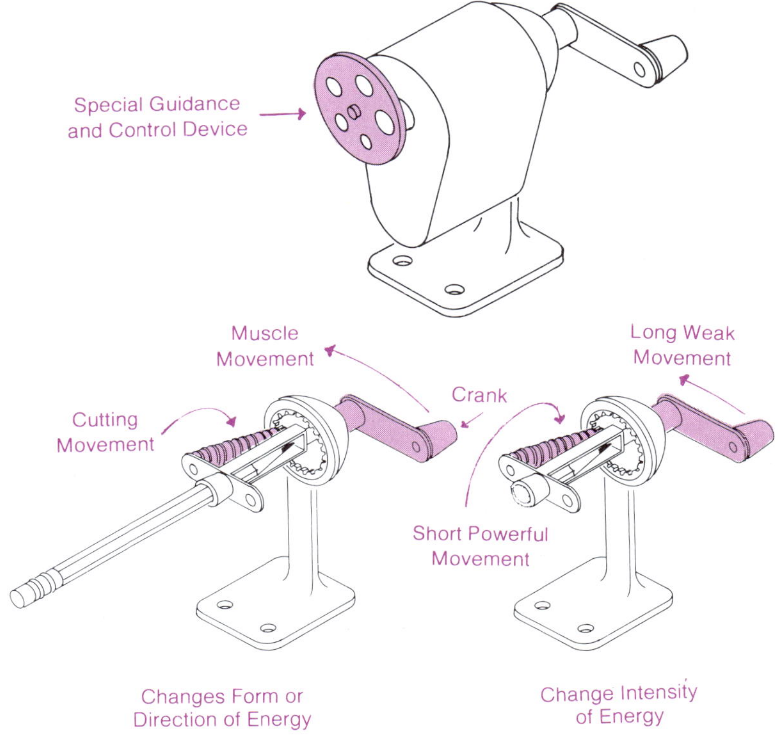

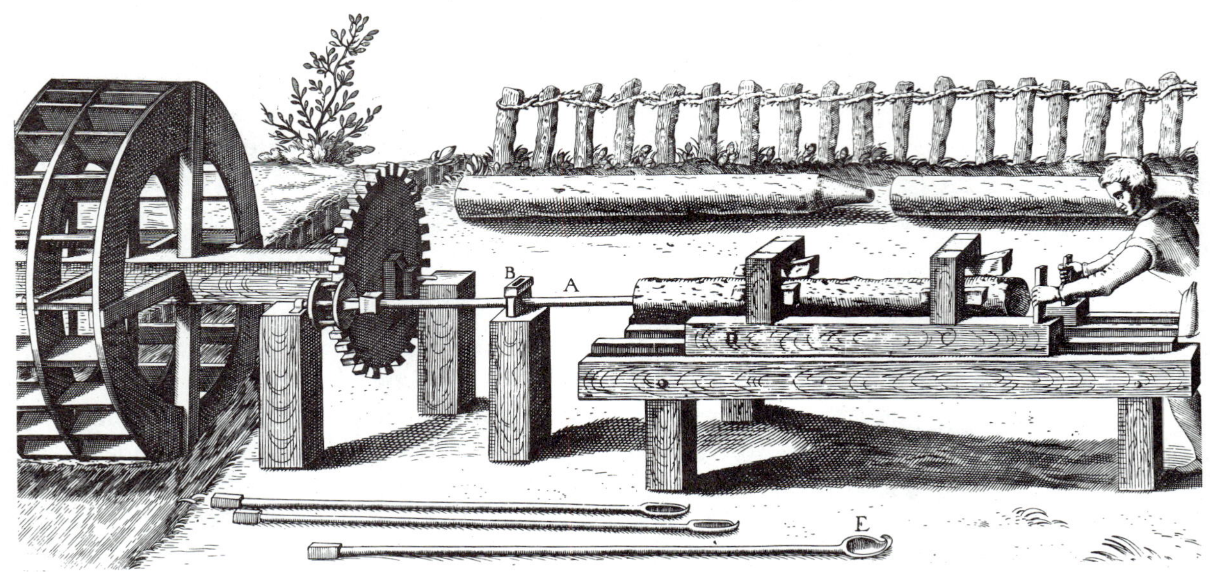

Machines like these were used to make water pipes until about fifty years ago. This machine worked like a pencil sharpener, but the material is removed from the inside rather than the outside.

You can see all four kinds of energy changes in the work of a pencil sharpener. The handle transmits the energy which you supply. This energy goes through a series of gears to the cutting area. The gears in the sharpener *change the form* of energy from muscle power to mechanical energy, making the blades cut the pencil. At the same time, a *change of direction* takes place as the cutter moves in a reverse direction compared with the handle's direction. Finally, the sharpener's crank provides leverage. This multiplies the power applied to the cutters, *changing the intensity* of the muscular energy you supplied. The cutters, by the way, are wedges. They focus the energy on the pencil.

Summary

All of the ideas covered in this chapter can be brought together in a diagram. The diagram describes what machines are and what they do. You can see that humans, materials, energy and information are all necessary for a machine to function and reach a desired goal or output. The machine goes through a process that changes (converts) the energy, materials and information into (1) some different form of material (such as a product, like a candy bar or a milk container), (2) a different form of energy (such as heat or light), *or* (3) a different form of information (such as a tape recording or a signal to another machine). The human element is involved in designing, operating and making use of the machine.

All machines are designed to help us reach goals; using materials, energy and information. Machines change these inputs through many different processes to meet the goals. The output of machines are new products, materials, energy and information. No machine is 100% efficient—they all generate some scrap or waste.

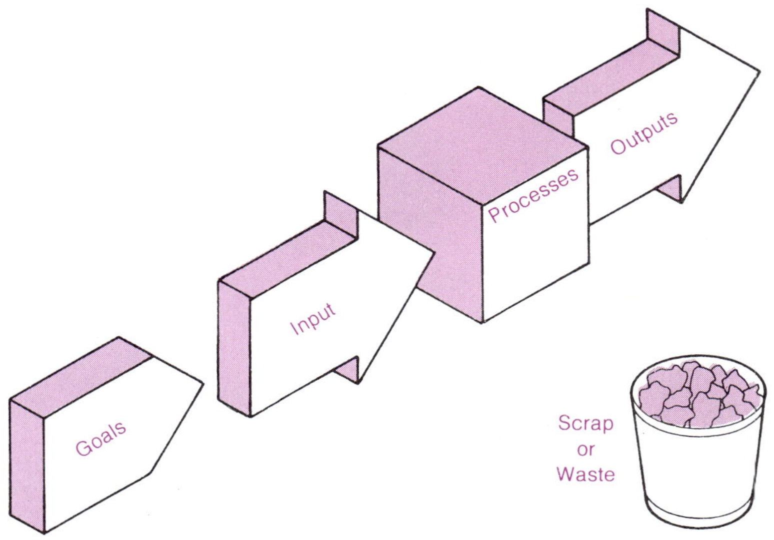

For any machine to do its work, it must contain all three systems: (1) a support and cover system to hold it up and protect the machine itself, (2) a energy transmission system to get the energy to where the work is to be done, and (3) a guidance and control system to make the machine operate. Every part of a machine, no matter how complex, fits into this simple scheme.

Key Concepts and Terms

Vocabulary

devices
energy conversion machine
force
fulcrum
information conversion machine
instruments
lever
load
machines
materials conversion machine
mechanical advantage
tools
wedge
wheel and axle

Machine Types

energy conversion machine
information conversion machine
materials conversion machine

Machine Uses of Energy

direct use, no change
change of direction
change in form
change in intensity

a
b
c
Fig. 2
d
e

Chapter 3 Materials

People use materials in a wide range of ways: (a) new machines for traveling in unusual environments, like on the moon, (b) grains that provide us with breads and cereals, (c) a silicon material for insulation, that even when hot can be held in your fingers, (d) old machines for extracting gold from its ores, and (e) microelectronic and micromechanical devices for building smaller, more efficient machines.

Materials are important for our survival and for realizing our dreams. People use materials to make things that are needed for safety and comfort. People use materials to feed themselves, build shelters, raise cities and journey across oceans and into space.

To understand technology, we must also understand materials. Materials are used to build our machines. Materials are then changed by these machines to make products. Materials are used to supply energy that we need for production, transportation and communication. Materials are the *basic ingredients* of technology. Sometimes materials are fairly simple and natural, such as soybeans, which provide oil and food. Materials may also be complex and manufactured, as is nylon.

Early humans had only natural materials to work with. Later, manufactured or synthetic materials were developed. Often, synthetics are combined with natural materials. T-shirts, for example, are sometimes made with a blend of polyester (synthetic) and cotton (natural).

synthetic

Many of the materials we use today are manufactured. To make these synthetic materials, energy and raw materials are necessary. Sometimes we get in the habit of using synthetic materials. This can have significant and irreversible effects on our lives. For example, most plastic is made from petrochemicals (oil). Great use of plastic has contributed to a shortage of gasoline, which is also made from oil. We are using coal, oil and other natural resources at a fast rate. These took millions

plastic

of years to form. Many scientists are wondering how long it will be before we entirely use up these resources.

There are many properties, forms, uses and varieties of materials. It helps us to group the almost endless variations of materials. Grouping helps us talk about materials and understand them more easily.

One of the purposes of this book is to help you develop a way of looking at categories of materials. This will help you understand and use materials wisely. We may also want to show how some materials take on different properties when two or more materials are combined together. Finally, we will explore what happens to materials when they are used. The three key concepts in this chapter are (1) the **characteristics** of materials, (2) their **structure,** and (3) their **function.**

characteristics
structure
function

Characteristics of Materials

There are thousands of materials available now that were unknown only fifty years ago. New ones are being developed each day. If a new job calls for a quality that existing materials do not have, researchers try to make a new material to fit the need. Astronauts' food, for example, looks very different from the ordinary food (materials) you might find in your refrigerator.

The product's function and special requirements determine what material should be used to make the product. Land vehicles such as automobiles must be strong and relatively light. The strength protects the passengers. The lightness is needed to make it easier to move the vehicle. A construction crane, on the other hand, must be very strong and it must be heavy enough to balance the weight of the load being lifted. An airplane must be both strong and light.

To make women's stockings, hosiery manufacturers may choose a fiber that can be heat-set to retain a given shape. To make athletic socks, the manufacturer may choose a different fiber, one that absorbs and cushions.

The function of a material depends upon the characteristics of that material. Often when you are talking about a boy or girl in your school, you might identify that person as being dark-haired, heavy, athletic, strong, noisy, or any of several other descriptive terms. You are speaking of that person's **characteristics.** Candidates for a space exploration program

Different materials have different characteristics that allows them to be used for making different kinds of products.

are selected partly because they have characteristics such as stamina, intelligence, technical knowledge and an ability to adapt to new conditions. When we talk of people, we often speak in terms of their characteristics.

Materials, too, have characteristics. Within any group of materials some can be identified as being soft, rough, heavy or brittle. Engineers call these characteristics of materials

properties

properties. Peanut brittle, salt water taffy and fudge all have some very different properties, yet each is sweet. Many of the types of woods that we use are relatively soft and easy to cut and polish. Metals are generally hard and will polish to a high gloss. Leather, however, is both soft and tough. Softness and hardness are properties of materials. Other examples of properties are brittleness, density, flavor, color, odor and ability to conduct heat.

The Structure of Materials

Materials often look very different from each other, yet in some ways all materials are quite similar. We need to understand what it is that makes materials similar. To do this, you must use your imagination.

Imagine you are stepping into an incredible shrinking machine. Each minute you are in the machine your size will be reduced to one half of what it was. After you are in the machine for a minute, those of you who were one and a half meters tall (4′11″) will be three quarters of a meter (2′5½″). We plan to keep you in this machine until you are small enough to walk around inside a sheet of aluminum. You will become so

atoms

small you will be able to stand between the atoms. If you continue to shrink at this rate it would take you nearly 33 minutes to be small enough to fit between the atoms of aluminum. If you had been growing at the same rate, you would be over two million miles tall and would bang your shins by tripping over the moon. Inside the aluminum, it would probably look much like the sky at night because of the great expanses of space in the materials.

Much of our information about materials as small as atoms has come from scientific investigation and theory. Some of the best scientific thinking compares atoms to small balls

electrons

shrouded in clouds of circling electrons. The atoms are held apart and in position by forces of attraction and repulsion, somewhat as the planets and stars are held in space. It can be compared to the way two magnets will attract each other or repel each other, depending on how you hold them.

molecules

To help you understand the nature of materials, we will give you a brief tour of the atoms and molecules of a few materials. We will consider how some atoms link together. Molecules are linked atoms. We will consider what shape molecules might take as they combine to form a material.

The atomic or molecular structure of a material determines the properties or characteristics of the material. If you were to buy a knife for wood carving you would not need one with the same molecular properties as a knife used by a surgeon for an operation. The wood you select to place in the ground as a fence post would be different in molecular structure from the wood used for constructing a fine violin.

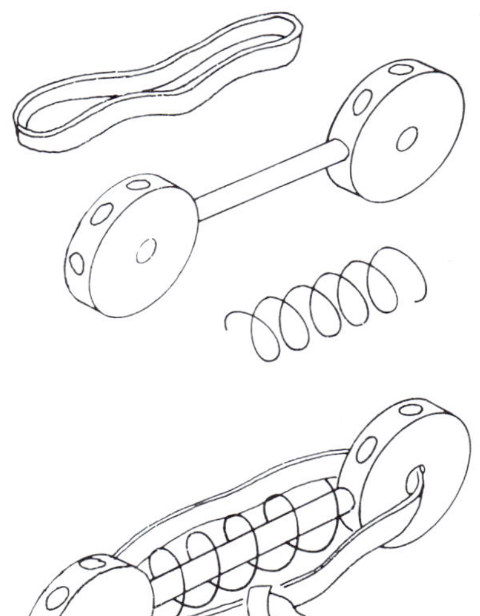

A simple model of the atom and atomic structure might help us. Here the atoms are represented by Tinkertoy spools. The rod represents the distance between the atoms. The rods would be invisible in a real set of atoms. The rubber band represents the attraction between atoms, while the spring represents the forces of repulsion between atoms.

Our model gets very complicated if all the lines of attraction and repulsion are shown. We will use the simplified model, knowing that each rod will represent attraction and repulsion as well as the distance between atoms.

The rods are important parts of our model. They represent the bonding energy that must be overcome to separate molecules of materials. When we cut metal or fabric, we break the bonds between molecules.

Most people use materials without knowing about the atomic structure. As people try to solve more difficult problems and synthesize more materials, they must examine materials in greater depth.

Compounds

Some materials, which we classify as chemical elements are made of only one type of atom. Copper is made only of copper atoms, for example. There are many chemical elements. A few other familiar elements are gold, silver, iron, carbon, oxygen, helium, sodium, neon, and uranium.

Most materials are combinations of elements. These are

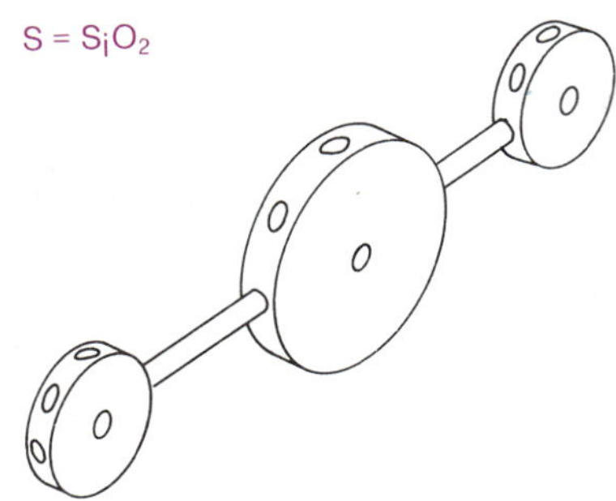

A molecule of sand combines two atoms of oxygen (O) and one of silicon (Si).

called compounds. Compounds are made up of many, many molecules. A molecule is made of atoms of different elements. Each molecule is exactly alike in any given compound.

Sand, for example, is a compound made of silicon and oxygen atoms. One silicon atom collects two oxygen atoms to form silicon dioxide, a molecule. Many molecules of silicon dioxide are in each grain of sand.

It is quite clumsy to spell out the names of the elements and compounds, so chemists have developed a shorthand system of symbols. Silicon is always designated as **Si.** Oxygen becomes **O.** The shorthand name of silicon dioxide is **SiO_2.** The **O_2** designates two atoms of oxygen. "Di" is the standard prefix which indicates the number *two.*

Common table salt is a compound made up of sodium **(Na)** and chlorine **(Cl).** Each molecule of salt contains the atoms **Na** and **Cl** combined in the same way. It is interesting to note that both Na and Cl are, by themselves, poisonous if eaten. When the NaCl molecule is formed, it becomes a material vital to human survival. "The salt of the earth" is a label used as a compliment by some people. Salt was so valuable it was used in earlier times as a medium of exchange (money).

Remember, all materials are made of atoms. Some materials have combinations of atoms, as do sand and salt. Other materials have only one kind of atom, as do copper and aluminum. Materials with only one kind of atom are called **elements.** Materials with combinations of elements are called **compounds.**

chemical elements

compounds

Copper atoms **(Cu)** on a copper surface often combine with the oxygen **(O)** in the air. This combination makes molecules that appear as a dark film on the copper object. Some people spend a lot of time polishing their copper goods to remove this film of oxidation. Copper is sometimes used for roofing. It is not cleaned, but allowed to build up a coating of oxidation. The coating will eventually turn green and become harder than the copper itself.

Atoms and molecules (material units) combine in relatively sensible ways. Some material units have almost no attraction (similar to magnetism) for each other. Some have tremendous forces of attraction. Other material units fall somewhere between these extremes. A simplified set of connections is shown here.

gas

In a **gas** such as air, each material unit (molecule) has almost no attraction for other units in the gas. There is more

Materials differ according to the attraction between their atoms and molecules. For example the cellolose molecule of wood grows to form a long chain-like structure.

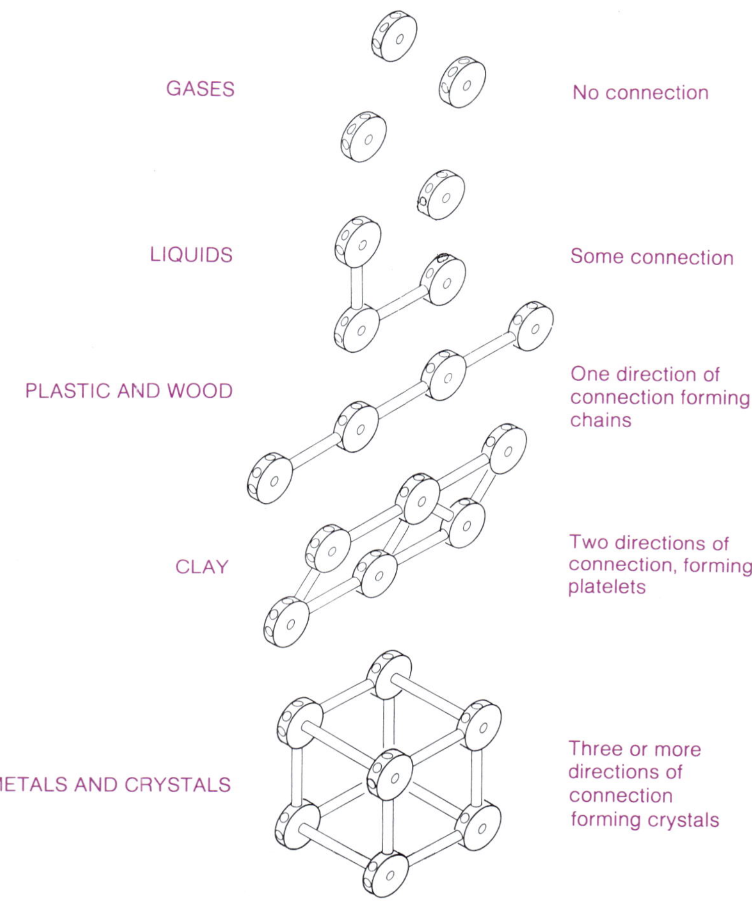

liquid — attraction between the material units in **liquids.** The attraction is greater in liquids that have greater density, such as mercury. Have you ever noticed how hard it is to separate a blob of mercury into smaller blobs? Generally, the molecules of solid — **solid** materials have the greatest attraction for one another.

Material units may combine in one, two, three or more directions. These connections and the strength of these bonds determine the properties of every material.

Chains of Molecules

The basic building units of materials can be combined in different ways. Some materials are made of units connected in chains of molecules

long chains. These chains are held together by the molecular forces of attraction.

Wood is one example of this. It is made largely of long cellulose molecules. The basic units of the giant cellulose molecule are glucose molecules (a kind of sugar). Several hundred of these glucose molecules may be joined to make a cellulose molecule. New molecules are formed each year as a tree grows.

The giant cellulose molecules line up and determine the direction of the wood grain. They are an important aspect of the properties of wood. Wood has long cells, as well as fibers of cellulose. The cells of wood and the long cellulose molecules are similar to a bundle of drinking straws held together by paste. Can you see why wood is strong if you cut across the grain, but can be split apart easily if you split in the direction of the grain? It is like our imagined bundle of straws. The straws can be bent, but they do not break easily. They can be separated lengthwise quite easily.

A second example of a material that has long chains of

A photomicrograph that magnified wood fibers many times shows the structure of the wood.

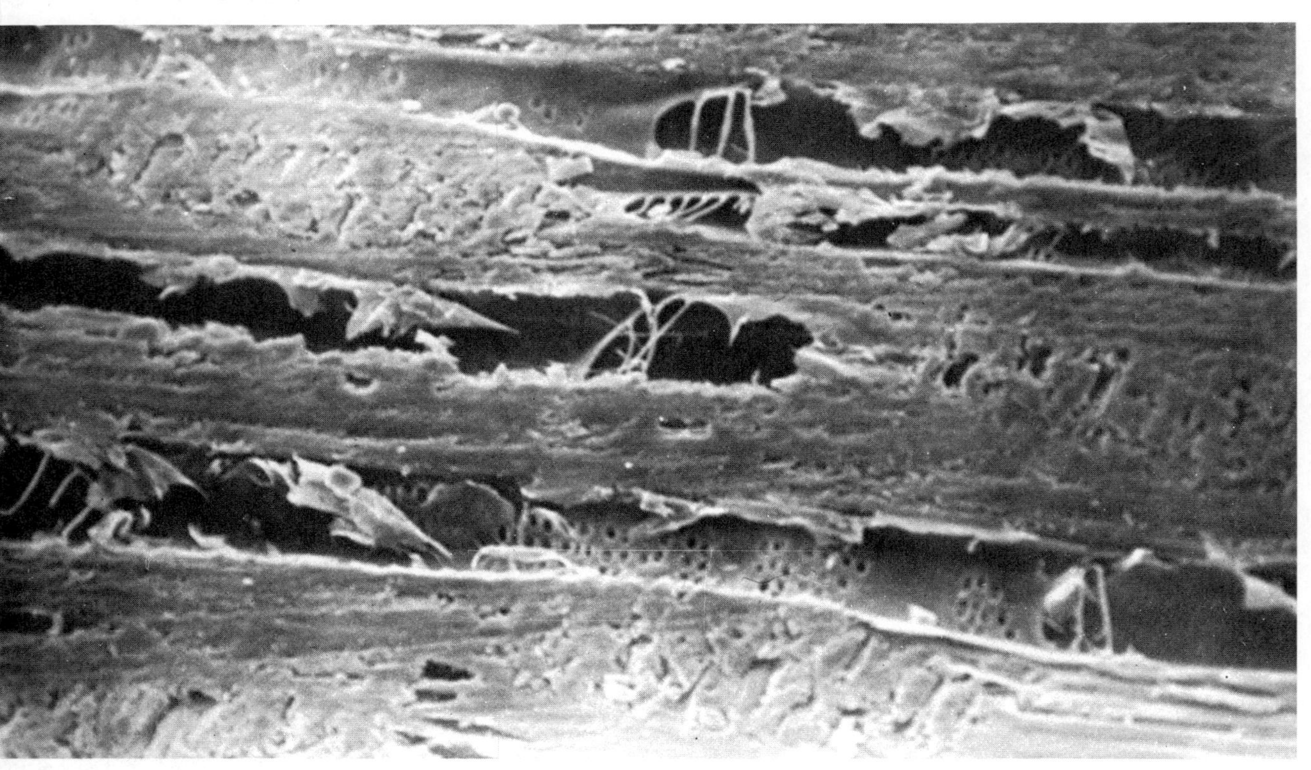

The cellolose molecules of wood cells are aligned with the cell walls. In what direction do you think the wood will split the easiest? In what direction is the wood the strongest?

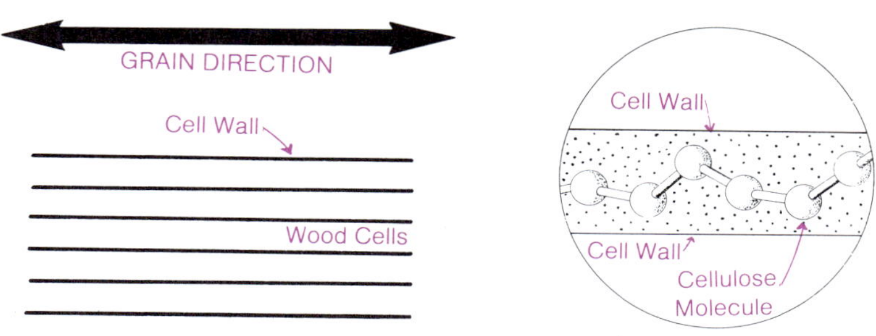

polymer

molecules is polymer plastic. The long chains of the synthetic material are very similar in form to the long molecules of cellulose in the natural material, wood.

The long chains of polymer plastic molecule are created by linking together many simple molecules over and over again. Polyvinyl chloride, for example, is a common plastic used in tires and cassette tapes. It is made of many vinyl chloride molecules. Our toy model may help again to understand the basic vinyl chloride units.

The basic building block of this plastic is the material unit CH_2 CHCl which is sometimes called a "mer." When many mers are connected to make a long chain, the term "poly," which means "many," is added. The result is the term "poly-mer."

Plastics vary in the way that the chains are formed and in the strength of the bonds that hold them together. Plastics

Vinyl chloride (CH CHC) is a type of plastic made up of four atoms of hydrogen, two of carbon and one of chloride. This plastic molecule is called a *mer*.

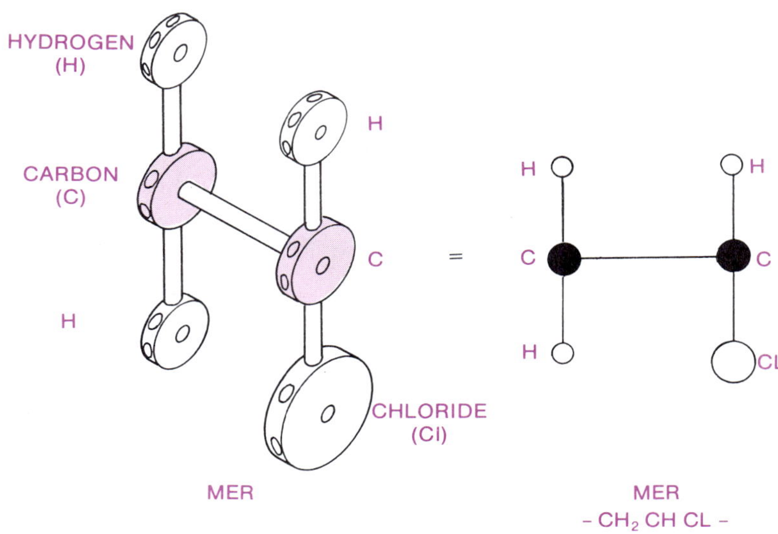

Many mers form in long chains to make *polymers*. A series of vinyl chloride mers form the polymer plastic called ployvinyl chloride (—Ch_2 CHCL—)$_N$.

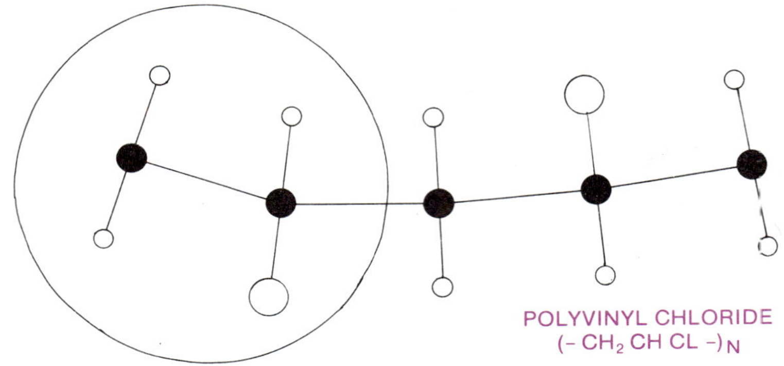

may be tough, fragile, fireproof, easily burned, light or dense. These qualities depend on the way the molecules are linked together.

These properties become important when choosing a material. You wouldn't use Styrofoam (often used in cups) if you wanted to make a crash helmet for a motorcyclist.

Because polymers are long molecules, they tend to be strong and to have some "give" when pulled lengthwise. These plastics are not strong if they are pulled so that the long chains separate from one another. You can prove this by taking a plastic bag and trying to tear it into pieces. Try it first in one direction and then try to tear it at ninety degrees to the first tear. This weakness is similar to the weakness of wood

The long molecules of the polymer plastics give them a grain structure that can stretch more easily in one direction than the other.

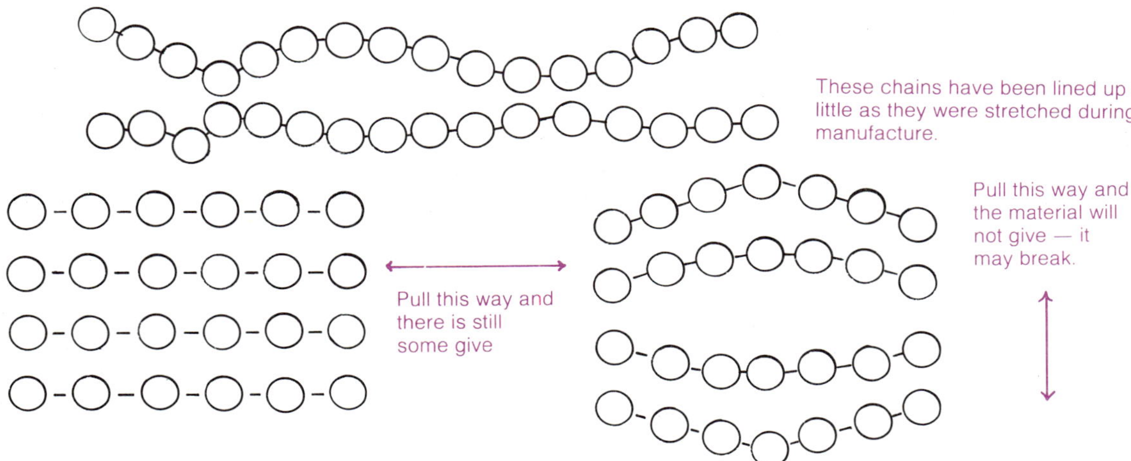

when you split it along the grain, between the cellulose molecules.

Other materials are formed by chains of molecules. Knowledge of molecular structure can help you use the best material for the desired purpose.

Platelets

platelets

attraction

Not all materials form chains or have a grain direction when they bond together. Some molecules bond into tiny **platelets** that slip and slide easily over each other. Clay, wool, graphite and blood are examples of such materials. The attraction between the molecules within a platelet is greater than the attraction between two whole platelets. (Think of a delicious pastry with layers and layers of dough that can be separated for slow eating.)

The molecular model of a platelet is basically two-dimensional.

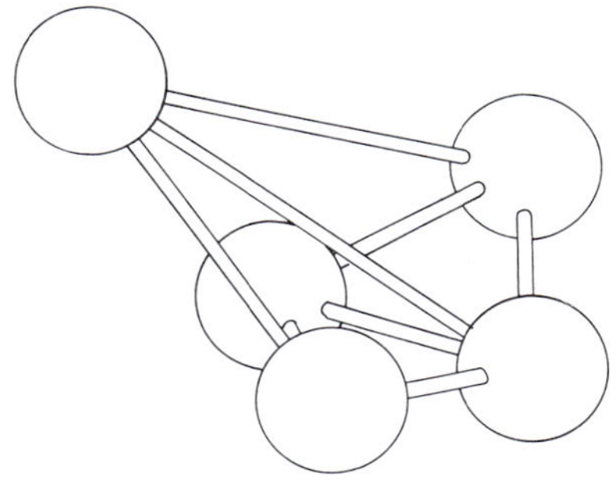

The properties of pliability (slipping of platelets over each other) and strength (the platelets maintain their shape) are valuable for some purposes. Graphite, for example, can be used for lubricating and can continue to slip and slide even under pressure. Some materials with platelets may be made more pliable by adding water or another solvent. A wool scarf stretches and changes shape when wet.

In most of these materials, the platelets can be changed by heating them. The wool scarf becomes very stiff and is no longer so good for keeping you warm if you iron it at too high a temperature. These changes can be used to advantage as well. Let us consider ceramic materials.

firing

green

Ceramics must be considered in two forms, the heated and the not yet heated. When ceramics are heated to a predetermined, high temperature, we say they are **fired.** Ceramic materials that have not been fired are called **green.** If greenware is mixed with water, it will become pliable and soft. In this state the ceramic molecules are formed by a weak attraction and take the shape of platelets. Platelets are very small and thin and look very much like microscopic, wet "corn flakes" of clay. These platelets slide over each other very easily. This accounts for the ease with which wet clay can be worked by hand or by machines to make many different products.

During firing of ceramics, however, changes occur in the materials. First, the water is driven off and the clay begins to

In this photomicrograph of unfired clay, the platelets look like a bowl of soggy cereal.

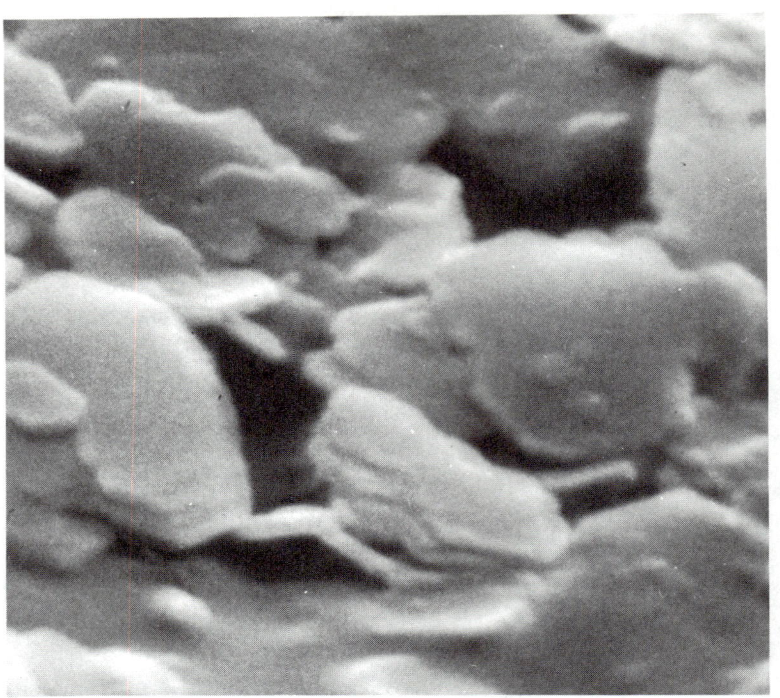

kiln

crystals

vitrification

dry. As the temperature rises in the **kiln** (an oven-like device capable of reaching temperatures of 2000°F and more), the platelets melt and fuse together. Crystals form and the material begins to look like stone. This point is called **vitrification.** The platelets are no longer evident. The clay has undergone a change that cannot be reversed. Ceramics that have been well fired will have a new molecular structure. This structure is more like the structure of metals than that of the slippery original material.

A model of a metal molecule crystal can be constructed by putting marbles of the same size close together in a technique called *close packing*.

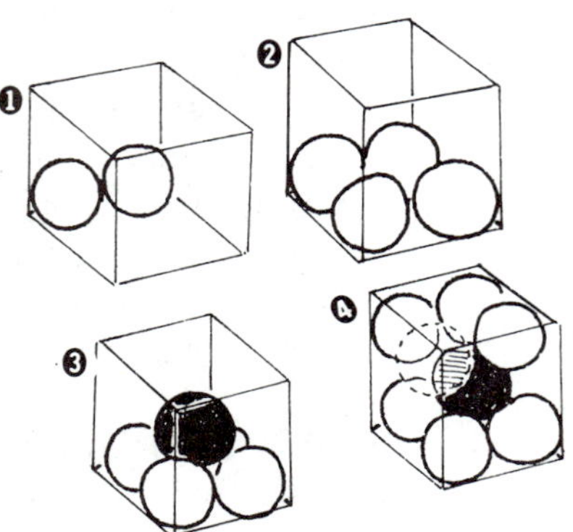

Crystals

When you think of a metal, you probably think of a material with a smooth, hard surface. If you could see the metal through a powerful microscope, you would find crystals packed together in regular ways.

When metal is obtained from the raw materials found in the earth, it is often heated to refine it. Chemicals are added to give the metal new properties. The properties of the metal depend upon the way the atoms of the metal link together to form new molecules as they cool.

In all metals the material units are packed close together. They form a relatively dense material. Engineers or technicians can decide how to cut or shape the metal if they understand the structure of the units that make it up. The structure of the metal also influences what type of product may be made from it.

hexagonal

body-centered cube

repulsion

In metals, atoms pack together to form molecules in a simple manner. You can build a model of a metal molecule by using marbles as the atoms. Some metals seem to have atoms that are stacked together in the shape of a cube. Other materials pack together in a six-sided, **hexagonal** shape.

The drawing of the body-centered cube indicates how it might appear under a microscope. Our model shows the marbles lying on each other. In the actual molecule, the basic units are a fixed distance from each other, not quite touching. Their internal attraction and repulsion makes this happen. In a metal, close packing makes the atoms very stable. The atoms will not separate from each other unless a very strong force is applied. Such a metal is very durable and resists bending.

The properties of the material depend upon the way the atoms combine to form molecules. The material's properties also depend upon the way individual molecules join to form

Four basic shapes are developed through close packing: the simple cube, the body-centered cube, the face-centered cube and the close-packed hexagon. The simple cube (not shown) would be the same as the body-centered cube, but without the dotted lines or the large black area in the center of the cube.

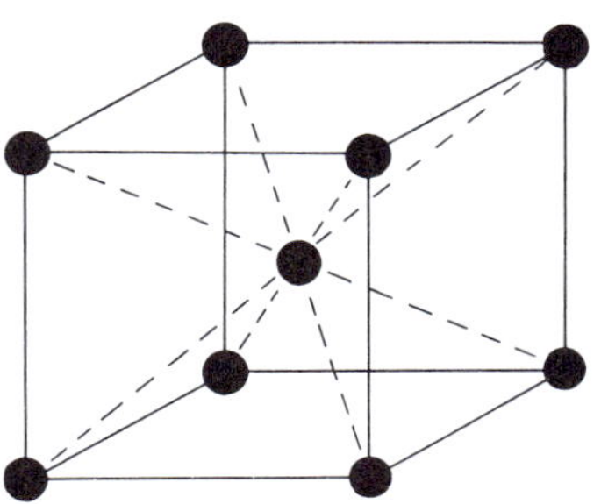

A BODY-CENTERED CUBE

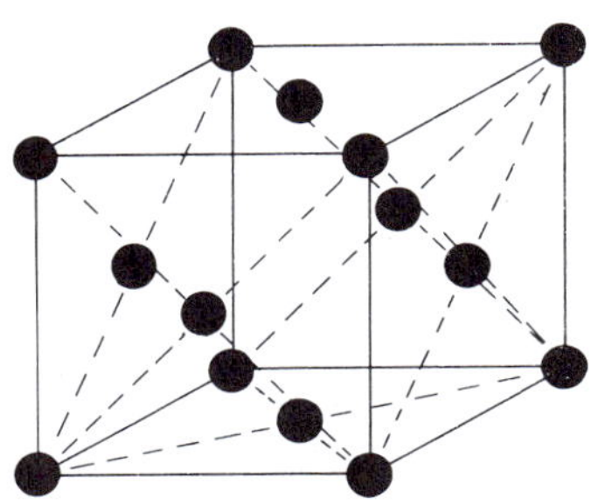

A FACE-CENTERED CUBE

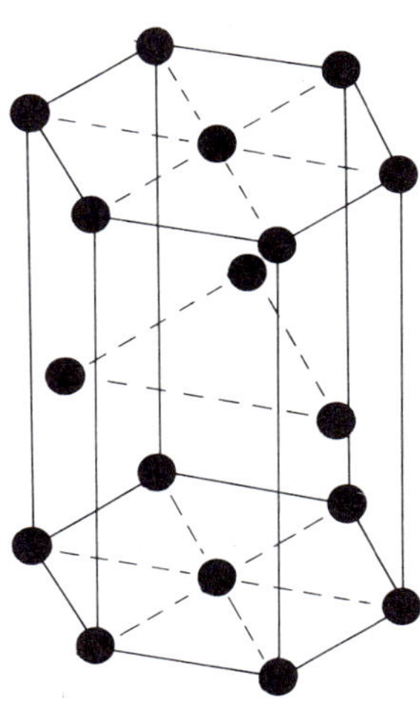

CLOSE-PACKED HEXAGON

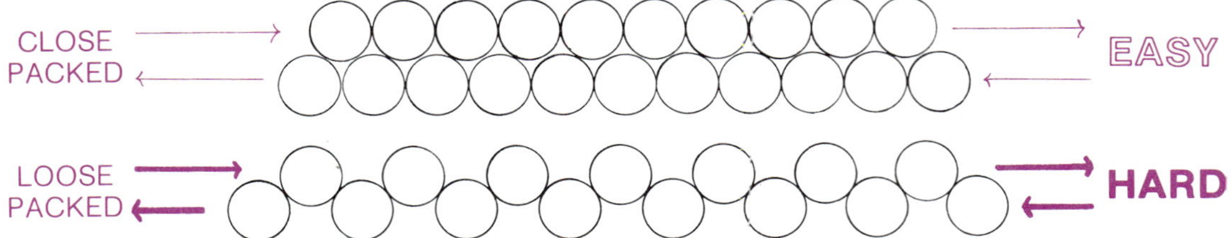

Because close-packed molecules slide over each other more easily, the materials can be shaped or molded more easily than loose-packed molecules.

the entire piece of material. Of all materials, only metals have both tightly packed atoms in each molecule and tightly packed molecules in the material as a whole. The best use of the metal, how it can be cut, and how easily it can be shaped all are determined by how closely the molecules pack together.

closely-packed molecules

In some metals the molecules are packed together in tight layers. In other metals there is more space between the molecules within the layers. In closely-packed metals, the metals move fairly easily over each other. If the molecules are packed together loosely, however, it becomes more difficult to move the layers past one another. These loose-packed materials can not be cut, molded or shaped easily.

loosely-packed molecules

Some materials other than metals are more easily identified as crystals. Diamonds, sugar, mica and salt are a few.

Not all materials form chains, crystals or platelets. New materials may be developed as materials are given new structures. The properties will change as the structure changes.

Functions of Materials

If you are to choose the best materials for a specific use, you will need to consider **(1)** some useful categories of materials, **(2)** some characteristics of materials in each category, and **(3)** how these affect the choice and use of materials.

Consider three categories of materials. These include:

monolithic materials

Monolithic Materials These are made of only one ingredient, element or compound. They do not change during use.

material systems

Material Systems These have two or more ingredients. The ingredients are combined for a specific function or to have a desired characteristic.

dynamic materials

Dynamic Materials These change as they are used.

Monolithic Materials

There are two kinds of monolithic materials: natural and basic. **Natural** materials are used as they are found in nature. **Basic** materials are natural materials that have been changed into different forms. The new forms make basic materials more useful for certain products. These materials are seen as *basic* to the making of certain products. *Basic* in this sense means "important to the whole product." Examples can be seen in the illustrations that open this chapter.

natural

basic

A raw material such as iron ore is not used in its original form to make products. First it is processed to extract the pure iron. The iron is then used to make many kinds of steel that go into bridges, automobiles, machinery and many other products. The raw material, iron *ore,* is called a **natural** material. *Pure* iron, because it has been processed from its natural state, is a **basic** material.

As you may realize, nearly all the products we use are made from basic rather than natural materials. Flour used in cakes, for example, is a basic material which is made by grinding (processing) the natural material, wheat. A diamond crystal may be admired in its natural state, but it is usually "cut" before being used in jewelry or in a bit for drilling or cutting.

A natural or basic material is called **monolithic** because it is used in one *(mono)* state, as found in nature. It is not combined with other materials at this point. When it is used, you

Two types of monolithic materials—some are found in nature; others are natural materials that are changed through a basic process for later use in a product.

MONOLITHIC MATERIALS

Natural	*Basic*
"produced or existing in nature"	"Important to the whole product"
Wood-Tree	Rawhide
Limestone	Pig Iron
Asbestos	Lumber
Iron ore	Leather
Cotton fiber	Furs
Wool fiber	Spun and woven fibers
Clay	
Gold	
Diamonds	

can expect it to stay about the same size and shape. You can also expect the material to stay relatively **inert** (inactive), not reacting to the materials around it.

Look around you and identify some materials that are used as they are found in nature. Also look for those materials that have been changed somewhat, but have not been combined with other materials. You can get a sense of the impact of technology when you do this. It can be hard to find monolithic materials. Many materials we use every day have been changed or combined with other materials.

Material Systems

More often than not, materials are combined into systems. Material systems are not really new. Clay brick reinforced with straw was used thousands of years ago to form a material system for constructing shelters. So, too, were mud, manure, straw and branches used in making walls of wattle (a material system still in use in some parts of the world).

Today, we use many material systems. Some examples are plywood, plasterboard, fiberglass, concrete, reinforced carbon fiber and fabrics backed with a thermal lining.

Material systems form the most-used types of materials. We define **material system** as two or more materials that differ in form, that become a new, more desirable material when they are combined. The resulting material system may be, for example, stronger, lighter or longer lasting than any one of the single materials used alone.

All material systems fall within two categories: those that are material **mixtures** and those that are material **layers.** These two terms (material mixtures and material layers) reflect the general ways that material systems are made.

material mixtures

A **material mixture** is a composite in which the ingredients of the system are mixed. They may be mixed mechanically, chemically or physically. The ingredients may form a random or an oriented pattern. Heat, pressure or both may be required to form a mixture. Many foods products are material mixtures. A recipe for mixing can be used to get consistently similar products.

material layers

A **material layer** is a composite in which materials that are different from each other are put together without actually blending the materials. Each material lends to the final product the special characteristics it possesses. Your home is probably made of material systems. If it is painted, the paint

MATERIALS SYSTEMS

MIXTURES

Composite: reinforced concrete, radial tires, fruit salad, particle board

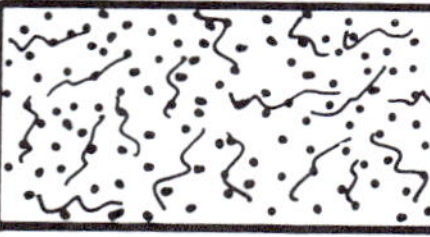

Diffused: carbonized steel, non-homogenized milk, tinted glass

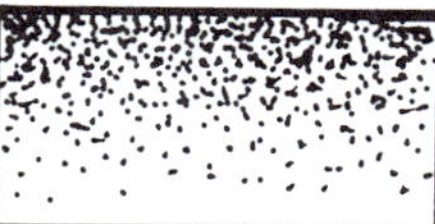

Dispersed: perfumed lotion, abrasive soap, pigments in paints, ink erasers

Fiber Reinforced: Fiberglass, linen paper, Masonite, chipboard

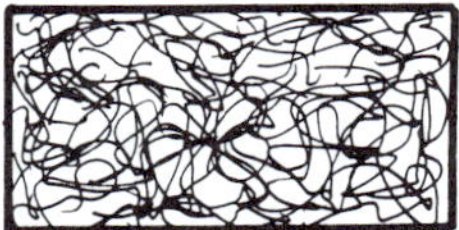

Alloys: solder, bronze, pewter, type metal, steel

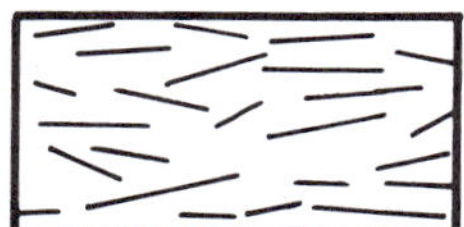

Powder Compacted: fuel screens, cattle salt cubes, fertilizer pellets, metal gears, plastic parts

LAYERS

Sandwiched: layer cake, corrugated cardboard, quilted vest, 25-cent piece

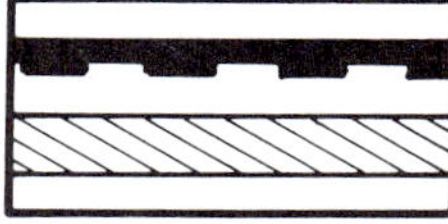

Claded: copper-bottomed pots, space visors, metalizing, spray welding

Bonded: galvanized steel, ceramics bonded to metal, iron-on patches

Coated: house paints, rust-proofing film, glazed donuts, enamel coating on bathroom fixtures

Laminated: countertops, plastic encasement of identification card, layered wooden beams, plywood

Material systems—those that are mixtures and those that are layers.

layer is bonded to the wood or other material, but most of the paint could be removed without destroying the base materials.

Dynamic Materials

Dynamic materials change during use. You can choose the material you want if you know that these changes will occur. There are three types of dynamic materials: **metamorphic, expandable,** and **functional.**

metamorphic materials

expandable materials

Metamorphic materials change slowly from their original characteristics to something different as they are used. The metal in the blade of a bulldozer is often made of a metamorphic metal. The blade gradually becomes harder as it is used to move dirt and rock. Cedar house siding, when left unpainted, gradually turns a beautiful gray color as it reacts to weather. We call these *metamorphic* changes.

Expandable materials are of two types, those that flex and those that foam. An example of a flexible material is a rubber balloon. Shaving cream from an *aerosol* can is an example of a foamable material. Both types of materials change more rapidly than in materials where composition or structural changes occur. Materials like the rubber balloon will change and repeat the process. Flexible materials can expand and contract. Flexible materials are used in tents, clothing, and as expandable containers for gases and liquids.

Expandable materials, like the shaving cream, expand only once. Some plastic and rubber materials use specially chosen ingredients that bubble up like a runaway dishwasher when mixed together. Depending upon the ingredients the new expanded material may become solid and hard or spongy and soft.

These foamable materials have many different applications. They can be used to provide a lightweight protective cushion for fragile parts that must be sent in the mail. Foamable materials have reduced the time and cost of insulating refrigerators. Because the bubbles retain air, expandable foams have been used to insulate older homes. Expandable materials can reduce the cost of heating in winter and cooling in summer.

Baking powder is another expandable material. When

Dynamic materials go through a process of change. Dynamic materials that change slowly over time are metamorphic. Others are expandable and change by growing and/or contracting. Others that change to perform different tasks are functional materials.

DYNAMIC MATERIALS

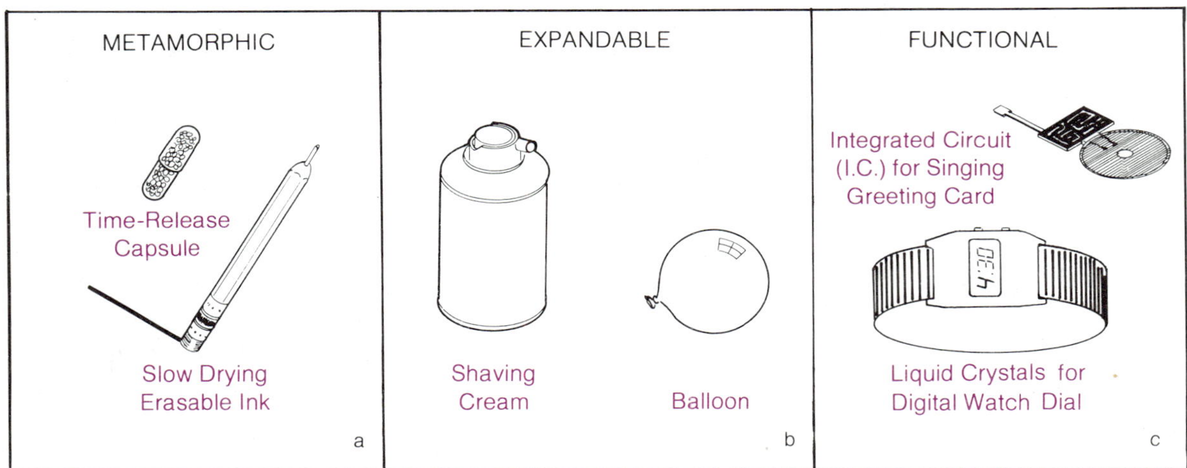

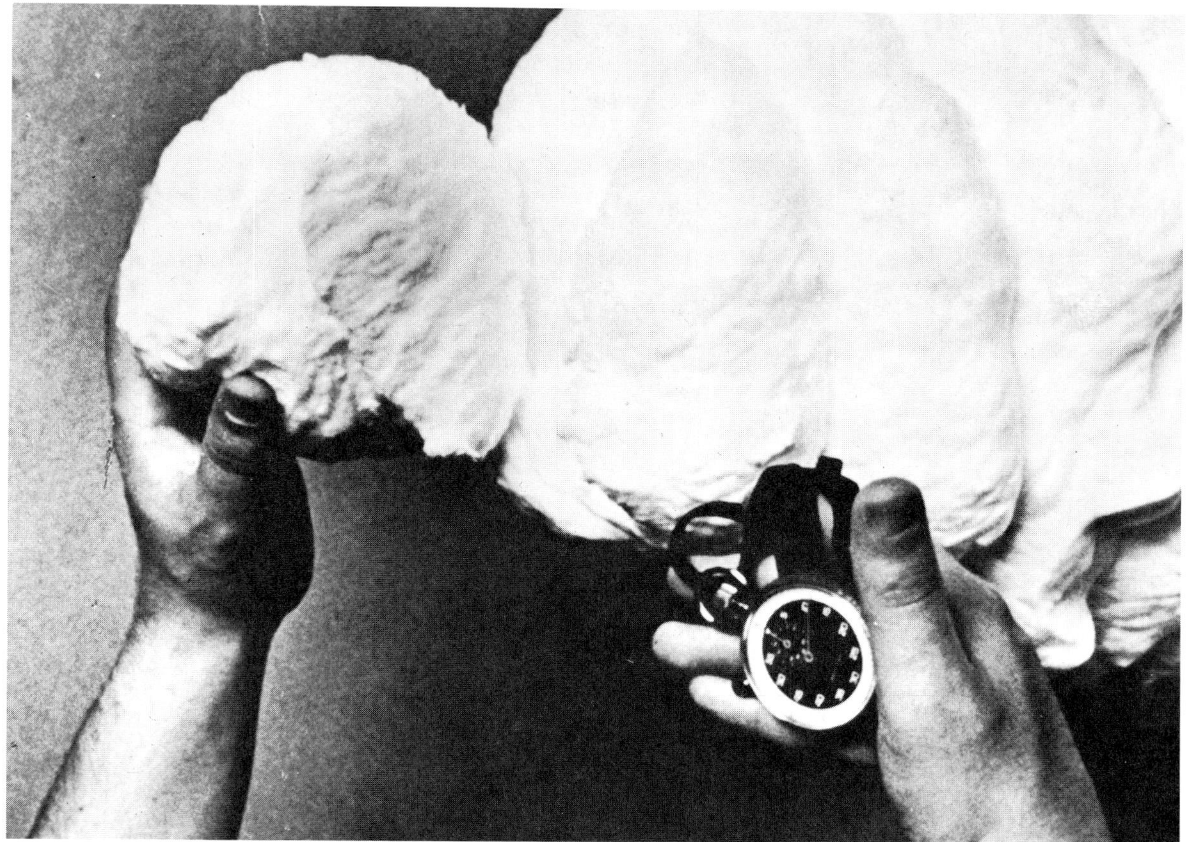

Expandable foams look like shaving cream during application. The ingredients used in a foam will determine its final form. The foam may be hard and rigid, as in some insulation materials, or soft and springy, as in pillows and packing materials.

mixed with water, baking powder changes chemically and releases a gas. This gas expands. If it is trapped in a cake flour mixture, it causes the cake to rise. This makes a light and fluffy cake.

Functional materials change back and forth from one state to another as an outside force is applied and then removed. Photosensitive eyeglasses are made with a functional material. The figures on a digital watch are made of liquid crystals which are made lighter or darker by electric currents. The action of sunlight causes the materials in the glasses to change and gradually become more opaque. Less light can get through. Once the sunlight is removed (when you go indoors), the glasses become more transparent and allow more light to reach your eyes.

A solid state refrigeration unit is another example of a dynamic functional material. When electrical energy is fed to the unit, one side of the unit absorbs heat from the other side. The second side becomes cooler. When placed in contact with the storage compartment of a refrigerator, the device takes heat out of the storage compartment, too. Heat is drawn from one side of the the solid state device to the other. There the heat is ventilated to the outside air. This process continues until it reaches the desired temperature (determined on the control/sensing switch). These changes take place with no moving parts. The functional materials in both the sunglasses and the refrigeration unit convert energy to new forms.

Families of Materials

families of materials

When we talk about a family, usually we refer to a group of people who are related to each other. Some materials are related and can be considered as family members. Many of the materials we have been studying are really part of families of materials. For example, the term "plastics" is a name for a family of related materials. This family includes many types of plastics such as vinyl, styrene, and acrylic.

metal
ceramics
textile
wood

We use such terms as **metal, ceramic, textile** and **wood** to indicate families of materials. We could name many individual members of these family groups. We may talk about the **food** family, for example. To help people understand the nutrients their bodies require, foods are often categorized into the Basic Four Food Groups. This groups together the many different foods that belong to each category. How many kinds of meat can you name? We could even try to name all the breeds of beef cattle.

ferrous metals

One of the largest material families is the **ferrous metals.** These are irons and steels. Iron is made from processing coal, iron ore and limestone, which are natural materials. Iron is seldom used in its basic state. Instead, it is mixed with small quantities of materials such as carbon, silicon, manganese, nickel, chromium, molybdenum, sulfur and phosphorus to form **alloys.** Within each group there may be dozens of individual steels.

alloys

We mix various ingredients in various amounts to make different types of cakes. This is much like making different types of steels. All cakes have flour, sugar and milk. All steels

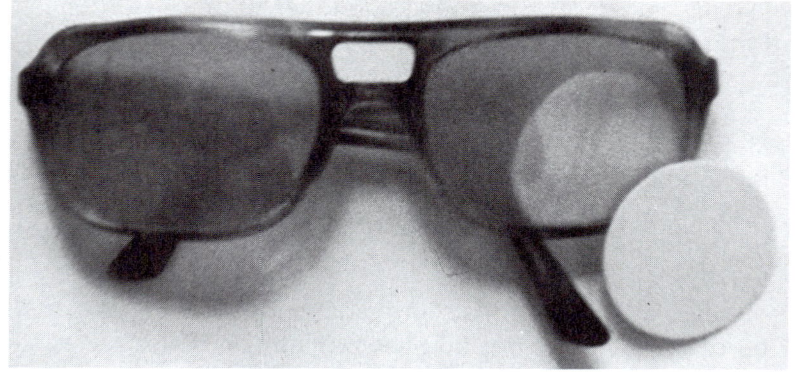

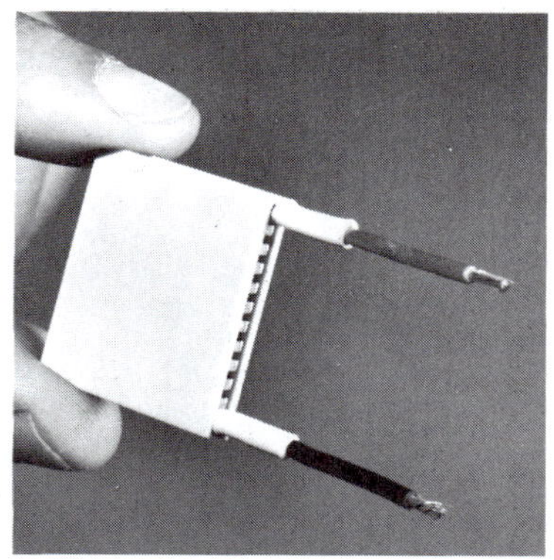

Functional materials change back and forth during use. The lenses of these sunglasses get darker when exposed to the sun and lighter when in the shade. When electricity is applied to this small solid state refrigeration device, it gets cold on one side and warm on the other. When the current stops, both sides become the same temperature.

have pure iron, carbon, and silicon. We can vary the amounts of the basic materials and add other materials to change the texture and lightness of a cake. The same is true for making steel.

Just as cookbooks contain hundreds of recipes for different types of cakes, so do the steel companies have formulas to make many different types of steel. These steels are designed to do different jobs and make different products.

This is true for other metals. There are more than 170 commercially-available alloys of aluminum, for example. The

One type of material may have many offsprings. These variations of a material are called a family. The Basic Four Food Group is a way to study food as a family of materials.

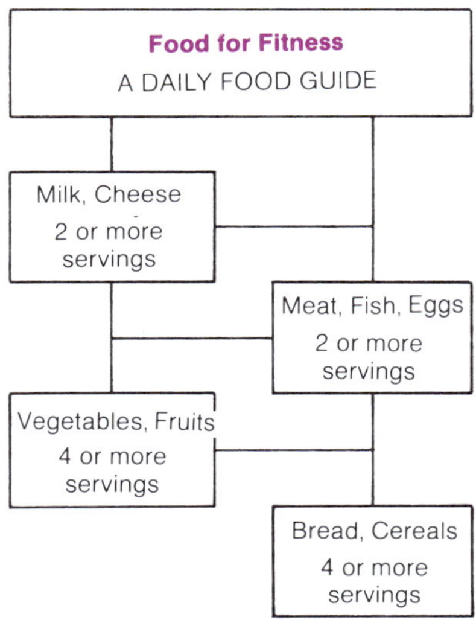

plastics family is large, too. It continues to grow as chemists formulate new plastic materials to do specific jobs.

Summary

Materials come in many natural and synthetic forms. Manufactured materials far outnumber the natural ones. Materials have quite predictable structure, characteristics and functions.

Materials have identifiable internal structures that form regular patterns. Materials have characteristics that can often be changed to fit specific needs. Synthetic materials, especially, can be formulated and produced to take on desired characteristics.

Tools, machines and materials are related in some very obvious ways. Tools or machines could not be built without suitable materials. The sophistication of the tools that we can make is directly related to the materials that are discovered. We name the phases of historic development after materials (the Stone Age, the Bronze Age, the Iron Age). This is an indication of the importance these materials have played in the development of humankind.

There is another important relationship here. Materials are converted into useful objects through the processes that the tools and machines perform on those materials. Tools and materials interact through one or more different **processes.** Processes are the topic of the next chapter.

Key Concepts and Terms

alloys
atoms
attraction
basic
body-centered cube
ceramics
chains of molecules
characteristics
chemical elements
closely-packed molecules
compounds
crystals
dynamic materials
electrons
expandable materials
families of materials
ferrous metals
firing
functional materials
gas
green
kiln
liquid
loosely-packed molecules
material layers
material mixtures
material systems
metal

metamorphic materials
molecules
monolithic materials
natural
plastic
platelets
polymer
properties
repulsion
solid
structure
synthetic
textile
wood

This advertisement appeared in 1889. It describes the processes involved in using the *new* Kodak camera.

THE KODAK

Is the smallest, lightest, and simplest of all Detective Cameras—for the ten operations necessary with most Cameras of this class to make one exposure, we have **only 3 simple movements.**

NO FOCUSSING. *NO FINDER REQUIRED.*

Size 3¼ by 3¾ by 6½ inches. **MAKES 100 EXPOSURES.** Weight 35 ounces.

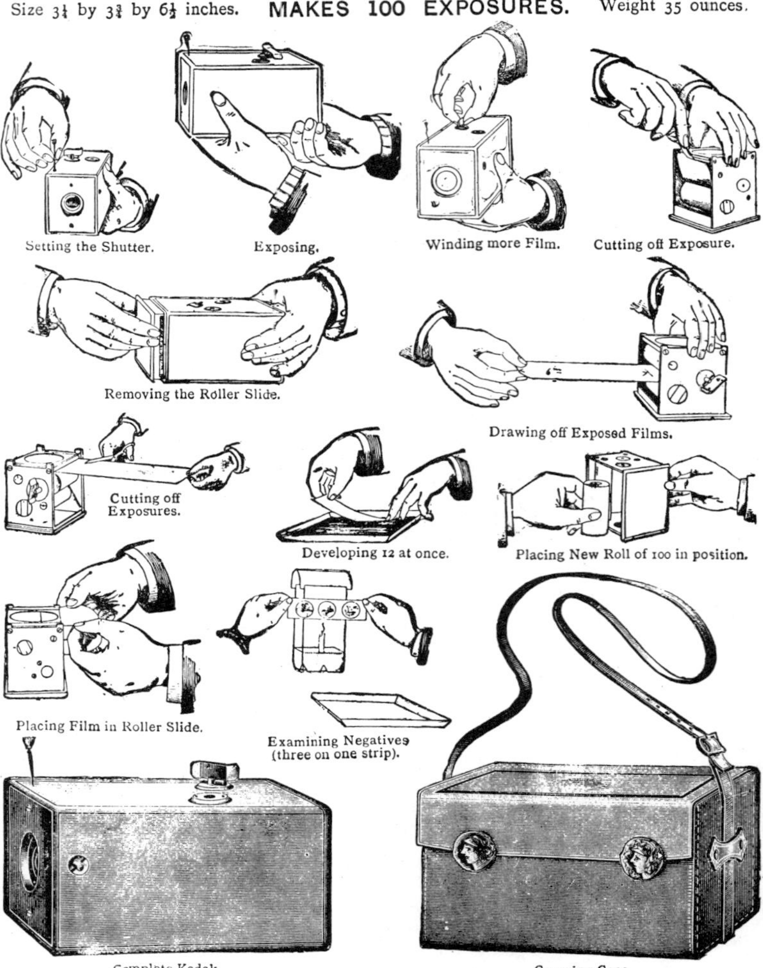

FULL INFORMATION FURNISHED BY

THE EASTMAN DRY PLATE & FILM Co., 115, Oxford St., London, W.

Chapter 4 Processes

Processes can be a little difficult to understand because they are not always visible. If you watch a pizza baking in an oven you do not actually see the chemical action (the baking) taking place. What you do see is the sequence of changes that the pizza goes through during baking. After the process of baking, you see the results of the changes that occurred.

sequence
action
changes
results

The term *process* as used here simply refers to a sequence of actions that lead to some result. As indicated in the example of baking a pizza, the key ideas are **sequence, action, changes** and **results.** In the chapter on machines, we discussed changing or converting materials, energy and information. The ways these conversions occur are called **processes.**

process

The Conversion Process

input
output
energy

Engineers have found that an input/output model is useful in thinking about processes. Engineers talk about a process occurring in a "black box." You can see in the diagram that if something is fed into one side, something else will come out the other side. The three stages of the black box are the **input,** the **process** and the **output.**

To make a phonograph work, energy is provided in the form of electricity. A material is added in the form of a record. Information input comes in the form of the switch turned to "on" and the speed set for 33⅓ or 45 RPM.

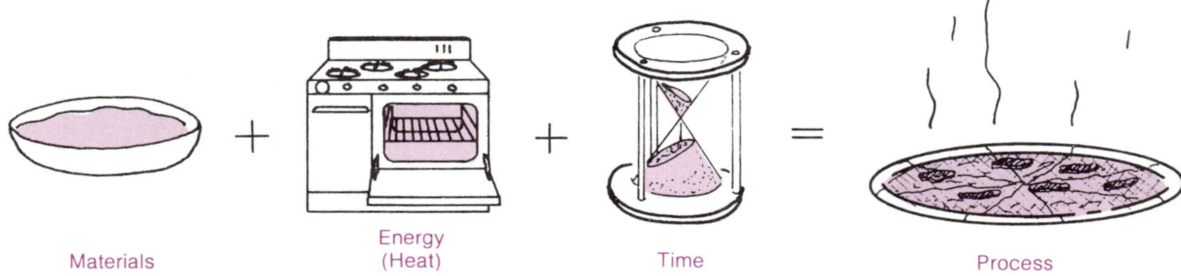

In a pizza baking process, heat (over a period of time) causes chemical reactions within the materials.

All of these variables are necessary before the machine can operate and the process of playing can begin. The change in electrical energy in the record player causes the turntable to spin. This, in turn, allows the needle to pick up the vibrations (as mechanical energy) and turn that energy into sound (a different form of mechanical energy). This is the process.

The phonograph is a good example of an information machine. The record player takes stored information "frozen" in the record and uses energy to convert that information to a more useable form (so you can hear it). The event would not be possible unless you knew the processes to follow. This is an example of an information conversion process.

The black box concept is one way to look at processes. Inside the imaginary black box some type of change takes place on materials, energy and information.

Goals
Inputs
Process
Outputs

A phonograph can be described using the black box concept. Inputs of materials (the record), energy (electricity) and information (the setting of the playing speed) result in an output of a new energy form of sound (music).

We can produce many different forms of energy by using other processes. By combining the right materials, we are able to make a battery which can produce electrical energy. In the battery, two paste solutions work together to release extra electrons and move them to the zinc case. The zinc thus has many free electrons to give up. The central carbon rod has a deficiency of electrons. Electrical energy is available as soon as the battery is linked into a circuit. The **electrical energy** is generated chemically in the battery.

electrical energy

The common flashlight battery produces energy (electricity) through a process of chemical changes among materials.

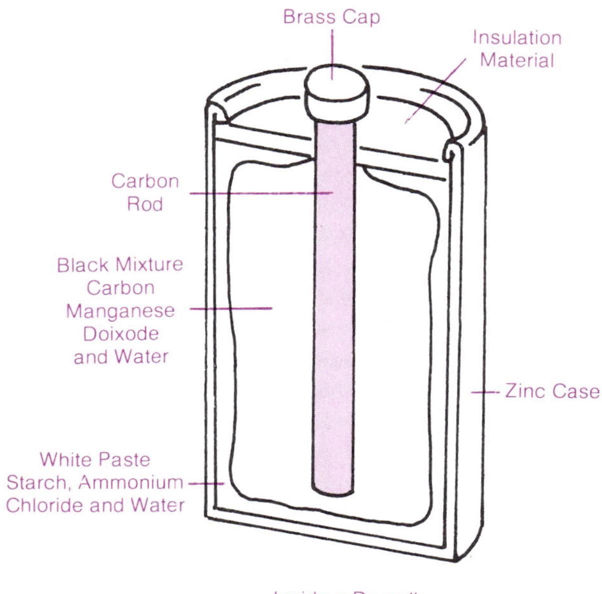

Inside a Drycell

The **sequence of actions** involves placing the right materials together in the proper places and connecting these into a circuit. The **change** is the exchange of electrons that takes place. The **result** is the production of new energy—electrical energy.

Coal can be burned. This is another example of energy conversion processes. When burning takes place, a chemical reaction releases energy from carbon-containing molecules. The initial heat allows those molecules to break some molecular bonds and combine rapidly with oxygen, an action we call **combustion.** By changing coal through burning, we can produce **thermal energy** (heat). This energy is much greater than the small amount of heat used to start the process.

combustion
thermal energy

In the examples of the flashlight battery and the coal, the materials are combined and changed to release the energy which exists in the materials (the zinc and carbon of the battery and the coal, which is relatively pure carbon).

Other processes are used to change materials into new materials and into products. If we have an adequate supply of thermal energy (such as that from burning coal) we can take a sample of tin and a sample of brass and melt them together by applying heat (thermal energy). A new and different material (bronze) will be produced. This process is called **alloying.**

alloying

Other **materials conversion processes** takes energy and new materials and changes them into useful products. These processes always require some form of tool or machine. There are many processes that can be used to create an almost endless number of products.

Popcorn offers a simple example of a useful material conversion process. Place the popcorn and oil (the material) in a pot on the stove. Now we are ready to start the process.

On a gas stove, burning converts gas energy into heat (thermal energy). The heat converts the water in the popcorn; it becomes steam. The steam expands and pushes out from within each popcorn kernel. When enough pressure builds up inside the kernel, it explodes with a pop. The oil helps to keep the corn from burning and helps to transfer the heat more efficiently from the pot to the corn. After popping, we can finish the process by adding butter and salt and have a useful product. The popcorn, now an expanded material, has been put through a process or series of processes to change it into its desired form.

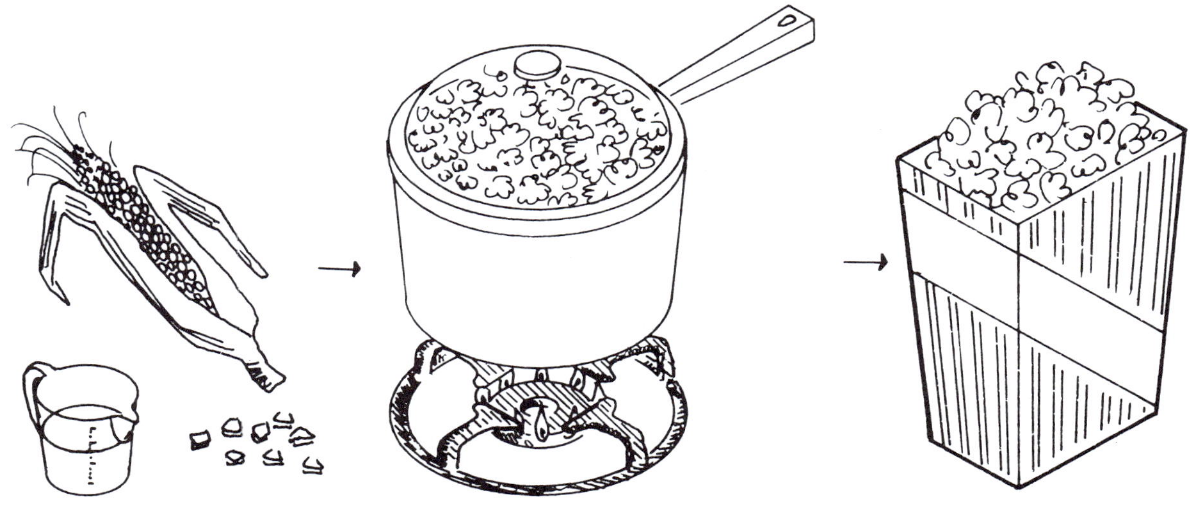

Making popcorn is a material conversion process.

Materials Conversion Process

The rest of this chapter will focus on processes that can be used to change materials. In later chapters we will apply these ideas to illustrate how energy and information may also be changed. The processes for changing energy, information and materials are basically the same. Luckily for us, the thousands of different processes used to change materials fall into four categories. Materials can be changed by (1) **adding** materials to each other, (2) **separating** materials, (3) making **contour changes,** or (4) making **internal changes.**

The "black box" idea of processes can be applied to a specific example of materials conversion. The raw materials (along with energy, in some cases) are the input. The changed material is output in the form of a finished product. We can divide each of the four materials conversion processes into smaller categories. Perhaps this will help you identify similarities and differences among the various processes. Knowledge about these processes can make you a wiser consumer of products. It can also allow you to choose the outcome you want when you construct or repair products yourself.

products

Have you ever wondered why bread must be kneaded before it is baked? Or why muffins should be mixed only until the ingredients are moistened and no longer? Why are cake

Raw materials can be changed into finished products through any of four conversion processes.

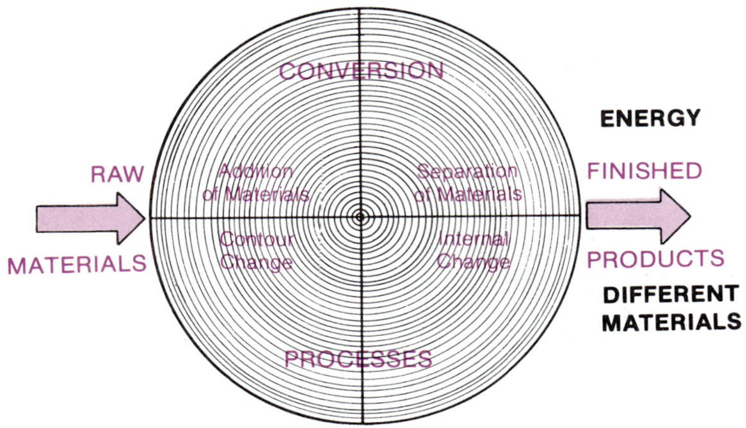

mixes beaten at high speed for several minutes while pastry is to be handled as little as possible? All this has to do with the **sequence of actions** in the mixing process and the desired **outcome.** Recipes have been developed to identify the sequence of actions as well as the duration and conditions under which the actions are to be applied. Because of this, the outcome (food) is more likely to be consistent and desirable.

Materials Conversion Through Addition

Materials are added to each other in many different ways. In the previous chapters we discussed materials systems. Materials systems are the product of the processes identified here.

mixing joining fastening coating weaving interlacing

It is possible to group the addition of materials into four categories. These include (1) **mixing** or formulating, (2) **joining** or **fastening,** (3) **coating** and (4) **weaving** or **interlacing.**

Mixing or Formulating

Perhaps the most obvious way of combining materials is to mix them together. You have done it many times. Water is mixed with water-based paints to make them thin enough to use. Sugar, flour and other ingredients are mixed to make cookies. Sand, gravel and cement are mixed together to make concrete.

These are all examples of combining materials through mixing. In some cases, materials that are mixed interact with

Several ingredients have been added to the water in this vat as an example of addition by mixing. The woman is capturing a layer of the mixture on a fine screen to make paper, one sheet at a time.

each other so that the original materials cannot be retrieved. In other cases, the original materials may be recovered for later use.

Joining or Fastening

Processes in this grouping are of three general types: adhesion, cohesion and mechanical. These processes result in the joining together of materials for structural or decorative purposes.

adhesion

In an **adhesion** process, the connection formed between materials is a surface bond only. The atoms and molecules of

Printing this book is an example of addition through adhesion—ink on paper. The above is an early example on the printing process.

the two surfaces remain separate. Adhesive processes include gluing, soldering, brazing, printing, painting and putting together a peanut butter sandwich.

Cohesion is a joining process activated by heat, pressure or both. A bond is formed by the attracting forces between atoms at the surface of each material. Cohesion processes include oxyacetylene welding, thermal welding of plastics and making a hamburger patty.

In **mechanical fastening** processes, materials are joined with fasteners such as screws, bolts, zippers, rivets, staples, clamps, tapes, thread and pins.

Coating

Coating is an addition process used to cover a base material for protection or decoration. We coat materials by adding a layer of a second material. Painting is one example. The paint has little penetration and sticks to the surface by adhesion.

These lengths of track are being welded together using addition by fusion. The ends of the track are covered with magnesium that burns, giving off an intense heat and melting the metal. The metal fuses as it cools.

The assembly of a skateboard is an example of addition through mechanical linkage.

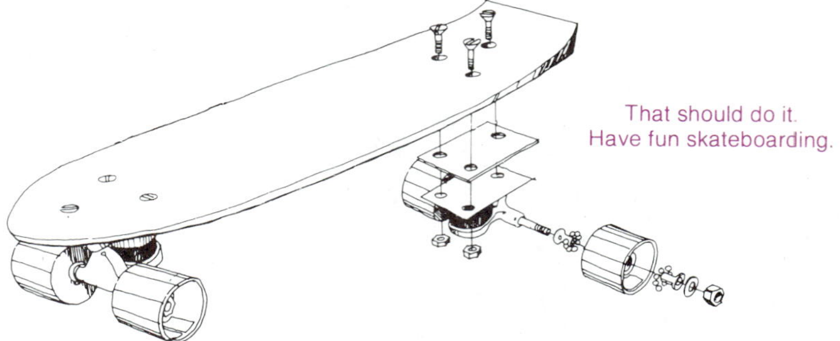

That should do it. Have fun skateboarding.

Glazing doughnuts and waxing a car are other examples of adding a coating.

There are coating processes that alter the surface of a material. In these instances, the second material penetrates by physical or chemical means and separation of the coating and the base material is less distinct. Examples of this type of coating are staining of wood (by physical absorption) and anodizing of aluminum (by chemical reaction).

layering

penetration

Coating may be used for protection and decoration. Can you identify any other processes that involve coating by **layering**? How about some others that are coated by **penetration**?

Weaving and Interlacing

In weaving or interlacing, the materials are added so they are mechanically locked together. Obvious examples are threads

Painting a porch swing to protect it from the weather is an example of addition through coating.

Automatically producing a trade-mark on a Jaquard loom is an example of addition through weaving and interlacing.

combined to make cloth, and wires combined to make screen. Weaving and interlacing are very old processes. They were used by people long ago.

Addition processes that have an intended and identifiable pattern we call **weaving. Interlacing** refers to materials added in random or non-patterned ways. We can take glass fibers and weave them together to make fiberglass cloth, for example. We can combine the glass fibers in a random pattern to produce fiberglass insulation.

Processes for Separating Materials

shearing

Shearing and sawing are two very common separation processes. Both of these processes separate material by use of a wedge. They are different because the saw separates the material by using a blade with cutting teeth. This produces waste in the form of chips. Shearing, however, uses two smooth blades (a pair of scissors, for example) that work together to separate the materials without producing a chip.

Shearing can be done two ways. Two cutting edges may cross one another, as scissors cut hair. A single cutting edge may go through the material, as a knife slices meat. Other examples of shearing tools include paper cutters, cheese graters, tin snips, bench shears, squaring shears, solid punches, hollow punches, wood chisels and cold chisels.

sawing

Sawing is done using a tool that has pointed, wedge-shaped teeth evenly spaced along the edge of a blade. Exam-

Cutting pizza is an example of separation with a wedge but without a chip.

ples include hacksaws, crosscut saws, rip saws, band saws, jig saws and jewelers' saws. Other processes similar to sawing include grinding, filing, shaping, drilling, milling, turning, sanding and polishing.

zero force

Shearing and sawing require some form of mechanical force. We can separate materials through non-mechanical processes as well. Chemical or thermal processes are examples. These are called "zero-force" processes because they do not use mechanical force. Bleaching, etching, dry cleaning, oxyacetylene torch cutting, carbon-arc cutting, chemical milling, laser cutting and ultrasonic cleaning are examples of zero-force processes.

Using solar energy to distill clean water is an example of a zero-force process. This causes a separation without a wedge and without a chip.

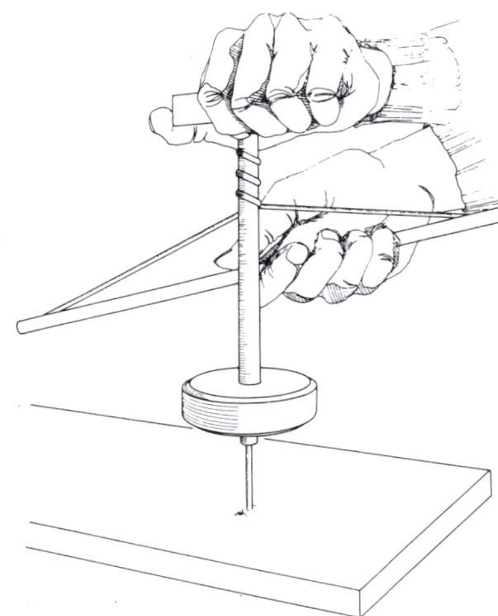

This is an example of changing materials by separation with a wedge and with a chip.

Processes for Contour Change

Contour (surface) change processes alter the outside of materials without removing or adding any materials. We can pour warm gelatin into a mold to determine the shape or contour of the mixture as it cools. This change stays after the gelatin is removed from the mold. This is the process of casting, actually, not molding.

Stamping, bending and stretching are other examples of contour change processes. All of these processes change the shape of a material. There are five general types of contour change processes: **casting, extruding, forming, pressing** and **molding.**

casting

Casting is the process of forming a desired shape by liquifying material and pouring it into a mold. When the material cools, or sets, it retains the shape of the mold. You may have used the casting process to make candles.

extruding

Extruding involves forcing a material through an opening (a die). This controls the cross sectional area and the shape. Toothpaste is extruded from a tube. Cookie dough can be extruded through a press. Plastic or metal is often extruded, as when making parts of storm windows.

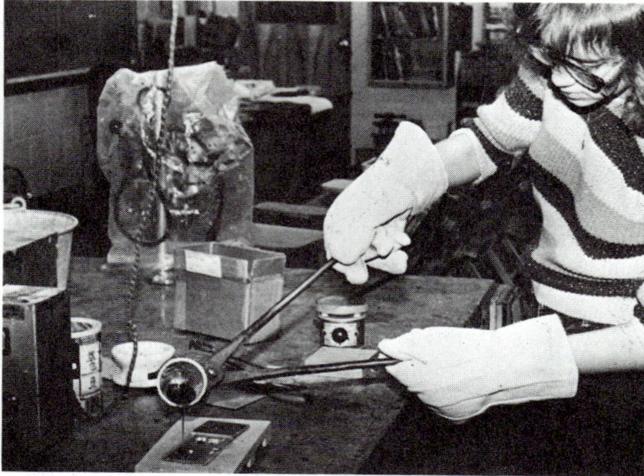

Melting and pouring molten glass is an example of processing materials by a contour change through casting.

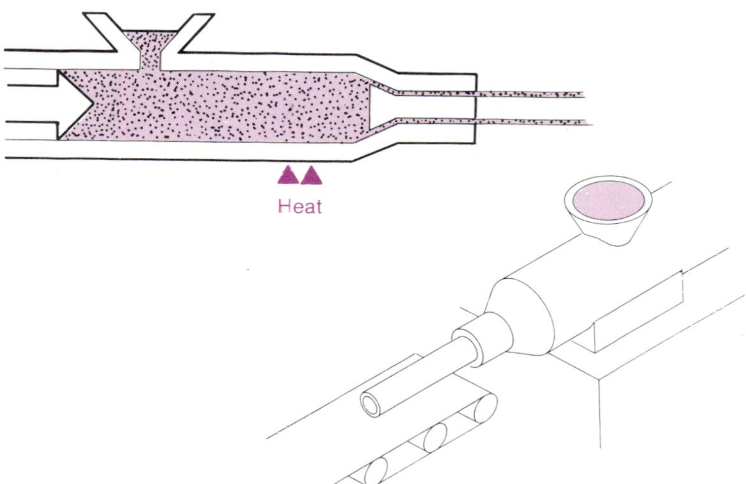

Heated pellets of plastic can be forced through a special hole (a die) to cause a contour change through extrusion.

forming

Forming is a process which shapes materials by impact or pressure. Forging is an example of impact forming. Forging is used to change the shape of a piece of metal while increasing the metal's structural strength. Horseshoes and very strong metal tool parts are forged. Bending, another example of forming, uses uniform pressure around a form. We use forming to shape a pie crust inside a pan.

pressing

Pressing shapes sheet material by squeezing it between two dies or textured surfaces. The surfaces force the material to take on a three-dimensional shape. Waffles are pressed in a waffle iron, metal bottle caps are pressed and hamburger patties are pressed.

Heated sheets of plastic can be shaped by vacuum and/or air pressure to cause a contour change through forming.

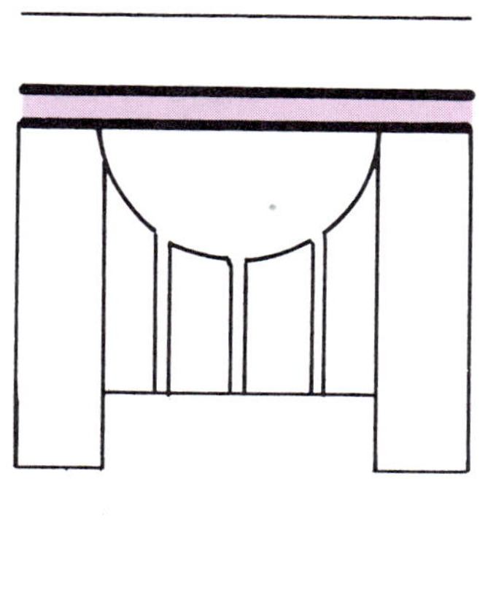

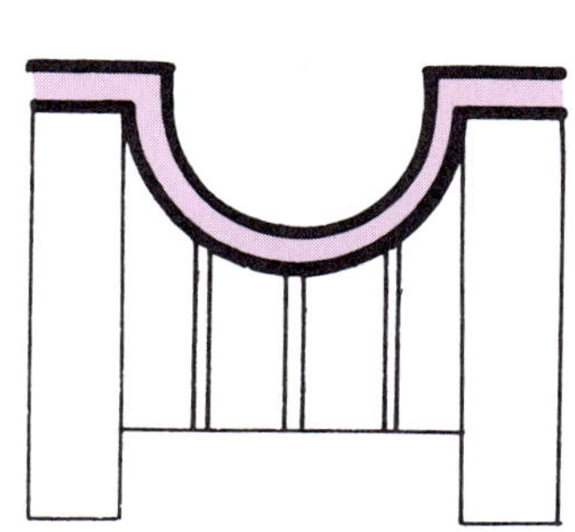

Powdered metal can be shaped into many different forms by a contour change through pressing.

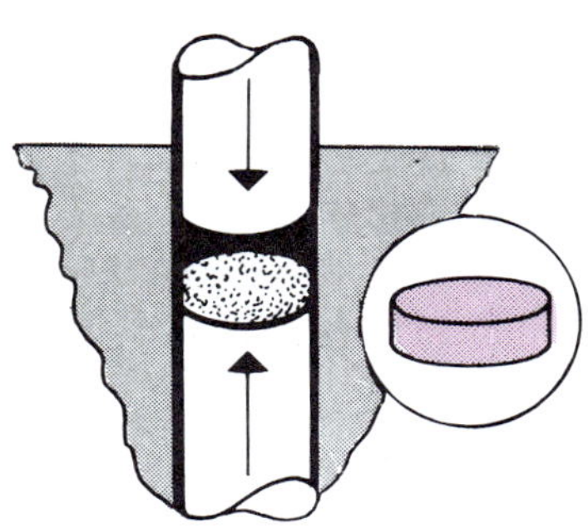

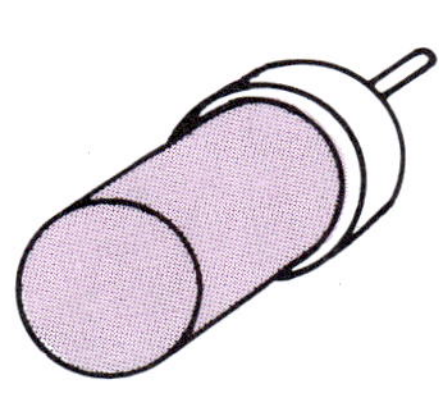

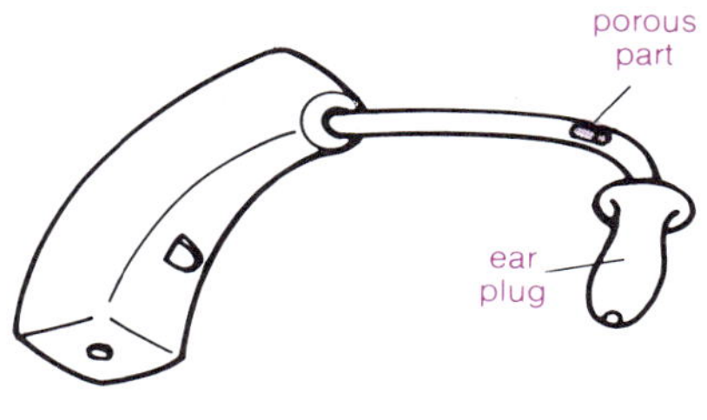

HEARING AID

molding

Molding is a combination of extrusion and casting. The material is forced into a mold (casting) under pressure (extrusion). Plastic cups and wastebaskets may be made through the molding process.

Look around for examples of products that have had their contour shaped to suit a given purpose. The handles on your scissors may be "form fitted" to be more comfortable. Plastic chairs may be molded to follow body lines. See how many other examples you can find.

Processes That Make Internal Change

Some processes are used only to make changes inside materials. These processes are not always visible to the eye.

thermal processes

Thermal processes such as baking, boiling and freezing produce internal changes. These processes are called thermal processes because they depend upon a change in temperature. A decrease in temperature is needed to freeze ice cream. Ice cream making, therefore, is a thermal process.

mechanical force

shock processes

Mechanical forces can make internal changes. Hammering, peening and explosive-forming are mechanical processes that use shock to cause the change in the materials.

magnetic processes

Magnetic processes can change some materials internally. A screw-driver, for example, can be made into a permanent magnet by passing it through a strong magnetic field. This

This ancient Welsh kiln shows how clay roofing tiles were fired and became glass-like by an internal change through a thermal process.

A drop hammer causes the piece of metal to become stronger through impact. This is an example of an internal change through a mechanical process.

polarizes the metal atoms. A needle can be magnetized to make a compass.

chemical processes

Chemical processes that make internal change include electrolysis, fermentation and film developing. The images on ordinary camera film are "developed" in the darkroom by chemical processes.

Acoustical processes also create internal changes in materials. Sound, in the range we can hear, causes changes in the molecular bond of specific materials. The most common acoustical processes use sound outside our range of hearing (ultrasonics). Ultrasonics can be used to increase the magnetic properties of selected materials. This is useful in materials testing. Recently, ultrasonics in health care and medicine has been used to supplement and sometimes replace X-rays.

The electrostatic spray painting process is very efficient. The paint is attracted only to the workpiece. In the spraying process, the paint and the piece to be painted are given opposite electrical charges that causes an internal change that is magnetic. The paint is drawn to the piece like a magnet.

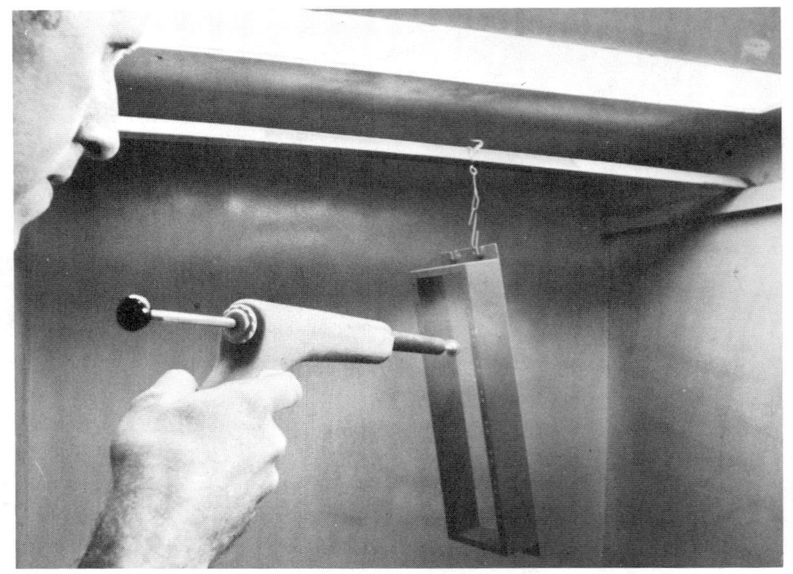

This electric arc light operated on batteries and was used in early movie projectors. The batteries are an example of internal change by a chemical process.

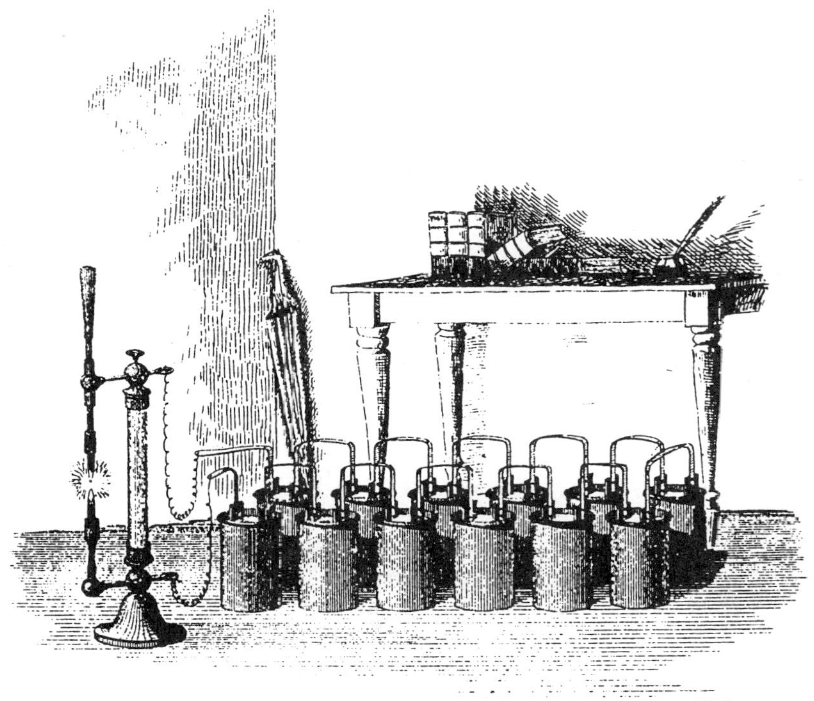

Kidney stones can be blasted apart by sound. This process uses an ultrasonic device to cause internal change by acoustic means.

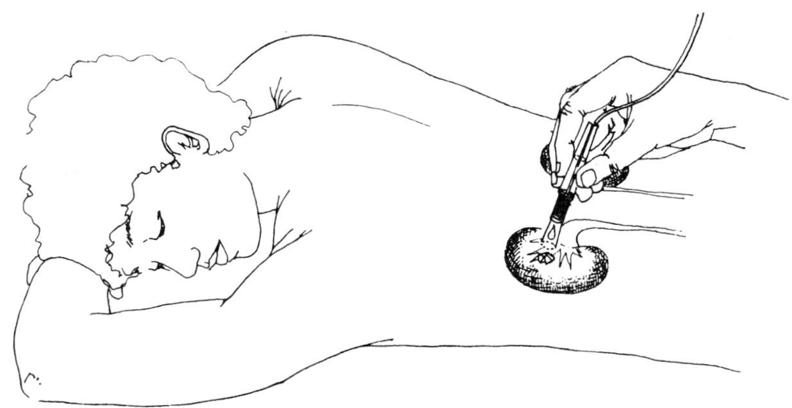

Optical processes refer to the change of materials that is caused by light. Some materials, such as plastics, change if left in sunlight for long periods of time. Photography is an example of using the effects of light on light-sensitive materials. Light also causes changes in materials that are important for electronics. These include the photoelectric, photoemissive and the photoconductive changes.

Electrical processes cause changes in materials by the passing of a stream of electrons through the materials. A flow of electrons can cause materials to change internally and vary

In typesetting, letters and other symbols are projected onto light sensitive paper that goes through internal change by an optic process.

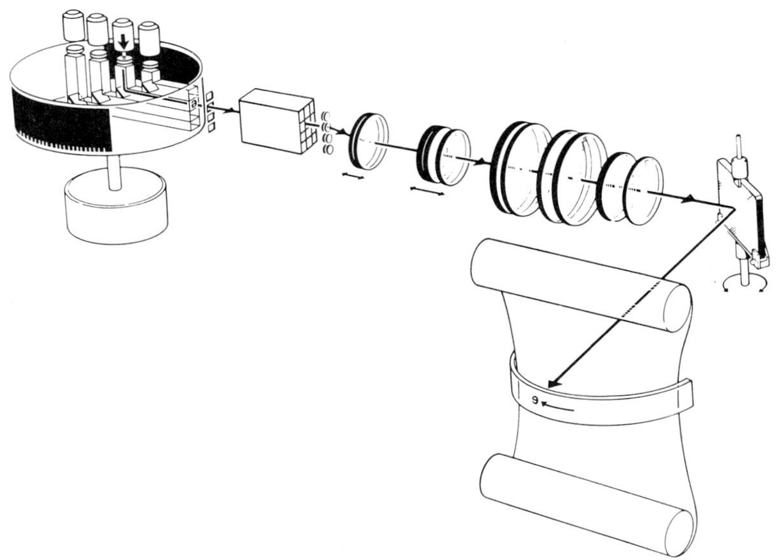

This prototype of an electric-discharge machining (EDM) device removes metal by causing internal change through an electic process.

their properties. Materials can be made to change their magnetic, conductive, resistive, light transmission and light emission properties.

Similarities in Materials Conversion Processes

The materials conversion processes presented in this chapter may appear to be different from each other. Yet, all of them are doing something quite similar. The processes change the way the materials are bonded together. We use these processes to create new bonds, break other bonds, or modify existing bonds in some way.

mechanical linking

Perhaps this is easiest to explain if we again consider materials at the molecular level. The molecules are shown as very small balls with invisible connections (lines of force) that hold them together, yet keep them apart. When we separate materials we are attempting to break these lines of force.

We can use mechanical energy to separate the bonds when we cut paper with a pair of scissors. We can use acid to etch or take away a small amount of metal. We can combine thermal and chemical processes for separating oxygen from metal ores to produce pure metal.

All conversion processes change inputs to outputs to reach a desired goal.

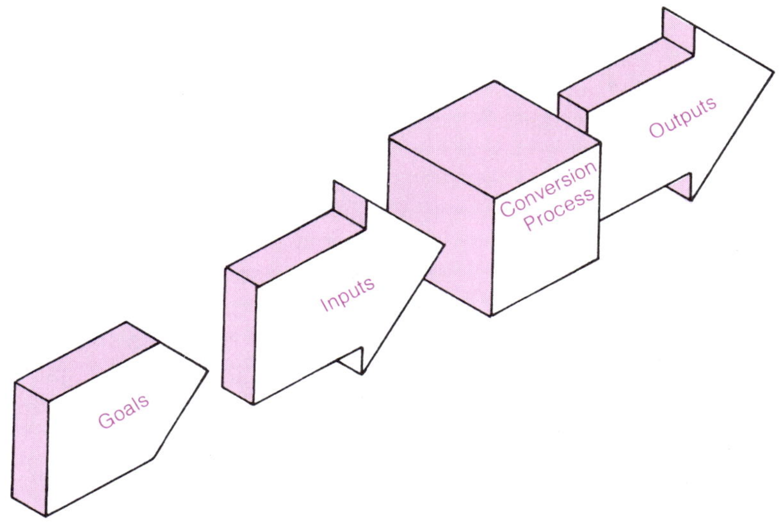

As we change materials through addition we are trying to create new bonds. We do this when we add a protective or decorative coat of paint. The paint forms new bonds between itself and the base materials during the *drying* process. If we paint a dirty surface, we may get a poor bond. We also create new bonds when we melt two or more metals together and allow them to cool, thus forming an alloy.

During contour changes of materials we usually rearrange bonds. We may pack the molecules more tightly, and we may shorten and strengthen the bonds. For this reason some auto parts are forged rather than cast.

We may also stretch materials apart, sometimes weakening them. A plastic trash can, for example, may tear at the corners, where the plastic was stretched during the forming process.

We often use contouring processes to take advantage of these changes. A good example of contour change of materials is the compacting of powdered metal in a press. The same process is used to put eye shadow or face powder into a makeup compact, or to make aspirin tablets.

Modification of materials by contour change usually means no loss of material. In the example above, we can pack finely powdered metal together to form a useable shape. The pressure developed in the press causes the granules of metal powder to hold together until they are heated and fused

fusion

(called "sintering"). Or, the powder can be loosely packed and then fused to make a very fine metal screen. A screen of this type can be used for separating dirt from gasoline in an automobile fuel line. We can pack the metal even more tightly and fuse it to form very strong bonds. This material can be used to make gears for automobiles.

Materials can be changed internally by creating new bonds or modifying old bonds. A chemical action is used to make yeast-raised bread. As the yeast reacts with sugar in the bread mixture, a gas is released and trapped in the dough. Kneading develops the protein to make strong walls which can contain the fine gas bubbles. The bubble walls are stabilized during the baking, which produces several other internal changes.

Internal changes can also be caused by pressure and impact. Such internal changes often cause a material to have different characteristics. For example, metal that needs to be tough and strong as for a crankshaft or is strengthened by impact in a process called *drop forging.* Now you know why some hand tools are stamped "drop forged." The manufacturer wants you to know the tool is a strong one.

Any process we select will make some kind of change in the materials we are using. It will be important to know what those changes are so we can take advantage of them. If we want to make an object that requires a lot of strength, we need to know what processes should be used. When we buy a product, we want to avoid one made by processes that could weaken the material.

Summary

This chapter focuses on processes used in materials conversion. It is important, however, to remember that processes are an element of *any* technological event. Processes are the sequence of actions that lead to some result.

Processes can be used to reach a desired goal or output, such as new *energy, information,* or *materials.* Processes *change* materials by adding or separating materials and by making contour or internal changes. Some of these materials conversion processes change the structure and, thus, the characteristics of materials.

All of these processes require energy. Many processes are used to convert energy, as well. The following chapter will focus on energy as an element of technology and on some of the processes, materials, and machines used to convert energy.

Key Concepts and Terms

action
adhesion
alloying
casting
changes
chemical processes
coating
combustion
electrical energy
energy
extruding
fastening
forming
fusion
input
interlacing
joining
layering
magnetic processes
mechanical force
mechanical linking
mixing
molding
output
penetration
pressing
process
products
results
sawing
sequence
shearing
shock processes
thermal energy
thermal processes
weaving
zero force

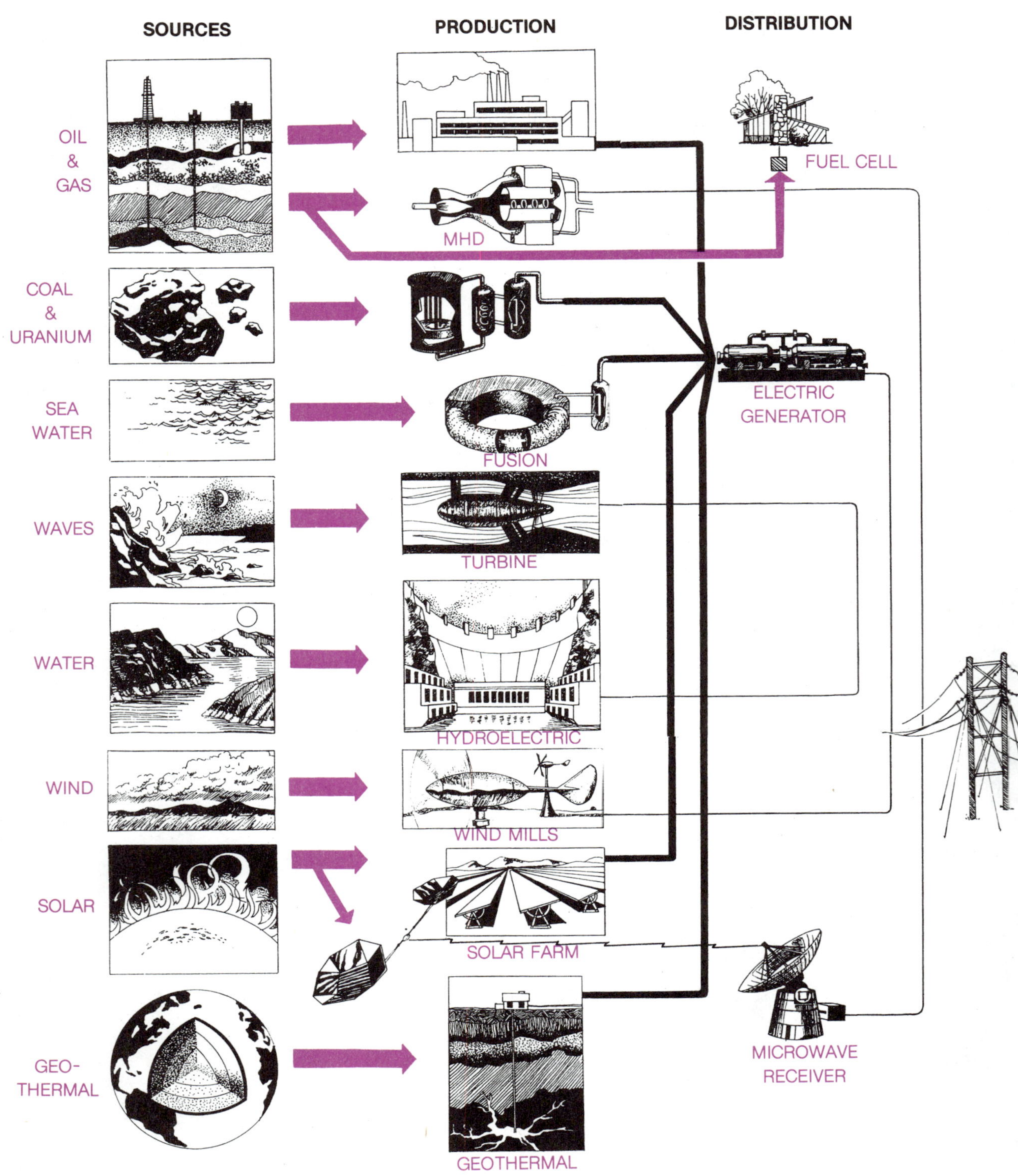
SOURCES
PRODUCTION
DISTRIBUTION
OIL
&
GAS
MHD
FUEL CELL
COAL
&
URANIUM
ELECTRIC
GENERATOR
SEA
WATER
FUSION
WAVES
TURBINE
WATER
HYDROELECTRIC
WIND
WIND MILLS
SOLAR
SOLAR FARM
MICROWAVE
RECEIVER
GEO-
THERMAL
GEOTHERMAL
WELL

Chapter 5 Energy

energy conversion

thermal energy

Energy is available to us from several sources and can be converted into many different forms. These three concepts—sources, forms and conversion—represent the major sections in this chapter. The intent of energy conversion processes is to put raw energy into practical use.

It can be difficult to understand energy because its source is sometimes hidden and because energy often is converted from one form to another. For example, energy from coal may be released through the chemical process of burning to produce heat (thermal) energy. This energy can be used to heat water to form steam. The pressure from the steam can then be used to drive a generator. The generator, in turn, transforms mechanical energy into movement. This movement uses magnetic energy to produce electrical energy. The electrical energy is then distributed through service lines to homes, industries and other energy users. Toasters change the electrical energy into heat. Electric lamps change it into light. Motors change it into motion.

The list could go on. Energy comes in many forms. It can be changed from one form to another, sometimes easily, sometimes with difficulty.

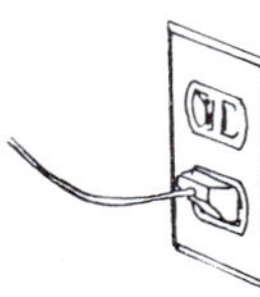

◀ Humans have found many ways of converting the prime energy sources found in nature to energy forms more suitable for their use. Electricity is widely used because of ease of distribution.

Primary Sources of Energy

If we trace our energy sources, from the electricity we use, the food we eat and the fuel we burn, we end up at the sun. The

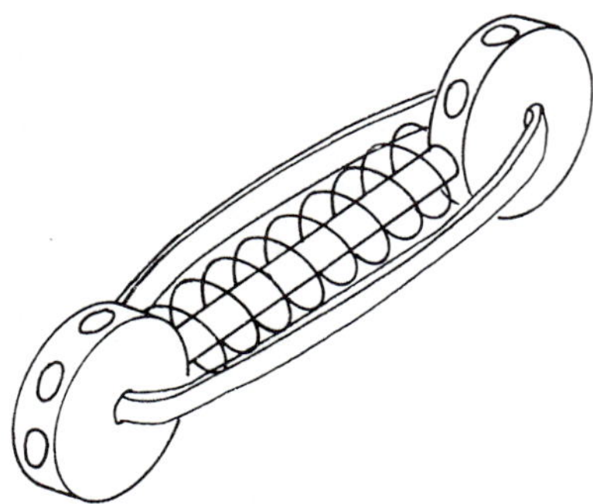

The basic source of energy is in the bonds within and between atoms and molecules.

sun is our biggest primary energy source. But the sun is just a giant energy "machine." The actual sources of energy are the bonds within and between atoms and molecules. Energy from the sun comes from nuclear reactions in the sun. These reactions free the energy in the particles of material (the atoms and molecules) that make up the sun.

Much of the energy we use on Earth is given off as bonds are broken between atoms or molecules of materials. Most of the time these bonds are broken by burning. Often our energy sources are fuels rich in carbon, such as coal, gas, food and oil.

Energy is released from a fuel in a fairly straightforward way. Fuels have material units with high energy bonds. During burning, these bonds are broken and new, low energy bonds are formed. The difference in the energy is released as (converted to) heat.

bond energy

For example, as a bond of a carbon fuel is broken, new bonds between carbon and oxygen are formed. This process results in a gas of carbon dioxide and carbon monoxide. Neither of these gases require as much energy to exist as did the original fuel. Energy is released in the form of heat.

prime energy sources

Energy goes through many changes of form, but the basic source of all energy is the bond between material units. Large usable reservoirs of energy are called **prime sources.** We generally use prime sources as the starting points for energy conversion. Originally, however, these prime sources came from an even more basic level.

Wind is one of our prime energy sources. Movement of the earth and changes in temperature actually create the winds. We can use windmills and sails to harness this prime energy source, which is a form of natural energy. But we seldom try to harness or change the conditions that originally produced the wind (the spinning of the earth and the warming and cooling of the atmosphere). In fact, we may not want to tamper with these natural conditions. We will discuss each of the prime energy sources.

Energy from the Sun

solar energy

Solar energy is radiated from the sun. Solar energy includes visible light, infrared waves, microwaves, ultraviolet and several other energy wave forms, and atomic particles. This en-

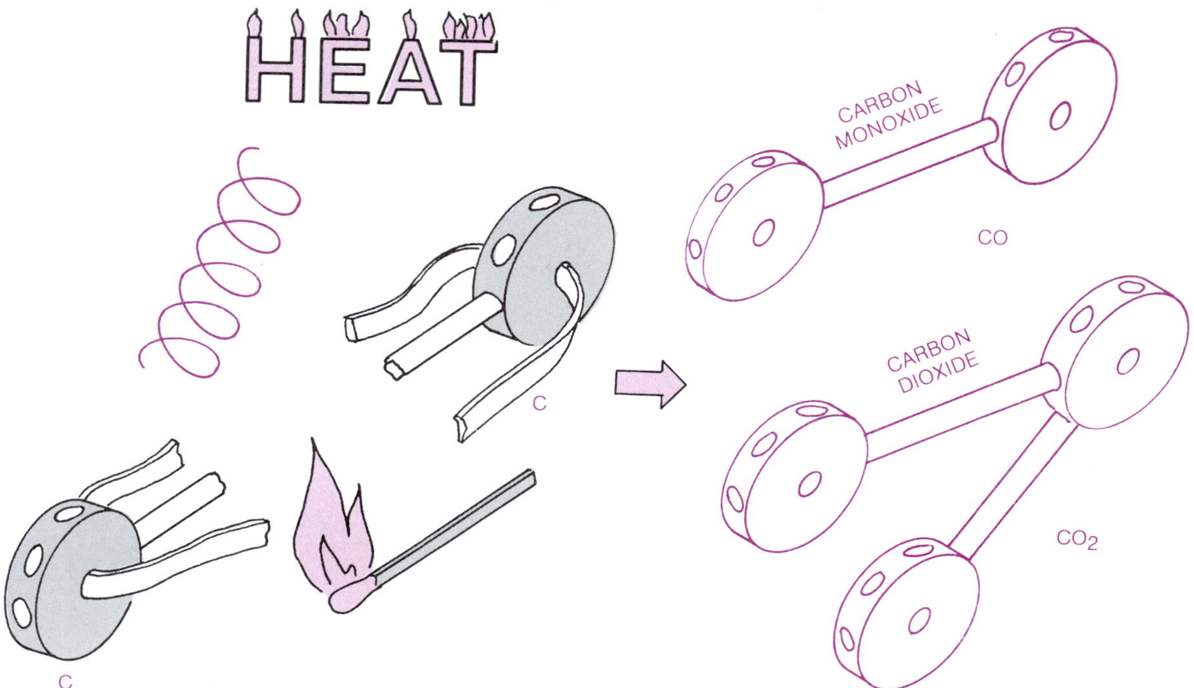

This simplified model shows how a fuel releases energy. For example, through burning, coal breaks down into new compounds of carbon dioxide and carbon monoxide. Since these require less energy in their molecular bonds, the excess energy is converted to heat.

ergy heats the oceans and land surfaces and helps create air movement (wind) around our planet. Solar energy also supplies plants with the necessary light energy for photosynthesis. Solar energy keeps our planet warm enough to maintain life.

The sun provides all our energy except what comes from nuclear and geothermal sources. Wind power, most water power, fossil fuels and fuel from plants or animals are sources of energy derived from solar energy.

The sun operates like a gigantic hydrogen furnace. Deep inside the sun millions of tons of hydrogen are converted to nearly the same amount of helium. A large portion of the hydrogen is converted to energy which eventually emerges on the sun's surface as heat and light.

There is no energy source available for human use that comes even close to the reservoir represented by the sun. The amount of solar energy delivered to the earth annually is 5,000 times the energy of all the fossil sources combined.

Solar energy provides a unique solution for operating a signal bouy.

Energy from within the Earth

magma

geothermal energy

Geothermal energy is a vast reservoir of heat that comes from the molten rock, called **magma.** Magma makes up almost all of the volume of the earth. The earth's surface is a hard crust of rock only a few miles thick. This crust floats on a molten ball of liquid rock. You have probably seen pictures of that rock when it has surfaced as lava, coming from a volcano. We use heat from this geothermal source. The heat comes from large domes of the molten rock that have been pushed within six miles of the earth's surface. Water from the surface of the earth trickles down through cracks in the rocky crust. The water is heated to very high temperatures, as high as 500 degrees Fahrenheit. Because the water is under tremendous pressure, it remains a liquid instead of turning to a gas. As it is driven to the earth's surface, however, it boils and gives off large amounts of steam (water in gas form). Working like a giant percolator without coffee, these geothermal domes continue to provide steam and hot water year after year. Old Faithful, a geyser in Yellowstone National Park, is one example.

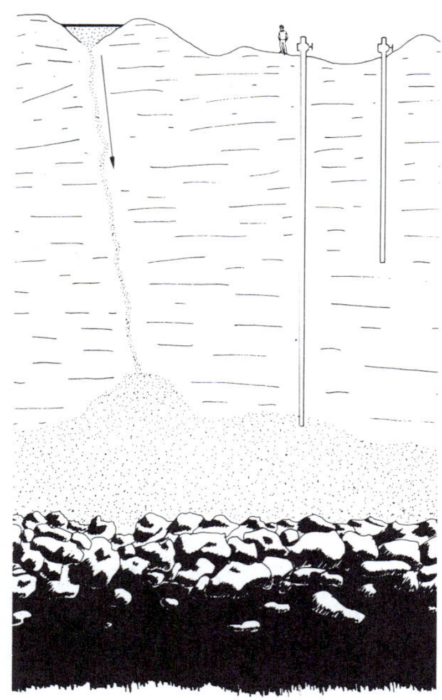

Water is converted to steam by heat from the geothermal energy present in the molten rock beneath the Earth's crust.

There is a huge amount of energy stored in magma. There are relatively few places on Earth that provide convenient geothermal domes from which energy can be used. Geothermal energy now can be used only in areas that were blessed by natural occurrence of magma domes. Someday we may develop techniques, such as a nuclear "mole" to extract usable thermal energy even in places where the magma is more than six miles down.

Energy from Movement of Air

Wind energy is created by differences between zones of high pressure and low pressure in the air. These zones move around. They are caused by the circulation of heat and air on the surface of the Earth. Air from high pressure zones is drawn to low pressure zones. The larger the pressure difference between the high and low zones, the greater is the speed of the winds. Violent storms, such as hurricanes and tornadoes, are created when unusually low pressure zones are generated.

The sun provides the Earth with heat. The air in our atmosphere rises and falls as the air, land and sea are heated. This

Windmills of different designs catch the energy of the wind.

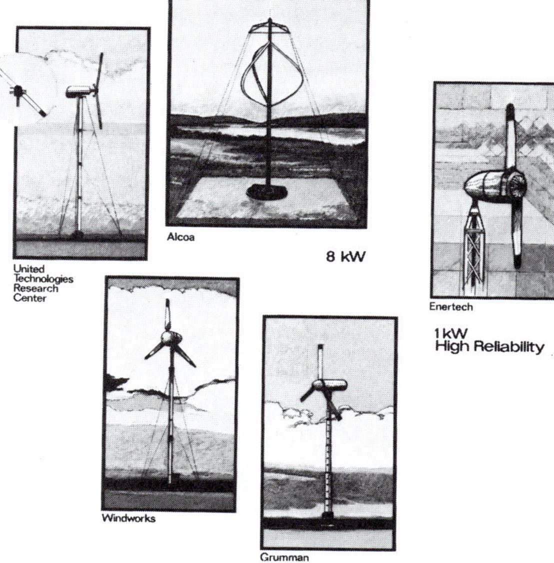

Sails are being added to some modern ships to harness wind power.

rising and falling helps create winds. The rotation of the earth on its axis is another movement that helps create the winds. The rotation of the earth causes the air to swirl in predictable patterns and directions.

The wind has been used for centuries as a source of energy. Sails on ships and windmills capture small amounts of wind power. Many believe that the wind and the windmill provided the energy that started the Industrial Revolution. Use of steam as an energy source came along after the course of change had already begun. Wind is being considered again as a potential source of energy to drive large ships.

Energy from Movement of Water

gravitational energy

Tidal energy is unique. It is one of the few forms of energy that is drawn from sources other than solar. The tides draw their energy from the gravitational forces of the sun and moon. The gravity of the sun and the moon serve to make oceans pulse in long, steady counts.

The intensity of the pulses depends on the position of the sun, moon, and earth. If they all fall in a generally straight line, the tides are extra high and low. At these times the tides are extra powerful. If the moon and sun form an angle with the earth, the water is pulled in more than one direction. The tides are then less extreme and less powerful.

Tidal power takes two general forms. The most common is in the form of waves. These can be seen and heard breaking on coastlines around the world. This movement has continued over millions of years since the seas were first formed.

The second form of tidal power is obtained by the gradual filling and emptying of a bay or estuary. A dam can be built across the opening, to hold water at high tide. At low tide the water can be released. The dropping of the water creates energy which can be used to drive machines. Tidal energy represents a very small fraction of the total power reservoir available for human use. The difficulties of harnessing the tides have made tidal energy low on the list of practical research efforts.

tidal energy

Water power is also available as a result of the Earth's gravitational force, which causes water to fall to a lower level when unhindered. The water energy reservoir is replenished by the natural weather cycle. Surface water evaporates be-

TYPICAL PRESSURIZED WATER REACTOR

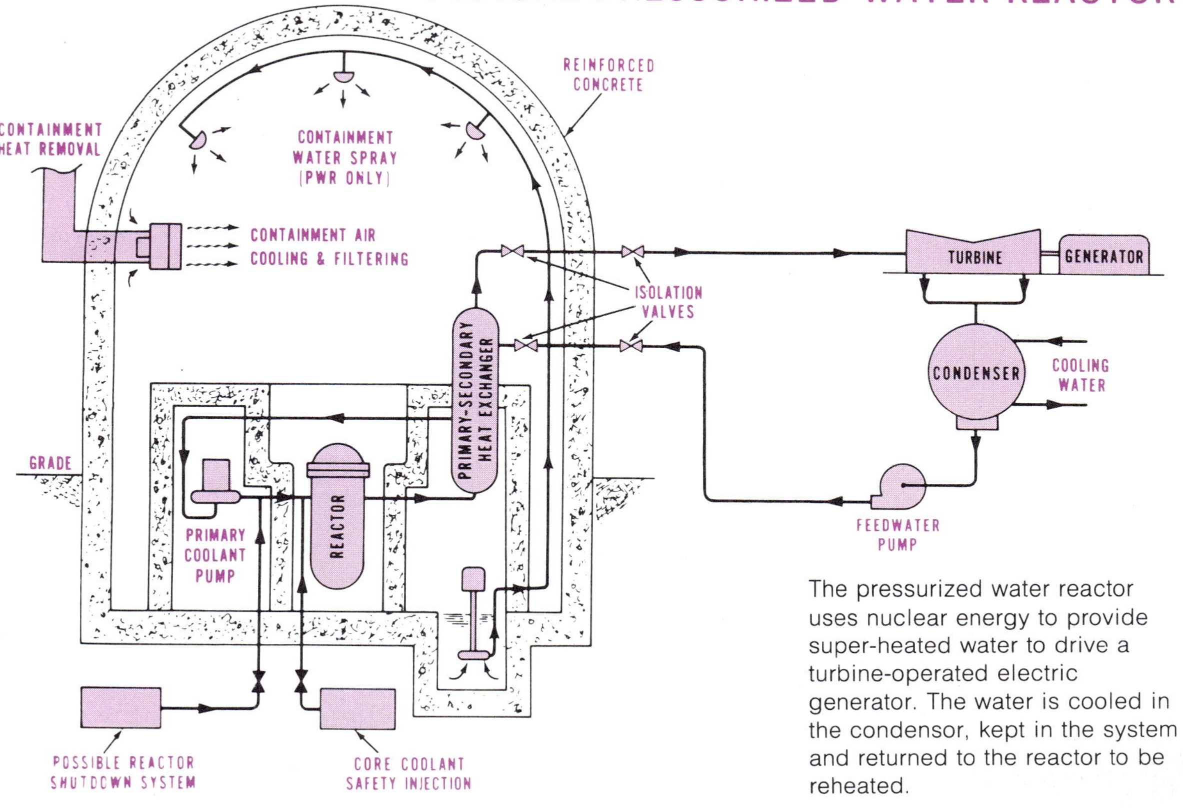

The pressurized water reactor uses nuclear energy to provide super-heated water to drive a turbine-operated electric generator. The water is cooled in the condensor, kept in the system and returned to the reactor to be reheated.

cause of the sun's heat. The water is redeposited in the form of rain, snow or dew. These simple mechanics renew the water energy source and make this a continuous source of energy.

The water stored in a mountain lake or a mill pond represents *potential* energy. This energy has potential for doing work. As the water flows from the lake or mill pond, it falls to a lower level. Falling water has energy from movement similar to the energy of a falling brick. The moving energy of the falling water is called **kinetic energy.** A falling brick has energy to break open the shell of a walnut. The falling water has the energy to turn a water wheel or a turbine, which can drive an electric generator. In this way the kinetic energy can be converted to mechanical energy.

kinetic energy

turbine

The greater the fall of the water, the greater is the force that can be developed. This height is called the **waterhead.** Tall

waterhead

This machine may have been the first known turbine. It operates in a manner very similar to the turbine used in the nuclear reactor above. It was powered by wood or coal, however, and could not recover the steam.

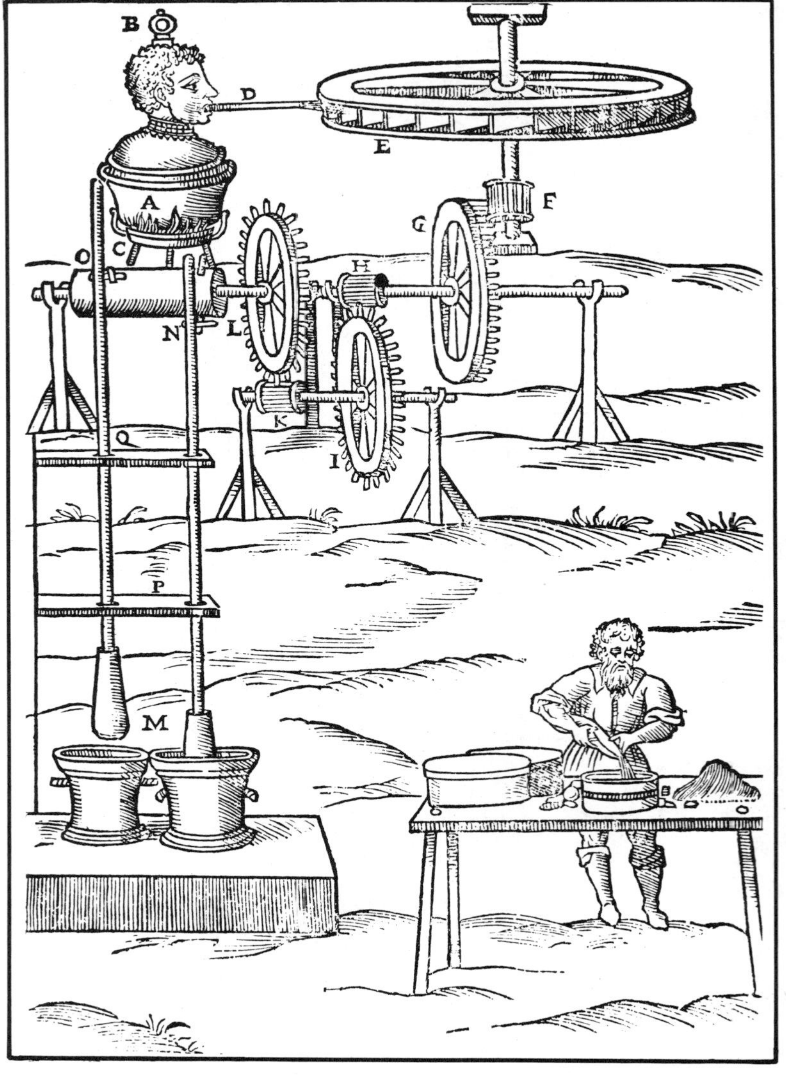

dams allow the waterhead to get very high. The great fall of the water can turn a turbine at great speeds and with great power. A moving river has little waterhead. Even though a river's energy can be harnessed by water wheels, the degree of efficiency is much lower. This does not mean that the potential of moderately moving streams and rivers should be neglected. Applications of these "low-head" sources of water power can be very practical.

Water, even from a low waterhead (the height of the water in the dam), can be used to drive a turbine at high speeds.

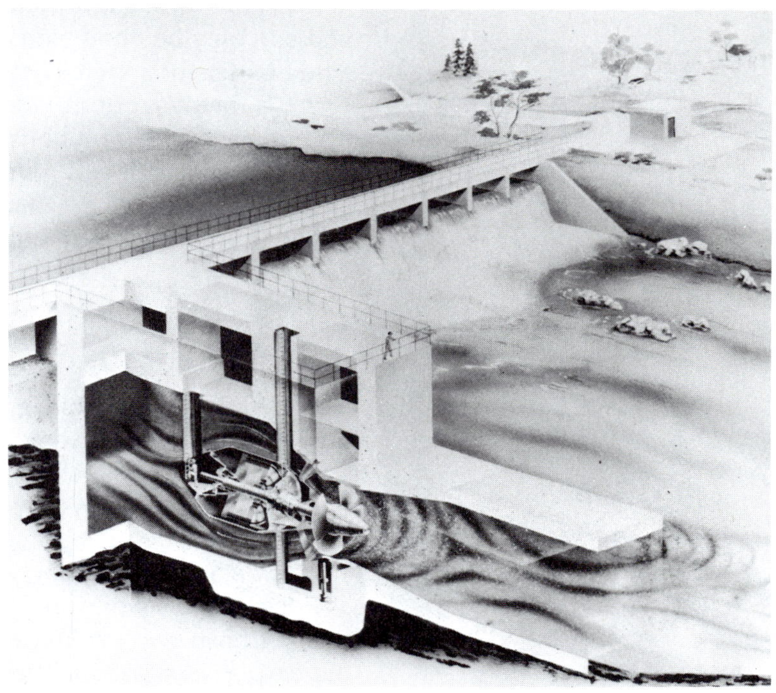

Energy from within the Atom

nucleus
fission
fusion

Nuclear energy is one of our newest forms of harnessed energy, available only since the 1940s. It is energy from the nucleus of the atom. Nuclear energy results from tearing the nucleus of the atom apart **(fission)** or from combining the nucleus of one atom with the nucleus of another **(fusion).** The forces that hold the nucleus of an atom together are very complex and extremely powerful.

Initially, nuclear energy was used for destructive purposes, in atomic bombs. Today, nuclear energy can be controlled and used as a heat source to make steam for generating electricity and for many other uses. There are two sources of nuclear energy.

1. In nuclear **fission,** the nucleus of atoms is split apart. Tremendous amounts of energy are released.

2. In nuclear **fusion,** the nuclei of two different types of hydrogen atoms unite or fuse together. This produces a helium nucleus, a neutron, and a fantastic amount of energy.

The fusion reaction requires extremely high temperatures and can only be done with a special nuclear fuel. This fuel is made by fission reactors called **breeders.**

It is difficult to control the vast amounts of energy required for nuclear fusion reactions. The waste products from fission remain dangerous for many years. These concerns have created many arguments about the use of nuclear power as a major source of energy.

Energy from Fossils

fossil fuels

Fossil fuel is vegetable and animal matter that has been buried for millions of years under a certain set of conditions. Usually, vegetable and animal materials decay—in the presence of oxygen, which is one of the gases in air. The decaying material releases carbon in the form of carbon dioxide gas. This is called **oxidation.**

oxidation

Fossils, however, are formed when decaying material doesn't have enough oxygen to decay entirely. Only part of the carbon is released. The incomplete oxidation and incomplete decay result in high-carbon deposits. These may take the form of peat, coal, oil, shale, petroleum or natural gas.

photosynthesis

The sun begins the sequence of events which create fossil fuel. Green cells in plants capture small amounts of the solar energy that reaches the earth. The plants carry out photosynthesis to convert light energy to chemical energy. In a way this is similar to our bodies digesting food to get energy. Plants live and grow on the energy they get from the sun. Animals live and grow on the energy stored in plants and in other animals.

biological energy

Through the normal life and death cycles, these chemical and biological stores of energy are released at about the same rate as they are produced. Over millions of years, only a tiny portion of the chemical energy in plants and animals has remained unused. This unused chemical energy is stored as fossil fuel.

The process of fossilization is still going on at a very slow rate. Peat is an example of a fossil fuel that is still forming. It would take a million years to produce less than one percent of the fossil fuels already formed and being extracted now.

Yet, because it is so easy to burn these fossil fuels, they have been our major source of energy for the last 200 years.

Coal remains an important fuel. Each barge in this fleet near Morgantown, West Virginia contains about 900 tons of coal and is waiting shipment to steel mills and power utilities.

Based on estimates of our current use, the fossil fuels could be depleted in only a few more centuries. Oil and natural gas will be used up first. Coal reserves may last a century or two longer. In other words, limited amounts of fossil fuel now exist on earth. Because little new fossil fuels are being generated through natural means, it is inevitable that we will someday exhaust the supply of this convenient type of fuel.

Energy from Life Forms

Biological energy comes from living things. Living plants and animals create food, fuel and muscle power. To do this they

Organic materials may be converted into usable fuels of solids, liquids and gases through a biomass process.

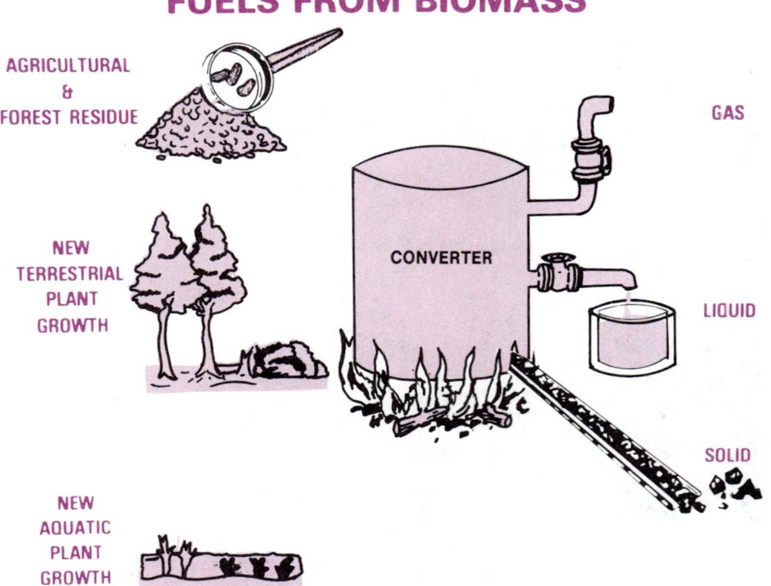

use minerals, organic compounds and in the case of plants, sunlight.

Green plants are the primary food producers. They are the only organisms that can use light energy to make chemical energy. Through photosynthesis, plants produce glucose. They use glucose to live and grow.

glucose

Glucose is the basic energy ingredient of the foods we eat. The energy is in fats, carbohydrates and proteins. When animals consume more glucose than they use, it is stored in the body primarily as fat. The fat is a reservoir of energy.

efficiency

chemical level

There is great variation in the **efficiency** of plants and animals that use and provide biological energy. Of the solar energy that falls on plants, only about one percent is converted into chemical energy. About ten percent of this chemical energy in green plants can be digested and used by animals, including humans. Of that ten percent, only about one quarter is used to power work by our muscles. Most of the remaining energy is dissipated as heat (normal for warm blooded animals). Humans are about this efficient at converting energy that is eaten as animal **protein** (in meat, milk, eggs). Protein is used for building and repairing muscles, as well as for energy.

protein

Biological energy is based on solar energy and chemical energy. These are important steps in the food chain. Use of

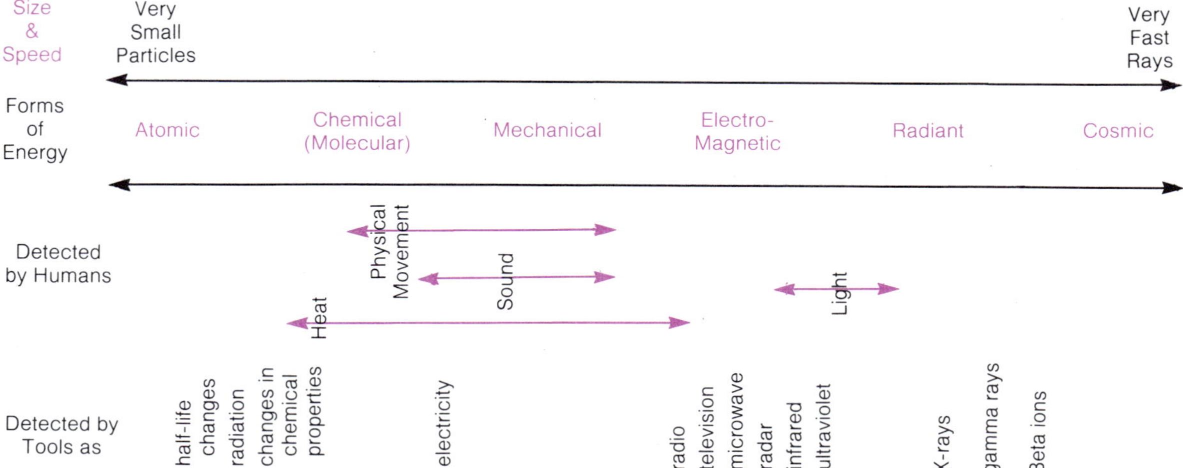

This matter-to-energy spectrum ranges from the very small atomic particles on the left to the very fast cosmic rays on the right. All forms of usable energy fit between these two extremes.

these energies makes muscle power, another form of biological energy.

Less and less biological energy is used in technological events each year. Muscle energy from humans and animals is still the main source of technological energy in many countries. Muscle power in more industrialized countries is used primarily to control other, more powerful sources of energy.

For many years, people burned dried body wastes of animals. This uses some of the biological energy that is eliminated by humans and animals. Recently, researchers have attempted to convert many kinds of waste into a usable source of energy, methane gas.

Forms of Energy

Energy sources are reservoirs from which we draw power to move ourselves, our tools and our machines. The energy sources must be converted into usable forms. There is little practical value in merely knowing the sources of energy. An understanding of the forms of energy may lead to **practical application,** and that is valuable. Three related ideas will help you understand energy forms.

First, all energy is similar. Second, in theory, any energy form can be changed into any other form. And third, energy cannot be made or destroyed. Energy can be only changed, or lost during the change process.

energy spectrum

One way to study the range and relationships of energy is to arrange energy forms into a theoretical spectrum. The energy spectrum provides a "clothesline" on which we can hang the major categories of energy forms. The spectrum ranges from energy as extremely small units of matter (atomic particles), to energy of incredibly fast vibrations (cosmic rays).

Close to the center of the spectrum is the energy activity that we can observe through our senses. The energy particles are large and travel slowly enough to be detected. The gaps occur because our unaided senses (eyes, ears) cannot perceive all energy forms. Energy forms in the middle range (heat, light and sound) have been used for a long time. Use of energy in the outer ranges proceeds as new tools are developed.

The spectrum has six major bands, ranging from matter in one direction to energy in the other direction. The bands are **atomic, chemical, mechanical, electromagnetic, radiant** and **cosmic.** The energy changes from one form to another along the spectrum. At some point the energy may have characteristics of two energy bands.

Atomic Energy

Chapter Three stresses the importance of the bonds within and between atoms and molecules. Remember that the bond

A geiger counter can detect atomic particles and amplify the minute energy charge in order to display it on a screen.

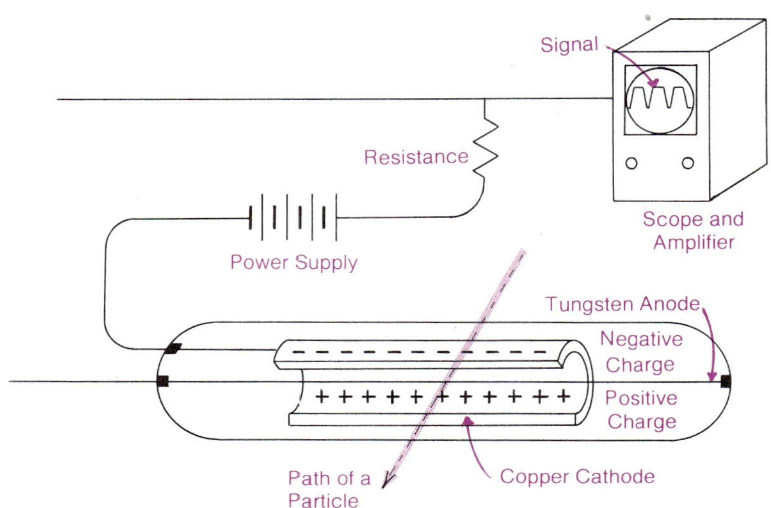

attraction and repulsion forces

is a balance between attraction and repulsion forces. Both the attraction and the repulsion forces are energy. Energy is

atomic energy

released at the **atomic level** when the balance within the nucleus of the atom is changed. This can happen if new bonds between atoms do not contain forces as powerful as previous bonds. During such a bond change, energy is released. The activity of combining and recombining goes on at the atomic level all the time. The process involves tiny particles that require special instruments to detect the activity.

In the process of fission or fusion, the bonds are deliberately changed. Much effort is then required to control the process so that only the desired amounts of energy are released. An uncontrolled reaction would produce energy at all ranges of the energy spectrum.

Chemical Energy

molecular level

At the **molecular level,** the bonds between molecules change. We are still concerned with the energy contained in attraction and repulsion. Chemical processes such as burning, photosynthesis and digestion release energy that is

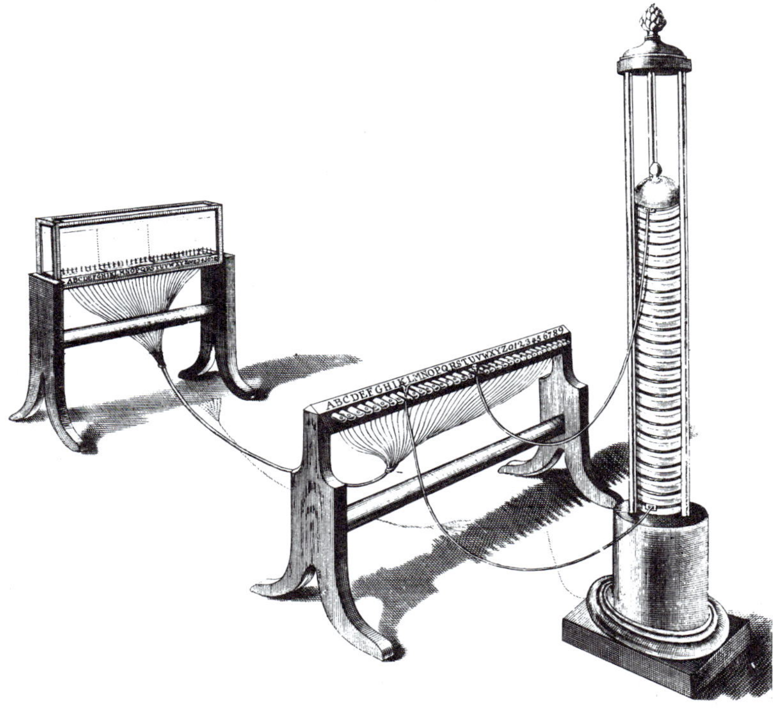

This unsuccessful telegraph used electricity from the Volta pile at the right to cause energy conversion through chemical means. The electricity was sent through selected wires to send a message. Bubbles of gas would form on the wires in the water and spell out the message.

stored in high-energy bonds within the molecule. As new molecules that contain lower-level energy bonds are formed, the excess energy is released. These changes can often be detected as heat.

Mechanical Energy

Our mechanical uses of energy involve **conversion** of energy rather than **release.** The materials are not changed but maintain their molecular structure. Processes are applied to *store* energy or to *control* and *focus* energy. Remember, energy cannot be made or destroyed, only changed.

compression
tension

The forces involved are push or pull actions between the molecules which make up the materials. These forces are called **compression** and **tension.** When you bounce a ball, the energy from your arm throws the ball to the floor. The material of the ball absorbs the impact with the floor and the molecules are *compressed* closer together. The molecules resist the change and attempt to push back to their original state. This causes stress. The stress is released in the bounce of the ball. If you catch the ball before all the energy in the stress is released, you will feel the impact on your hand.

You use *tension* forces when you pull back a rubber band to shoot a paper rocket. You apply tension to the band by pulling it. The molecules will resist the pull and try to return to the resting state. The stress will pull the molecules back together and at the same time propel your rocket toward its target. If your finger gets in the way, you feel the force as the rubber band snaps back.

This moose-killing device is an early example of energy conversion by mechanical means.

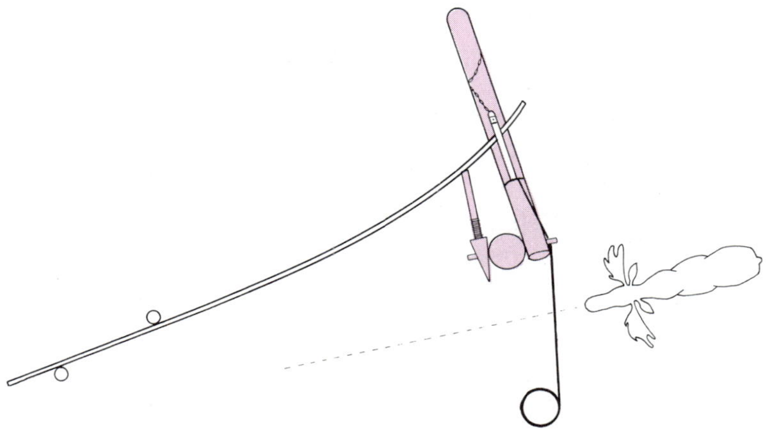

The working of glass requires heating of the materials through energy conversion by thermal means.

mechanical level

At the mechanical level, energy conversion processes can store energy from one source or form for later release. Energy may also be taken from one source or form and focused through the use of tools or machines. An example is the use of a turbine to convert the energy from moving water into electrical power. Another example is using steam produced by boiling water to do work for us.

Many technological events use energy at the mechanical level. We can detect this energy through heat, sound and light. All of these forms involve movement between molecules.

Suppose we first compress and then stretch a material. The change in the strain produces a rhythmic vibration. Compression and stretching is done by the muscles attached to your vocal chords when you speak. In industry, devices such as vibrators and oscillators can make similar rhythmic changes.

oscillator

If we stretch and compress something a little more rapidly, internal strain becomes detectable movement. We are actually creating a **frequency,** motion which is constantly changing with a regular tempo. A tree swaying back and forth in the wind has a frequency, just as a vibrating tuning fork has a back and forth motion and frequency.

frequency

Have you ever strummed a guitar? You set the string in motion with your pick or fingers. The string doesn't move in only one direction, however. It moves back and forth. The

This man is using an ultrasonic testing device that determines if there are any flaws or weak spots in the railroad rails. This device works on an energy conversion by acoustic means.

distances traveled each way are nearly equal. The movement continues to decrease until the guitar string is at rest. This back and forth movement is called **vibration.** The number of vibrations in one second is called **frequency.**

vibration

As we increase the frequency of the vibration, we reach a point on the spectrum in which the energy can be heard as sound. Continued increases in frequency send the pitch of the sound higher and higher. Eventually the frequencies are out of the audible range and into the area of ultrasonics.

Electromagnetic Energy

There is a limit to just how fast we can vibrate an object. This frequency limit is generally determined by the material. If you vibrate a piece of glass fast enough it will shatter. Metal will shatter, too. Even if the material can stand the strain of the increasing frequency of vibration, there is a point at which a given material just can't move fast enough. There is a physical limit to mechanical vibration. Beyond this point is the electromagnetic band of the spectrum.

electromagnetic band

The **electromagnetic band** is very similar to the mechanical energy band. The difference is in the frequency of the vibrations and how these vibrations are generated. In fact, the electromagnetic band overlaps the mechanical band. In the mechanical band we find alternating electric current. At the low end of the electromagnetic band, devices cause electrons to flow first in one direction and then the other. The movement is similar to the back-and-forth tension and compression of the mechanical energy band. In this case, the reversals of the electron current flow are important. Reversal

Large generators supply most of the electricity that we use. This one uses energy conversion by magnetic means.

of the magnetic attraction causes the push and pull, and the movement of the electrons. At this point, energy is no longer physically moving objects. This energy is in particles and waves.

As we increase the frequency of this motion, we change the direction of the flow, slowly at first, then faster. The faster we change the direction, the greater is the frequency.

generators

Most frequencies below 7000 cycles per second are generated by means of electrical or electromechanical techniques (generators). The most common **generator** consists of a coil rotating in a magnetic field. This coil or wire cuts magnetic lines of force and creates an electrical current. Each time the moving wire changes direction, it creates an electrical current. The faster we rotate the coil, the greater the frequency.

electromagnetic level

ultrasonics

Eventually, however, there will come a point at which the coil cannot rotate any faster. This point is the top frequency limit of the electromechanical band of the spectrum. Frequencies above 7000 cycles per second normally must be produced by other devices such as ultrasonic transducers. The most common instance of **ultrasonics** is sonar for underwater detection of submarines and fish. The higher frequencies of ultrasonics overlap the lower radio frequencies.

infra-red

The frequency of energy waves increases through the spectrum of radio waves, television waves and microwaves. After these we move into yet higher frequencies that are infrared energy. This is the beginning of the spectrum of visible

A prototype of the first television camera used energy conversion by means of light.

The microwave oven uses a special vacuum tube to cook food through the use of energy conversion by radiant means.

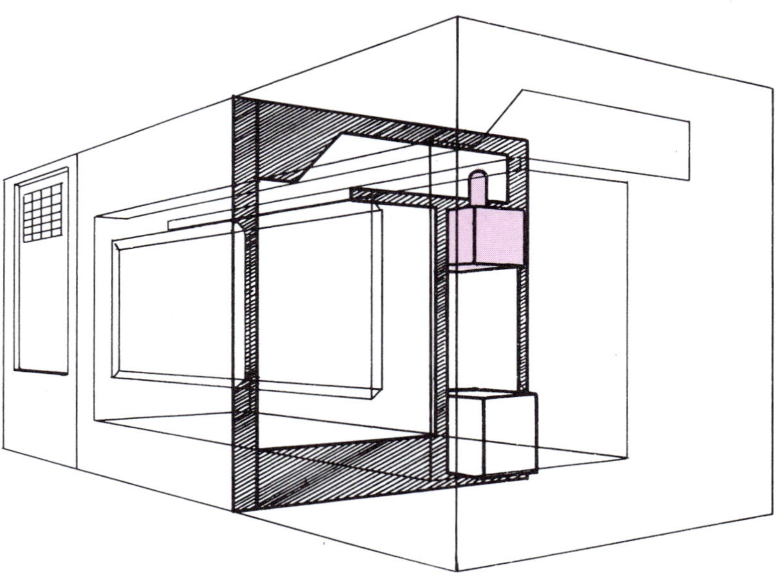

light. The spectrum changes in color as we continue to increase frequency. The light spectrum goes through the reds, yellows, blues and into the ultraviolet.

cosmic energy
radiant band
x-rays

Radiant and Cosmic Energies

Now we reach the energy level of the **radiant band.** X rays and gamma rays are in the radiant band. We have only a few tools that use or generate energy at this extreme level of frequency. Energy at the extreme end of the known spectrum is called cosmic energy. Just as we know very little about energy at the extremely small level of matter (subatomic), we also know very little about generating or using energy at the extremely fast level of rays (cosmic).

Conversion of Energy

entropy

Earlier in this chapter we mentioned that all forms of energy are similar. In theory, any energy form can be converted to any other form. Each time the energy is changed from one form to another some energy escapes. This loss of energy is called **entropy.** Much of the heat from an engine is not put to practical use and is considered lost. If a device converts one energy form to another with a small energy loss, the process

ENERGY CONVERSION

	CHEMICAL	HEAT	MECHANICAL	SOUND	ELECTRIC	LIGHT
CHEMICAL	foods plants	fire food steam boiler furnace	rocket gas engine animal muscle	explosion smoke detector	battery fuel cell	candle oil lamp phosphor-escence
HEAT	gasification vaporization distillation	heatpump heat exchanger	turbine diesel gas & steam engines	explosion flame tube	thermopile thermocouple	fire arc lamp light bulb
MECHANICAL	gunpowder	friction brake	flywheel pendulum waterwheel	voice musical instrument	generator alternator steam power plant	flint spark
SOUND	hearing	sound absorber	ultrasonic cleaner	megaphone	telephone microphone hearing	light amplifier color organ
ELECTRICAL	batteries electrolysis electro-plating	toaster heatlamp sparkplug	electric motor relay	horn thunder loudspeaker	diode rectifier transformer	television fluorescent light light emitting diode
LIGHT	plant photo-synthesis camera film	laser heatlamp	photoelectric door opener electroscope	movie sound track video disk	solar cell photoelectric cell	laser light wave repeater

This Energy Conversion Chart shows some examples of how energy can be changed. Energy is usually changed into different forms such as electricity to heat. Energy may be changed from one form of itself to another as in the transforming of electricity from a high voltage to a lower voltage. Energy may also be changed to allow us to measure the amount of an energy form such as determining the temperature of a room.

is considered efficient. A larger loss of energy is considered less efficient. Energy conversion often will go through several steps, losing useable energy at each step.

Energy is generally converted from one form to another through the use of two types of devices, converters and transducers.

converters
transformers

A **converter** uses one energy form to produce a new form or to change the nature of the original energy form. Converters include people, plants, generators, engines, motors and transformers. Any machine, animal or plant can be considered an energy converter. In converters, energy is the input *and* the major output of the device. A **transformer,** for example, is a converter that changes electrical energy and steps the power up or down.

transducers
thermocouple

Transducers change energy into information. Examples include thermocouples, proximity switches and pressure switches. The output of a transducer must be usable for communications, control or both. Transducers cause an action to take place. For communications, the information must be changed into useful forms.

A thermometer is useful only if we can understand the output. A standard thermometer does not cause an action, so it does not qualify as a transducer. But the thermostat on a furnace is a heat sensor that sends a signal. The signal causes the furnace to shut down when an appropriate heat level is reached. The thermostat is a transducer.

Converters and transducers also are helpful as categories to show how energy is related to other elements of technology. Converters use energy to convert materials and energy into new forms. All machines, then, can be described as converters because all machines require some form of energy to operate.

Transducers, in comparison, change energy into information. This information can be used to control machines by sending signals to them. A radio alarm can send a signal (a small bit of information) to the switch of a coffee pot to turn it on. The heat sensor of the coffee pot, when it becomes too hot, will send a signal to turn off the coffee pot.

Transducers will be important in the following chapter on Information. We will also include them in later discussions on automation and cybernetics. Transducers are important in helping machines communicate with humans. They are also used when machines communicate with other machines.

Transducers are unique energy conversion devices that change samples of energy into information signals.

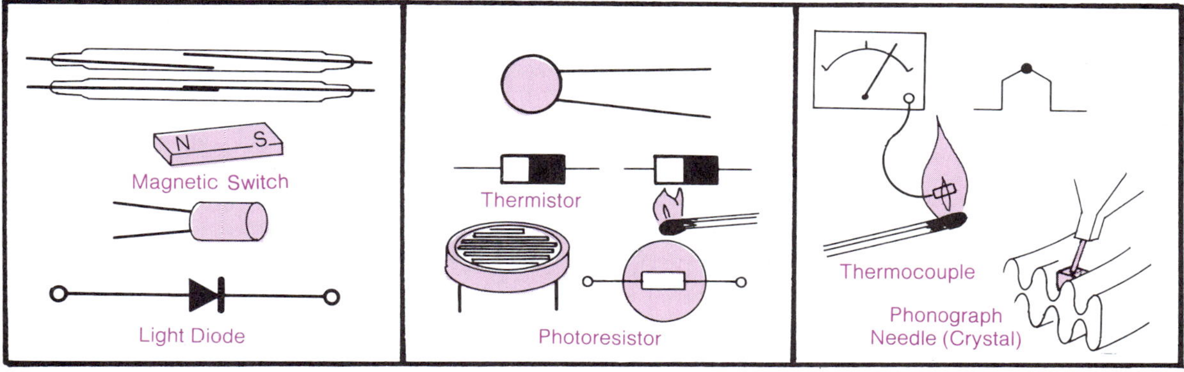

Summary

This chapter focuses on the **primary sources** of energy: solar, geothermal, wind, water, atomic, fossil and biological. These primary sources are related to a general energy spectrum with bands of energy which include: atomic, molecular, chemical, mechanical, electro-magnetic, radiant and cosmic. Two ideas are essential to understand the energy spectrum: **bonds** and **frequency.** Energy is captured in the bonds between and within elements and compounds. These are released by atomic, molecular and chemical processes. Energy is also found in the frequency of movement from very slow mechanical forms of energy to the very rapid cosmic forms. At extremely high frequencies, the energy is referred to as rays. Examples are X-rays and cosmic rays.

The usefulness of energy depends in part on how efficiently it can be converted from one form to another. Currently, we are limited in our knowledge of atomic and cosmic energy. We know very little about how to convert them to other energy forms. Most of the energy conversion processes we use are applied to the energy band within the middle of the energy spectrum. Converters are the devices that are used to change energy from one form to another. Transducers are unique devices used to change energy into information in the form of signals. Signals are necessary for the guidance and control of machines.

Key Concepts and Terms

atomic energy
attraction and repulsion forces
biological energy
bond energy
chemical level
compression
converters
cosmic energy
efficiency
electromagnetic bond
electromagnetic level
energy conversion
energy spectrum
entropy
fission
fossil fuels
frequency
fusion
generators
geothermal energy
glucose
gravitational energy
infra-red
kinetic energy
magma
mechanical level
missing
molecular level
nucleus
oscillator
oxidation
photoelectric cell
photosynthesis
prime energy sources

protein
radiant band
solar energy
tension
thermal energy
thermocouple
tidal energy
transducers
turbine
ultrasonics
vibration
waterhead
x-rays

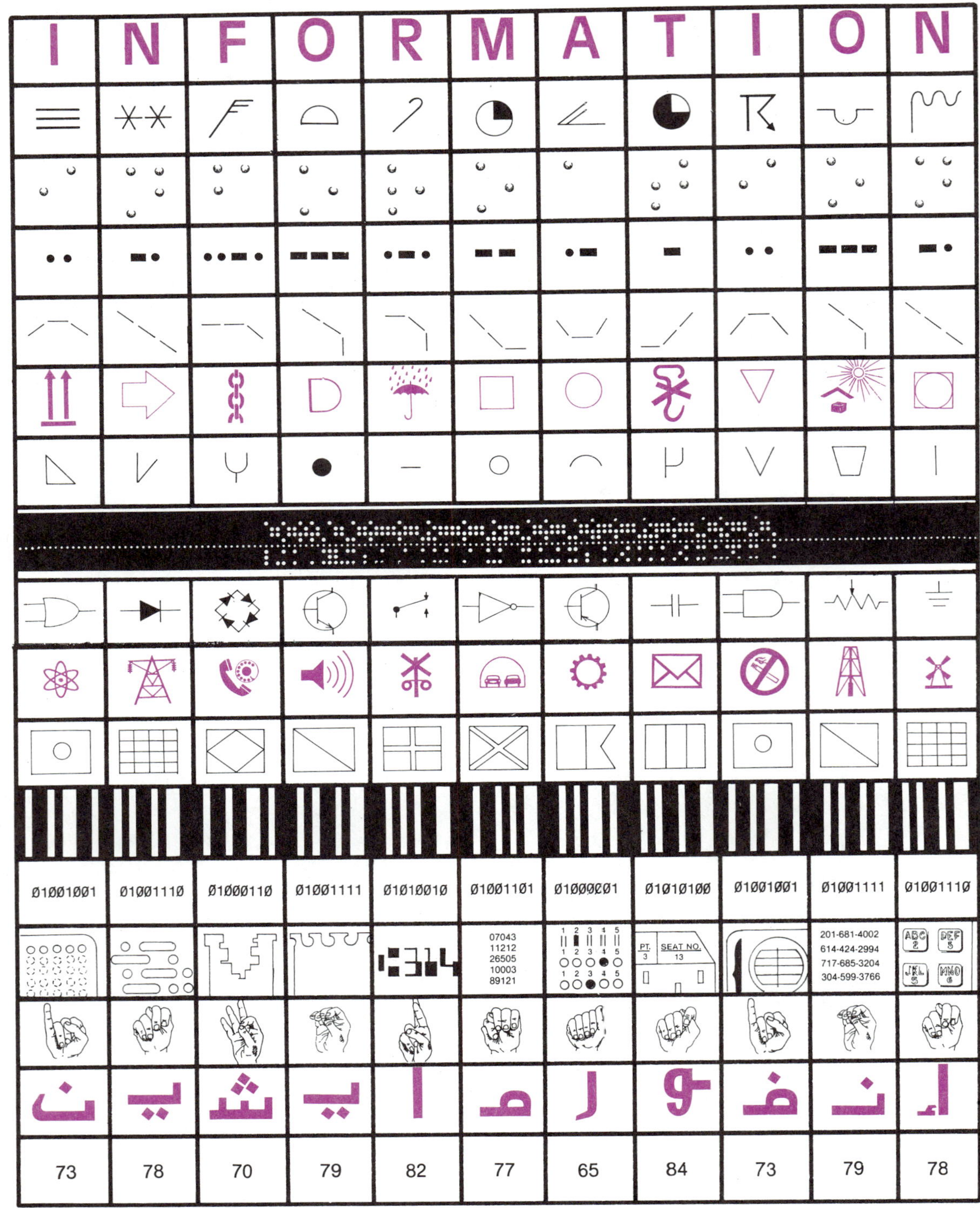

Symbols and signals of information provide codes by which humans and machines can interact. How many of these codes can you identify?

Chapter 6 Information

Humans and machines require information to do work. The study of information as a specific area is very new, less than fifty years old. It is important to understand what information is, what it does and how it is involved in technology.

A simple example is a common light switch that can turn lights on or off. When the switch is turned to *on,* the electrical circuit is complete. Electricity flows, and acts as information "telling" the bulb to emit light. A movement to the *off* position breaks the circuit. The switch sends the information for the bulb to become dark. On and off, light and dark, is all the information that the switch can provide. The code of two conditions, *off* and *on,* are similar to the conditions of other devices. Each of these examples really represents a *yes* or *no* condition that can be shown by the binary symbols of 0 and 1. The "0" symbol indicates that something is missing (off), while the symbol "1" indicates a condition where something is present (on).

binary

Let us consider another simple example of using information. Imagine that you are shipwrecked on a rocky deserted island. After you have been on your island for a long time you begin to forget the exact number of days you have been stranded. You decide, therefore, to keep track of the days by building a pile of stones. You place a new stone in the pile for each new day. Each stone is a symbol that represents a day, and the pile provides the information. Any time you want to know how long you have been there, you merely count the stones. You have built an information device.

information device

Many different devices can be used to generate and store information in both digital and analog forms.

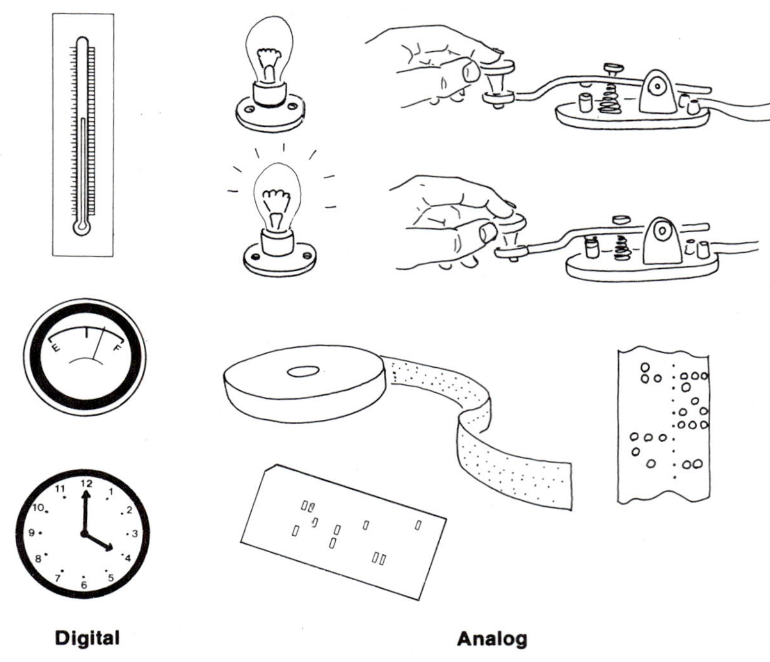

Digital **Analog**

Signals and Symbols

An understanding of signals and symbols helps to understand information, as well as communications and control. **Signals,** whether they are signs or impulses, are intended to *cause action.* When we flip a light switch, we send a signal to the light. Usually, the word *signal* is used when information is sent from a person to a machine or from one machine to another.

communication signals

symbols

Symbols, on the other hand, are signs that have inferred meaning. Symbols represent something. For example, some light switches have two buttons, one white and one black. The white button is a symbol representing a lighted room; the black button represents a darkened room.

The numbers in an elevator are symbols representing different floors. The number "3," obviously, is not the third floor itself, it only represents the third floor. It is a symbol. We interpret what the "3" means when we see it on a display of numbers in an elevator.

If you saw a "3" in the address of a building you would assume that the "3" indicated something other than the third

The most common form of information is digital. In this memory device, each stone serves as a digit and represents a day.

floor. A "3" as a symbol means something quite different on a street sign, a birthday cake or a three dollar bill. Symbols have meaning only when people bring that meaning from their previous learning. Examples of different symbols and signals are shown in the introductory illustration of this chapter.

Signals do not need to be complicated in order to cause machines to take some action. Suppose you are asked to count how many people pass through a doorway. That would be a traffic information problem. You could just sit and count everyone that passes by. But, an easier method would be to place a pad in the doorway with a switch built into it. The pad has to be big enough so that no one could step over it, and would allow only one person at a time to pass through the doorway. The switch is turned on by the weight of each person. Each time someone steps on the pad a small amount of electricity (the signal) is sent to a counter. This counter keeps track of the number of times that the switch is triggered. The pressure generated by the weight of the person is a form of mechanical energy. The switch converts part of that energy into a signal to the counter. Not all problems are as simple as this.

relay
photoelectric cell
digital signals

Suppose cars are being towed by a cable through an automatic car wash. We want to save money and water by designing the washing section of the machine to turn on only when a car is passing under the spray nozzles. The setup of this operation is fairly simple. A light source is directed toward a photoelectric cell. The photoelectric cell is attached to a sensitive relay or a digital device, both of which act as an electrical switch. When the light is turned on and shines on the photocell, the water turns off.

Why will this work for us? If the light can reach the photocell, there is no car to block the beam of light. When a car does move down the line, we want the photocell to sense the change (from the *light on* to the *light off*). If the light is blocked, the photocell will stop producing its small amount of electricity. This causes the sensitive relay to switch to *off* and the water pump to switch to *on.* The water is pumped into the spray nozzles as long as the car blocks the beam of light. Once the car has moved past the light, the photocell again sends electrical current to the relay. The relay turns *on,* which turns the pump *off.* The water stops until the next car comes through the car wash line. We have saved water by using a simple information system.

Automatic machines can use simple switches as the information devices necessary for controlling processes such as washing cars

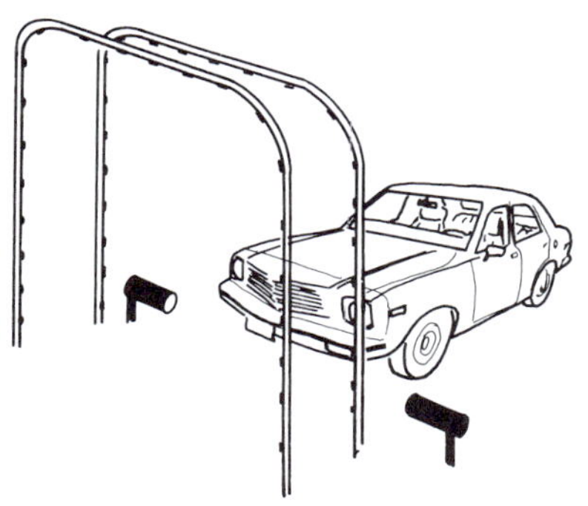

In some instances, it may be important to record the information that has been sent. In our first example, we were interested only in the total count of people who passed through the door. In the second example, we did not need to record the signals to the water pump. But sometime we may want to save the signals that a machine transmits.

telegraph

Morse code

A century or so ago, the major way of sending messages was by telegraph. A telegraph operator had to both send and receive messages in Morse code. It was not too difficult to send a message. The operator had it right there to read. But it was not as easy to receive a message. The operator had to listen to the code and write down the message. If the operator missed a word, it was necessary to ask the sender to send it again. A simple recording machine helped solve the problem. It made it unnecessary for the receiving operator to decode the message while it was coming in. The recording machine made a mark on the paper tape as it was drawn past a marking pen. Every time a signal was sent, an electromagnet drew the arm forward. The pen made a mark each time a

Samuel Morse, in addition to developing the telegraph and Morse code, invented this recording instrument. Electrical signals sent by wire cause an electromagnet to move a pendulum with a pencil attached to it. This pencil makes marks on a strip of moving paper. The marks on the paper serve as a record of the telegraph message.

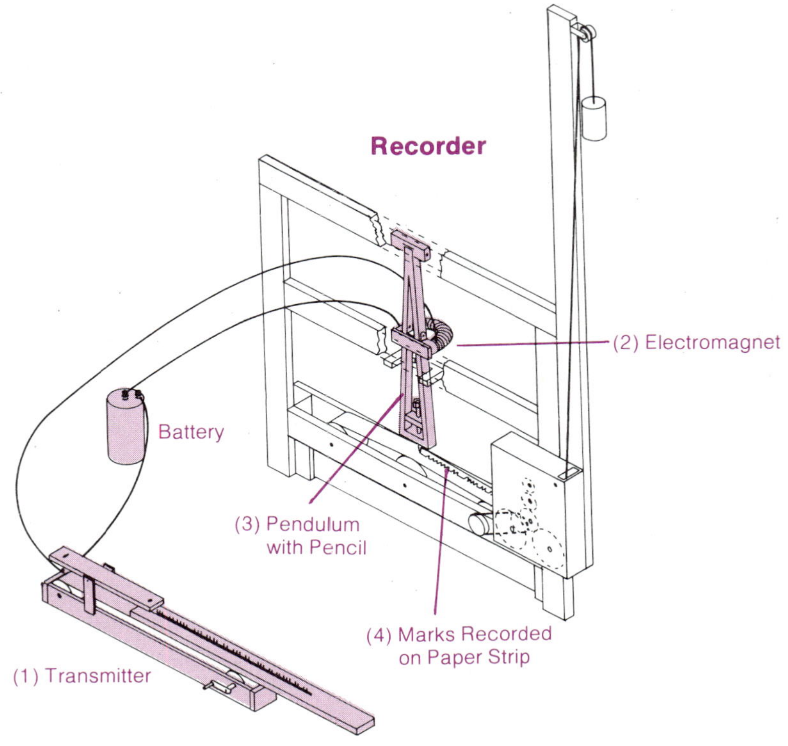

signal was transmitted. The record of the message was a long tape of these marks. The telegraph operator could decode during the transmission or later.

transmission

Information and Control

codes

guidance and control

Codes are used for communication and control. Machines are controlled through specific coded messages sent to the guidance and control subsystem of a machine. The guidance and control subsystem makes a machine operate as it should.

Every time we use a machine, we send it information in some form of prearranged code. The old-time player piano uses a roll of paper with rows of holes punched in it to store information (musical notes). Each row on the paper is for a specific key on the piano. A small vacuum develops behind the roll of paper. This vacuum comes from the bellows pumped by the operator's feet. When a hole for a given note passes in front of the vacuum, the outside air pressure trips the hammer for that note. The hammer then causes the note to be played. The player piano is an early example of a **"read only memory" (ROM) device** now quite common in com-

read only memory (ROM) device

The player piano served as a model for later more sophisticated tape-controlled machines. The information for the notes was recorded as holes in the paper roll.

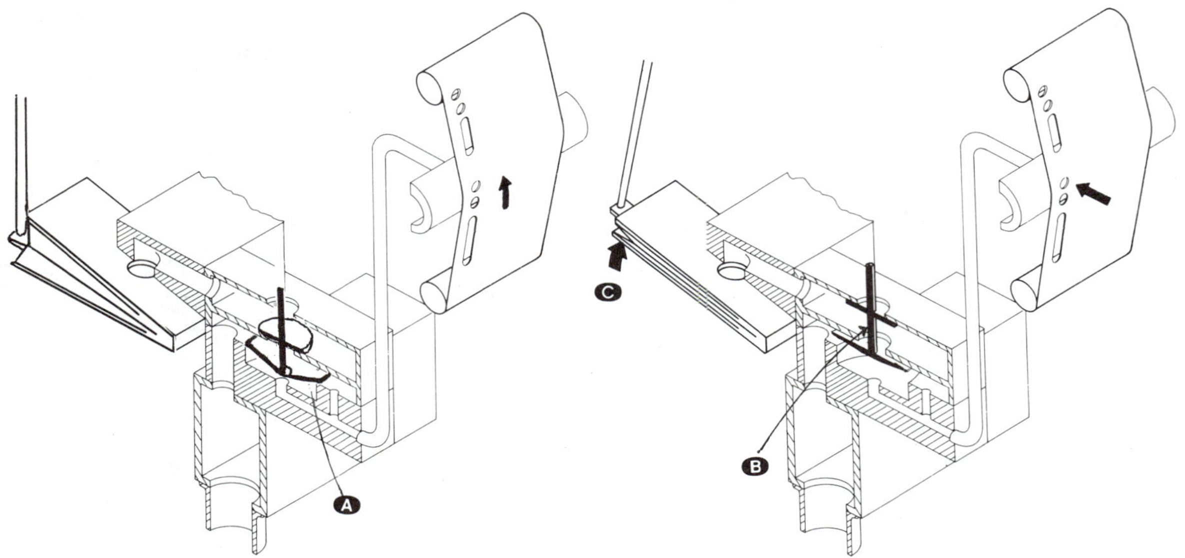

As a hole in the paper roll passes over the sensing hole, the vacuum collapses and causes the bellows to collapse and move the rod to hit the right note.

For the 1890 census, Herman Hollerith used some of the ideas from the player piano to design a device for punching information onto cards (A). He used other devices for reading the cards (B) and recording the information (C).

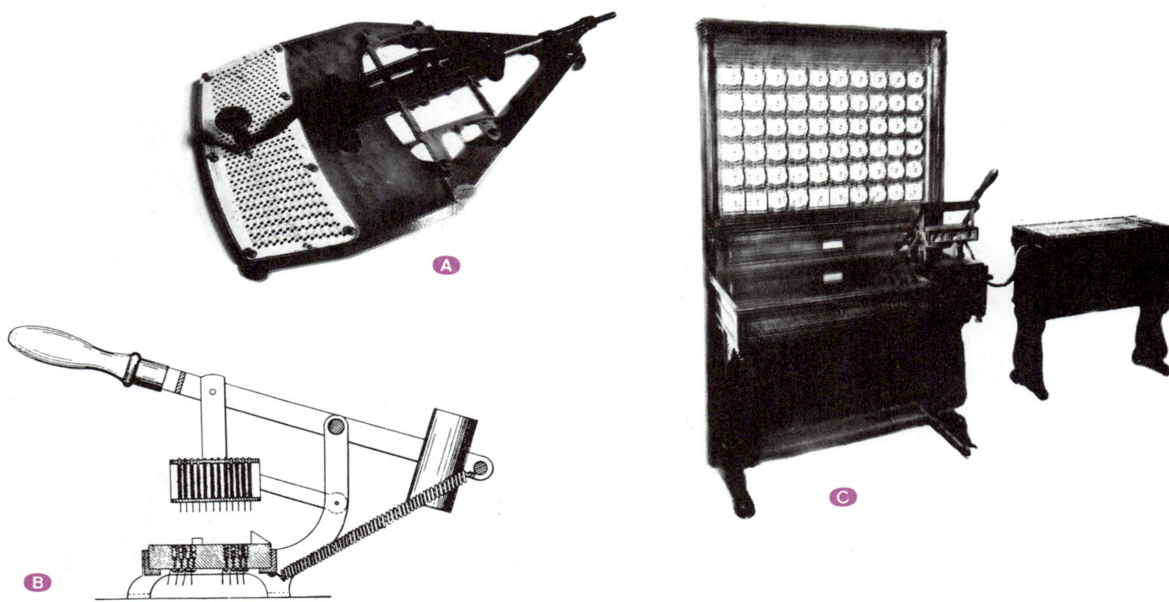

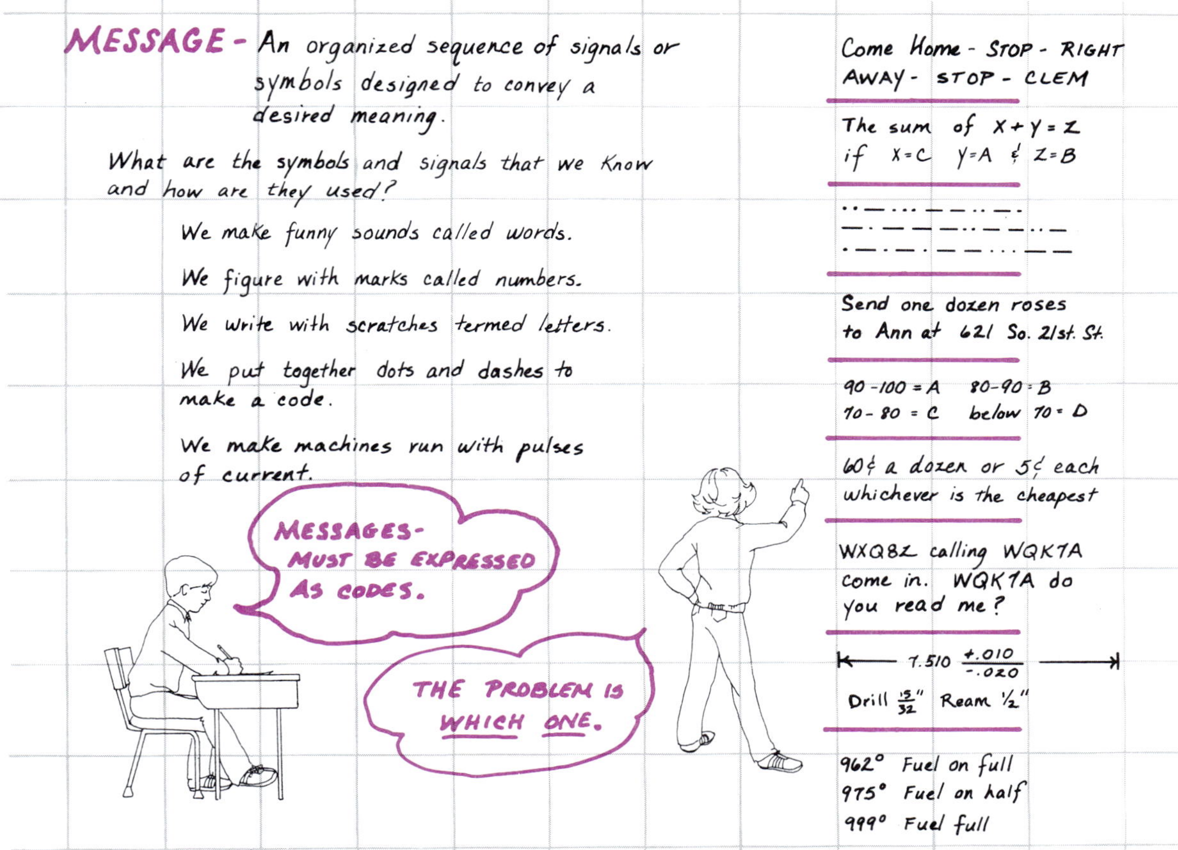

Examples of signals and symbols that we can use as codes to send messages.

puters. The piano reads only what is already coded on the punched roll of paper. The Jacquard weaving loom, the first machine to use punched tape, is another example of a ROM device.

message
receiver

A **message** is the information transmitted from the source to the receiver. The message consists of an organized sequence of symbols or signals. These are designed to convey a meaning. Meaningful messages are possible only if the symbols or signals are from an accepted code understood by the receiver.

Codes, Signals and Symbols Simplified

automatic control
numerically controlled (NC) machines

Machines that are controlled automatically through the use of punched or magnetic tape are called **numerically-controlled**

(N.C.) machines. These machines work much as does the player piano described earlier. Although most NC machines use electrical contacts to sense the holes in the tape, some machines use air, photo sensors or magnetic tape such as that used in a tape recorder.

subsystems

sensing energy

All N.C. machines work in the same general way. They must send a signal to the guidance and control subsystem of a machine each time a hole (or its counterpart) passes one of the contacts or sensors.

transducers

A very important device used to provide information for the guidance and control of machines is the **transducer.** Transducers take a small sample of energy and change that energy into a new form. The changed energy is information (a signal) for communicating with or controlling a machine. For example, we use a thermocouple as a transducer to control an oven.

As the oven gets hot, the thermocouple takes a small amount of that heat and converts it into electricity. When the heat in the oven becomes high enough to produce enough electricity, a switch controlling the heater is turned off. When the heat level drops, the contact is broken and the switch turns the oven on again. This cycle of *on* and *off* continues until a timer or a person turns off the oven's main switch.

Many machines are controlled through the use of transducers. Even the most sophisticated machines use transducers. They may use several transducers and integrate the signals through a computer or microprocessor.

The transducer is the major link between energy and information. If a signal is generated by sampling some form of

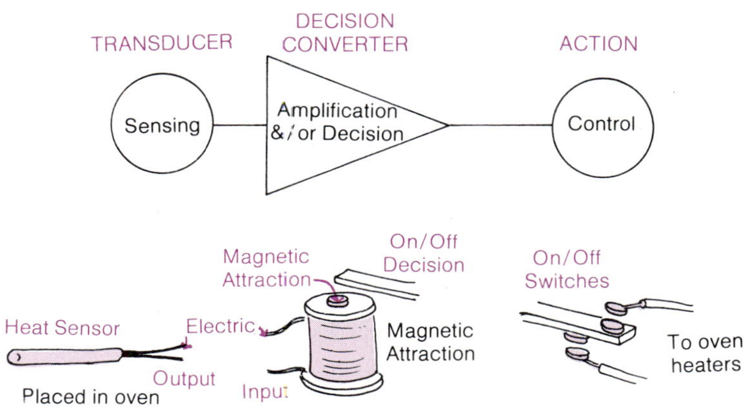

This simplified example of an oven's control system would work if the thermocoupler were very efficient. Only a small amount of electrical current is sent to the relay that controls the switch to the oven heaters.

Since most information devices must work on small samples of energy, they require transducers and amplifiers. The transducers send the signals to amplification/ decision units. These units control the desired action. In this example, a heat sensor sends an electrical signal to a sensitive coil, which serves as the one/off control for the oven.

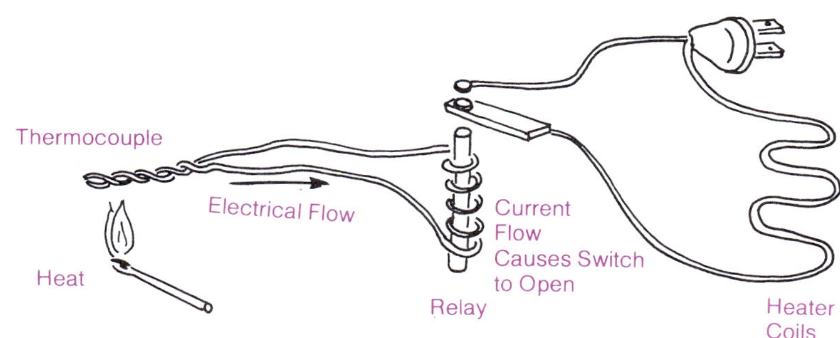

Information devices become much more usable if they can be programmed to perform a range of tasks. Even a toaster must be programmed to make light, medium and dark toast. In this way the toaster operates as a small primitive computer. When dark toast is wanted, more time is needed to move the contact arm (B). The opposite is true for light toast (A).

energy, whether heat, light, sound or movement, the sampling and the generating of that signal are done by a transducer. This is especially important in machines that are designed to operate automatically. Machines such as a furnace cannot be put on automatic control without transducers. Thermocouples, pressure switches and the like require information in the form of signals. The devices then send that signal to the guidance and control subsystem of the machine. Such a signal can be used to control a machine (turn it off, speed it up or make it hotter). The machine then becomes somewhat self-directed or automatic.

DARK TOAST SETTING

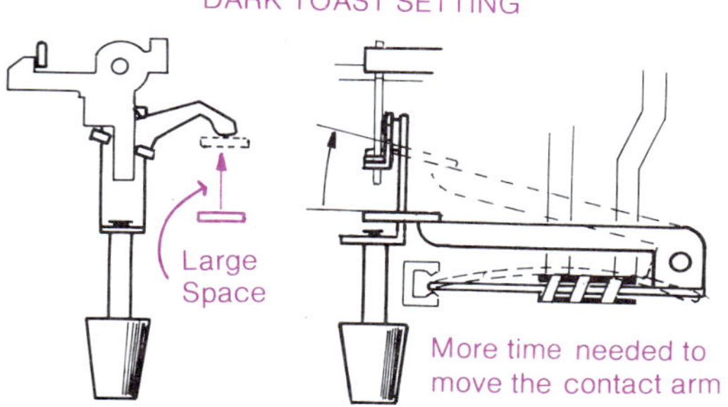

LIGHT TOAST SETTING

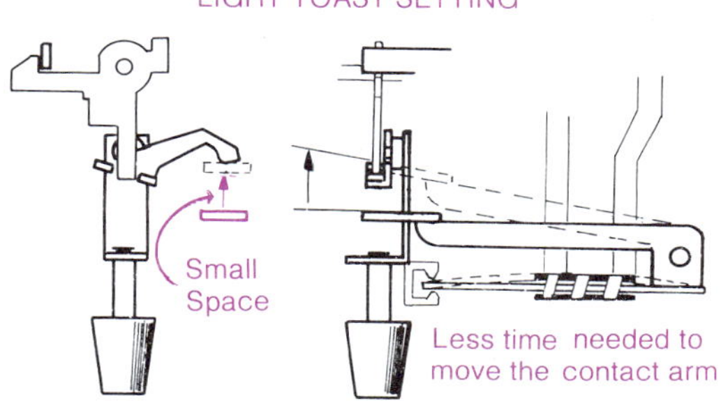

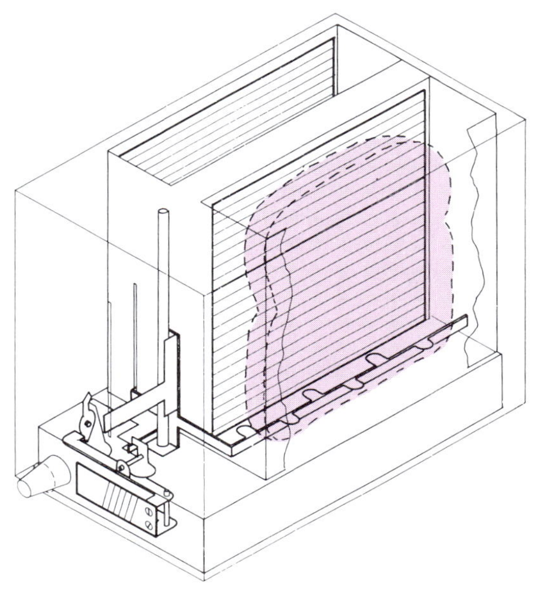

A tape of a reel to reel recorder can be punched to provide a numerical control (NC) program. The programmed tape provides the necessary signals for the pump to send water to the plant.

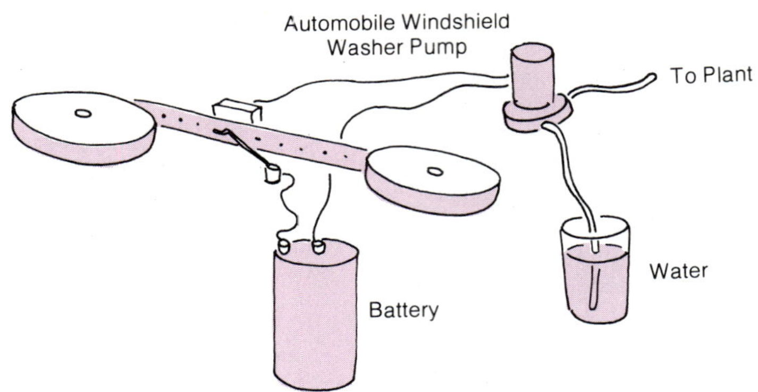

As indicated earlier, codes allow people to make sense out of signals. But the codes must be prearranged and cannot vary from the original form. For example, a toaster makes our breakfast toast light, medium or dark, depending on how we set the selector switch. The setting on the selector switch is the code that we can "program" into the machine. Selecting the code of "medium" mechanically transmits a signal to the heating element in the toaster to stay on a little longer than it would for "light." If, however, you use a piece of already toasted bread or a thicker slice of bread, your toast may not look "medium" when done. The code for the toaster controls only the time during which the heating element stays on. The program that is established when you set the toaster is quite similar to a program used with a computer. The difference is in the tremendous complexity of the computer program.

It is not very difficult to understand the use of information in a numerically controlled machine. A simplified example can involve an old-fashioned reel-to-reel tape recorder.

Suppose we plan a trip of some length. Because we will be away for some time, we are concerned that our houseplants might die from lack of water. To build our numerically controlled machine we will punch holes in the tape of the recorder. The holes will send a signal to a small switch. The signal is generated when the switch makes contact through the holes in the tape.

The contact will allow a battery to send electrical current to a windshield washer pump that we have taken from a car. As the tape recorder runs, the wiper switch is held away from the metal contact plate until a hole in the tape passes by. When the hole is sensed, the wiper switch touches the contact plate.

This sends a signal to the windshield washer pump. In turn, the pump gives the plant a small amount of water from the reservoir. Many holes punched together would cause the pump to operate for a longer period of time. A length of tape with no holes gives the plant time to dry out.

Although our machine would work, the tape might run out before we are back from our trip. We can prevent this by making a continuous loop of tape.

computers

All numerically controlled machines and computers, whether they are operated by punched tape or cards or by magnetic tape, are similar to our homemade machine. The machine can sense only "hole" or "no-hole" (0 or 1). We can make that mean whatever we want it to: yes/no, on/off, go/stop. Our machine has only one channel. Consequently it is restricted to one function—turning the pump on and off. If we increase our channels (rows) of holes, we can control more functions.

Standard numerically controlled machines use tape with eight rows of holes. One row is used to break up the information into blocks. The remaining seven are used to control the machine. At first glance, it appears that seven rows would control seven functions. But, because there are two states of the tape (a hole or no hole) and seven rows, you can combine them in 128 different ways. To calculate the variations, multiply two (the two states of the tape) by itself seven times, the number of rows in which a hole or no-hole could be found. The possibilities are 2 to the power of 7 or $2^7 = 128$ ($2 \times 2 \times 2 \times 2 \times 2 \times 2 \times 2 = 128$).

magnetic tape

Punched paper tape is still in use but magnetic tapes and magnetic discs are much more common for storing the digital

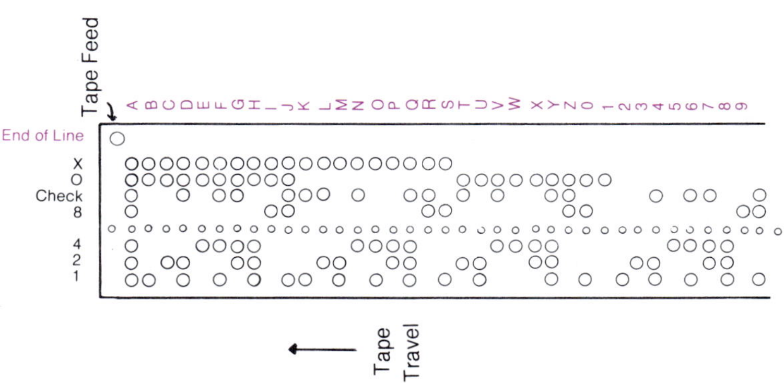

Numerical control (NC) tapes use several channels or rows of holes for storing information. The most common is the eight channel tape that provides four channels for information and four channels to move and control the tape movement and to check the information.

signals. Magnetic media require different devices for storing and retreiving the signals than are needed for punching and reading holes in the paper tape. Each magnetic blip is equivalent to a hole. Each space that is not magnetized is equivalent to a no-hole in the paper.

The major method of capturing digital information on tape or disc is by typing or, more accurately, "keyboarding." Home computers have a keyboard that is used for information input from humans. On the keyboard are letters of the alphabet and numbers (symbols) that you can understand and use. As the keys are tapped the computer converts these symbols into electronic signals it can understand and use.

A variety of devices and codes are available that make information input easier and faster. These include the punched cards used years ago in the control of weaving machines, the cards used in the 1898 national census, the cards

The keyboard is one of the most used methods for capturing information. These examples include one of the first typewriters, an early keyboard for punching words in braille, a unit for typing lettering on engineering drawings and a modern split-keyboard that is designed to fit the human hands.

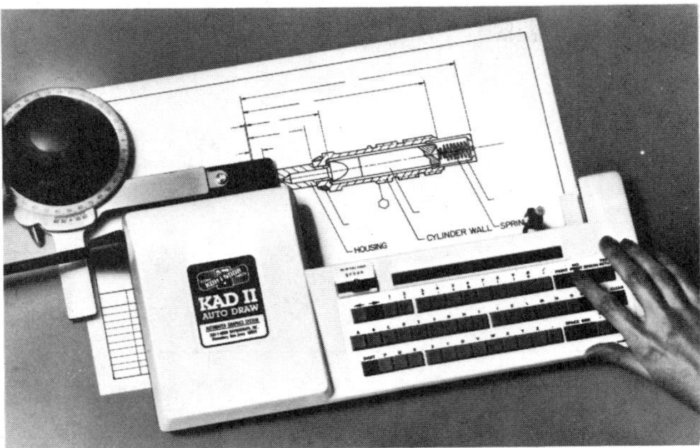

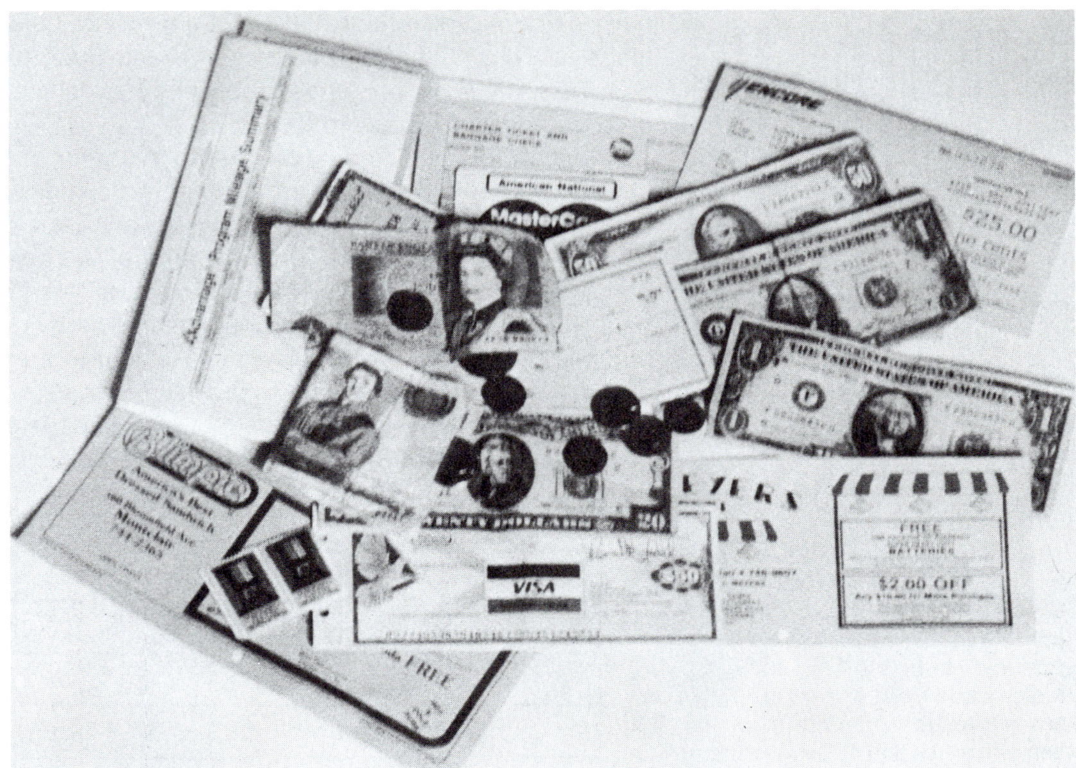

Money takes many different forms and is either symbolic or signal information. The material of the money itself is usually not valuable. The value is in the resource that the money represents. Many financial transactions are conducted by transmitting electronic blips or signals from one banking account to another. The electronic signals, like money, are valuable only for what they represent.

that standard keypunch machines still use and the zebra-stripe or bar codes. New input devices continue to appear on the market such as: joy sticks, mice, lightpens, pointers, rolling balls, digital tablets and touch sensitive screens on the video or cathode-ray-tube (CRT) monitor. There are some computer systems now available which recognize spoken words, but they have a limited vocabulary. Some programs will even recognize each person's "voiceprint."

Summary

Information exists as patterns of material and energy in the forms of signals and symbols. In the form of signals, information is used to cause some intended action. Information in the form of symbols is used to infer meaning.

Information is the basic building block from which communications and control are constructed. Both communica-

Bar codes are being used more often for the storing of information. The most common bar code is the Universal Product Code you see on food packages.

tions and control require that some sharing of meaning take place as people and/or machines interact. Such meaning is possible only when an understood code specifies what the units of information are to be.

The technology elements discussed in earlier chapters are closely interrelated with information. Energy and information are related when small samples of energy are used in generating information as signals. The sampling of the energy is accomplished by small devices called transducers. These signals are important connections between energy, processes and machines. Information as signals is sent to the guidance and control system of machines. These signals control the operation of the machines which in turn guide the machine's processes. The connections between materials and informa-

The different codes and symbols included in the chapter frontpiece are listed below. Line 1: Symbols from a weather map. Line 2: Braille code for the blind. Line 3: Morse code used first on the telegraph. Line 4: French semaphore code used for sending messages from hilltop to hilltop over long distances prior to electronic communication. Line 5: Symbols used in material handling to indicate how a part should be transported and handled. Line 6: Symbols used to indicated the different welding processes that are to be used during fabrication. Line 7: An eight channel punched tape with a hidden message. Line 8: Symbols used in electronics and computers. Line 9: International traffic symbols. Line 10: The International Maritime Code used with water craft. Line 11: Optical bar code. Line 12: The binary code used in computer. Line 13: Different devices that are used to store information for our use. Line 14: the American Sign Language for the deaf. Line 15: The Arabic alphabet. Line 16: The hexidecimal code used in computers.

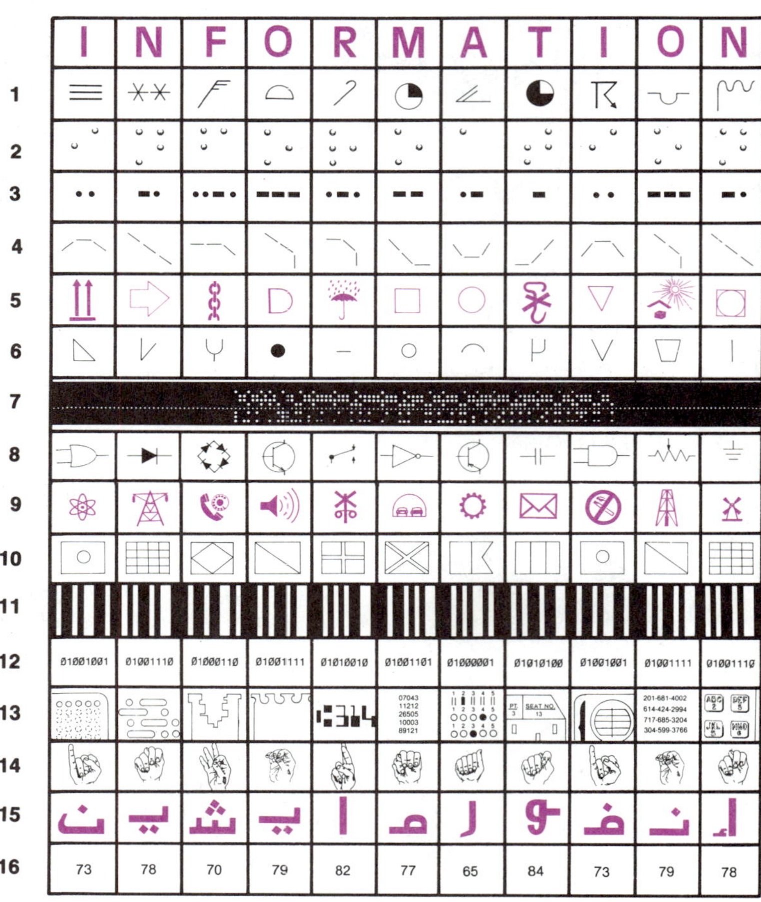

tion are somewhat removed. Materials can be converted through processes that are controlled by information in the form of symbols and signals. Materials and information also interact through on-off devices, such as solid-state switches. The switches are made of dynamic materials that are changed by a signal. The material changes (on) and then returns to its former state (off). This makes information processing possible with computers and microprocessors.

Information provides a major link between humans and machines. Symbols provide a form of information that humans use with each other and also allows machines to communicate with humans. Signals provide the information in a form for humans to communicate with machines and for machines to communicate with each other.

Key Concepts and Terms

automatic control
binary
codes
communication
computers
digital signals
guidance and control subsystems
information device
magnetic tape
message
Morse code
numerically controlled (NC) machines
read only memory (ROM) device
receiver
relay
sensing energy
signals
symbols
telegraph
transducers
transmission

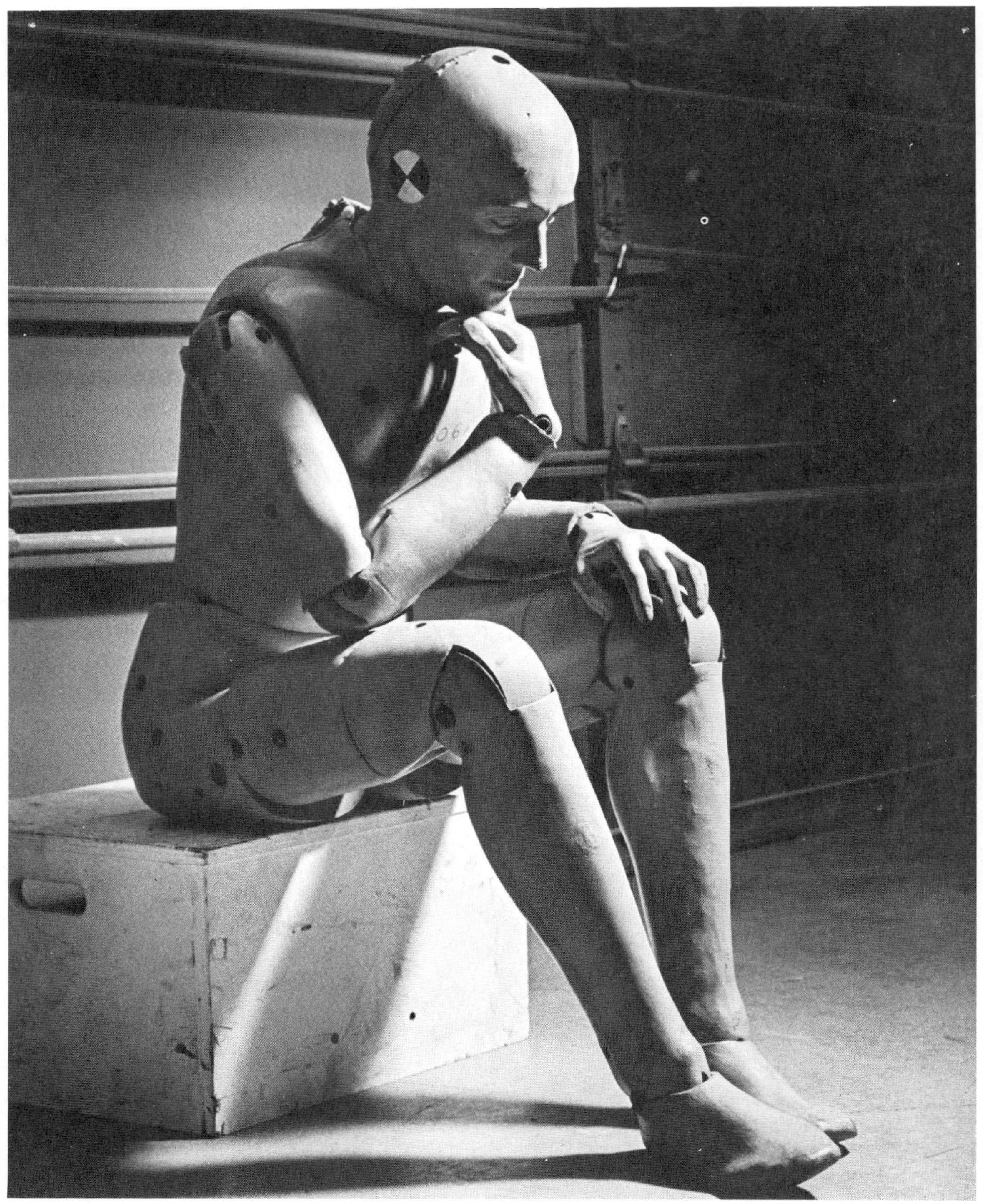

Chapter 7 Humans

In this chapter, we will look at technology in relation to humans. Technology was developed by humans to meet human needs. Thus, people's goals and needs are reflected in the applications of technology. Many human factors influence the development of technological systems. The human is one of the six elements in a technological event. The machines, materials, processes, energy and information all interact with humans. People influence technology; and technology, in turn, influences people.

What is a Person?

Perhaps we can best understand who we are if we look at ourselves as foreign creatures. Engineer/inventor Buckminster Fuller has described the human as:

> A self-balancing, 28 jointed adapter-based biped; an electro-chemical reduction-plant, integral with segregated stowages of special energy extracts in storage batteries, for subsequent actuation of thousands of hydraulic and pneumatic pumps, with motors attached; 62,200 miles of capillaries; millions of warning signal, railroad and conveyor systems; crushers and cranes (of which the arms are magnificent 23 jointed affairs with self-surfacing and lubricating systems), and a universally distributed telephone system needing no service for 70 years if well managed; the whole,

◀To understand how the human relates to technology, consider the human body as a structure and as a machine. With the understanding we can then turn our attention to more complex relationships between humans and technology.

> extraordinarily complex mechanism guided with exquisite precision from a turret in which are located telescopic and microscopic self-registering and recording range finders, a spectroscope, et cetera, the turret control being closely allied with an air conditioning intake-and-exhaust, and a main fuel intake.

Fuller describes humans in much the same way we might describe a machine. Although people are much more complex than machines, machines have been designed to imitate humans.

Your Body the Machine

skeleton

In Chapter 2, you learned that there are three basic components to all machines: support and cover, energy transmission and guidance and control. These components can also be found in the human body. The **skeleton** forms a framework of bones which supports your body. Some parts of the skeleton, such as the skull and ribs, play both a support and cover function. The bones in your arms and legs act as levers that are moved by pulling of muscles to move you from place to place. Your skeleton is made up of over two hundred bones and provides you with a frame capable of being loaded with well over a ton of weight. Yet, the bones in your skeleton probably weigh no more than twenty pounds! Did you realize that your skeleton is such a superbly designed structure?

skin

The major part of the human's cover system is the **skin.** It has an area of approximately twenty square feet. The skin provides a protective covering to keep harmful germs and dirt out of the body. It keeps itself moist and stretches to move as you move. It repairs itself when damaged. Skin works to keep the temperature of your body within safe limits. Skin is even waterproof.

tendons
cartilege

The human energy transmission system includes our muscles, tendons and cartilage. Since the human is self-propelled, the digestive system provides an inboard power plant to change food to energy and replacement material.

muscles
nerves

Muscles are attached to the bones by tendons. The nerves that control the muscles transmit signals to make the muscles contract. When the muscle contracts or shortens, it pulls the bone. The muscle cannot push the bone back in place, so another muscle paired with it contracts to pull the bone back.

We use muscle energy to move ourselves and our tools. Some muscles are much more powerful than others. For example, the large muscles in the legs provide much more power than do the muscles in the face and head.

joints

The **joints** are a unique part of your skeletal system. Some joints move only back and forth as a hinge. Other joints, such as in your shoulders and hips, work as a ball and socket hinge and move in several directions. These living hinges also always provide their own lubrication. A built-in sac encloses the joint to keep the oily liquid in and abrasive materials out.

lubrication
abrasive

The human guidance and control subsystem includes the nervous system and all of the senses—sight, sound, smell, taste and touch. Some parts of the body work automatically. For example, we do not need to think about it to breathe or blink our eyes. Other parts of our body require conscious guidance and control. We must decide to get up and walk.

The human skeleton is similar to the support and cover system of a machine. Sometimes when a joint or hinge of the body wears out it can be fixed with a replacement part, or *prosthesis*.

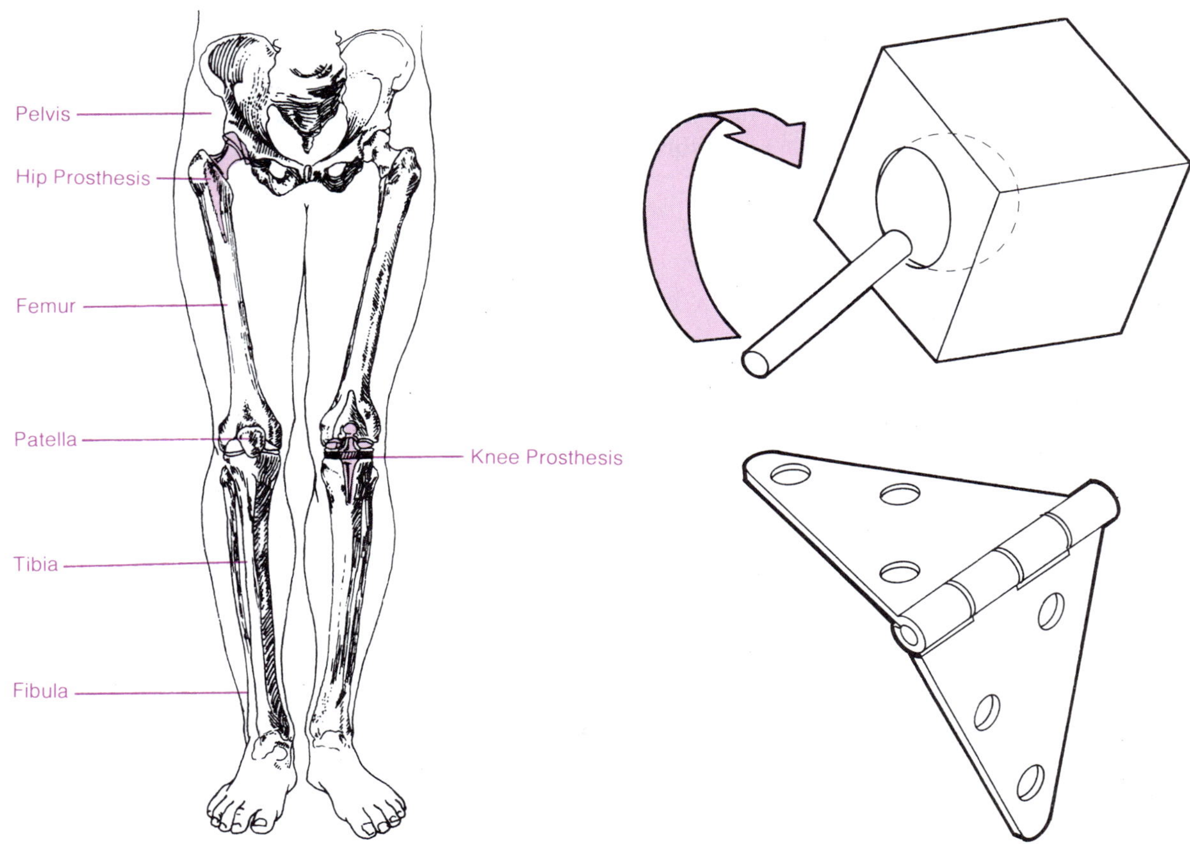

Our brain then sends signals to make our muscles move. The process of moving is very complicated. Babies must learn gradually to control parts of their bodies.

Some movements require fine coordination. Our guidance and control system usually takes data from several senses at once and coordinates it in a split second. Thought processes occur much faster than muscle processes. Think of all the times you have forgotten what you were going to say before you could get the words out. Scientists are just beginning to understand how the brain works. The more knowledge we obtain, the more respect we have for the extremely intricate processes of the human brain.

Knowledge of the body is important when industrial designers fit a product to a user. Tools and machines must have an appropriate size and shape to be useful to humans. The size and shape of many of the things we use came about without much planning. Most household scissors are made the way they are chiefly because the shape is easy to manufacture. Only recently have designers attempted to develop a new shape that will better fit the human hand. Many products have never really been designed, but just sort of "happened." The size of a pencil, a toothbrush, cups and dishes probably just evolved without much planning.

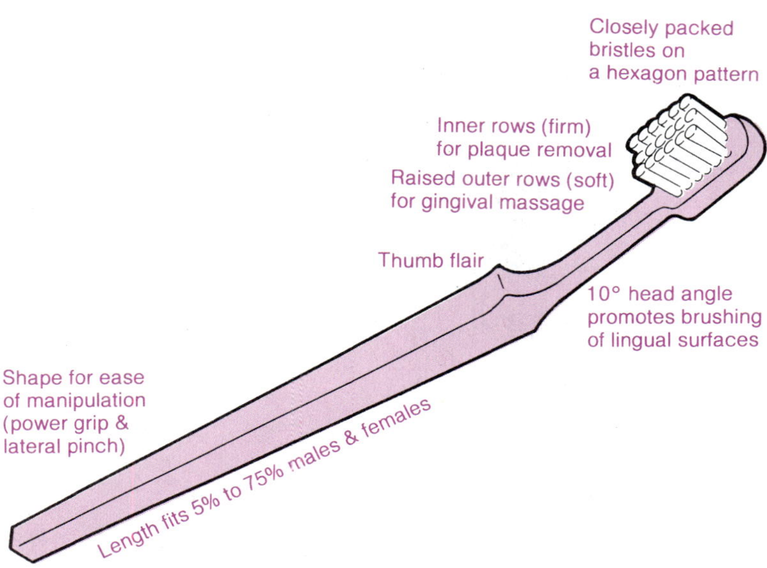

Products like this toothbrush, scientifically designed for use by humans, are relatively recent developments. Such work is done by industrial designers, a profession that is only about fifty years old.

In the past thirty to forty years, however, a new area of industrial design has emerged. Engineers and environmental designers are now beginning to pay more attention to how people use the products of technology. Only recently, the first toothbrush was designed according to a detailed analysis of how people clean their teeth.

strength

The size of the human body is only one type of information that is helpful in understanding how the human relates to the other elements of technology. Other physical characteristics include movement, reach and strength.

To understand movement of the body and its parts it will be

Detailed measurements of the human's size and performance are important to engineers and designers.

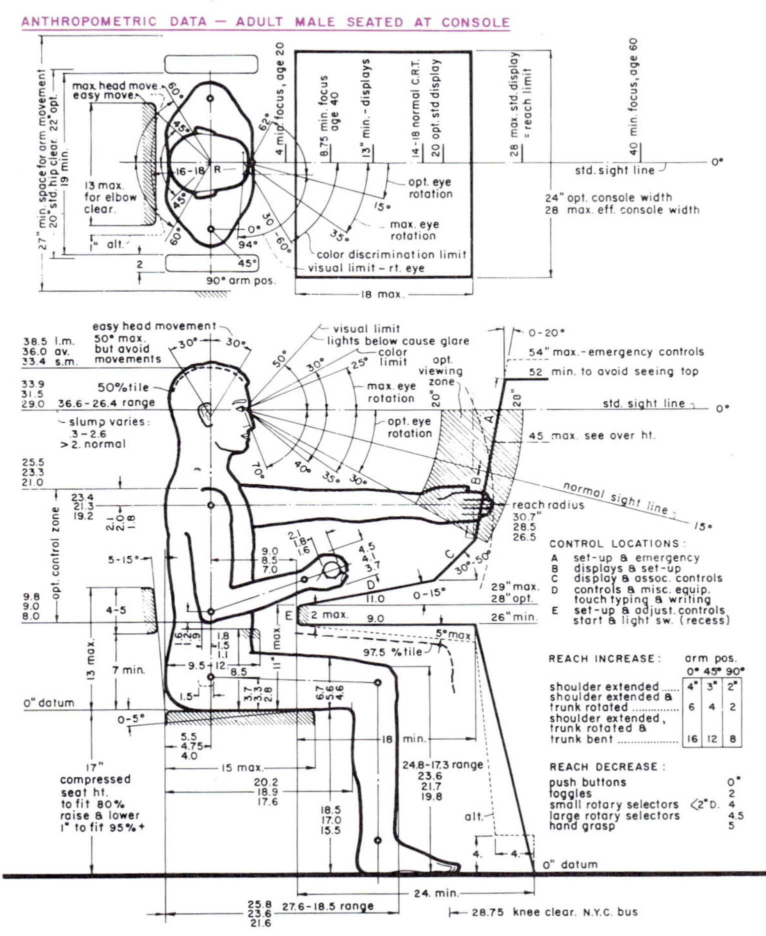

helpful to compare the body to machine parts. We can make the following comparisons:

1. **bones**—support
2. **skin**—cover
3. **joints**—bearing surfaces
4. **joint linings**—lubricant
5. **muscles**—motors, shock absorbers, locks
6. **tendons**—transmission cables
7. **nerves**—control and feedback circuits

These body parts work together to allow us to move about and to engage in many different activities. One important type of movement is reach. Reach becomes especially important if the subject cannot move from a given position, such as a pilot's seat or a wheelchair.

Strength

biceps
extensor muscles

Your strength varies, depending on the muscles you use and the movements you make. You can lift more weight with your biceps than you can push with your extensor muscles. There is also considerable difference in the strength of the wrist as

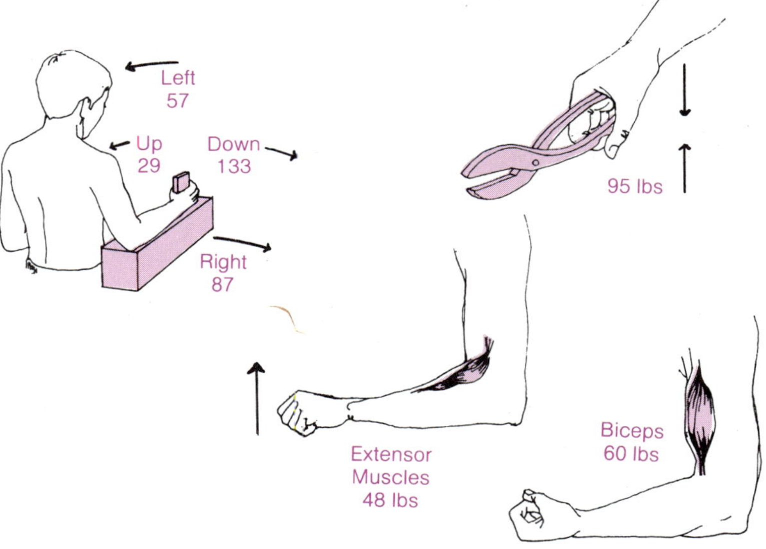

The strength of a human's muscles varies widely, depending on which muscles are being used and in what direction the movement is made.

The human body is affected by many environmental factors. We are comfortable and can survive within a narrow range of these factors.

ENVIRONMENTAL TOLERANCE ZONES

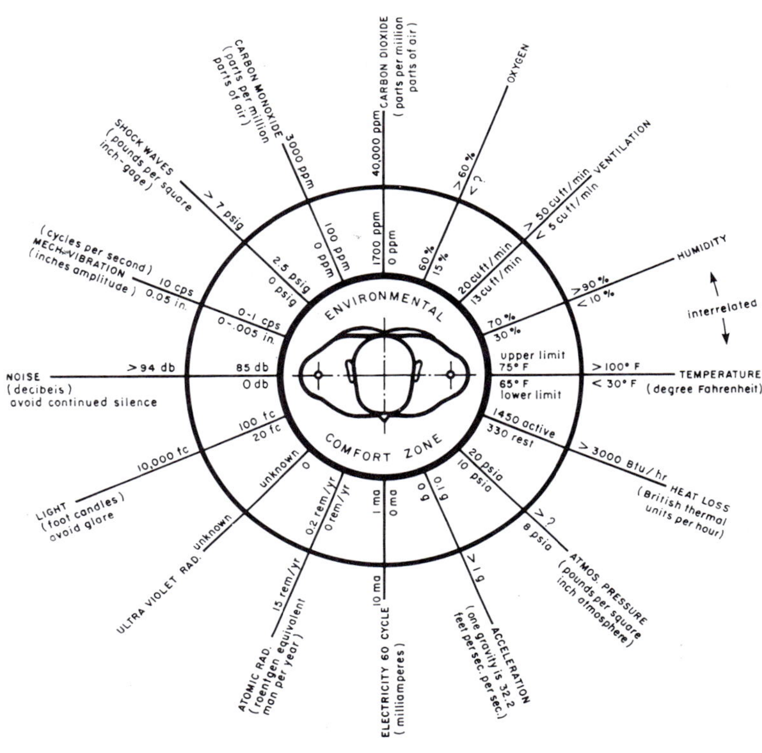

THE BAND BETWEEN THE CIRCLES INDICATES THE ZONE FROM COMFORT TO THE TOLERANCE LIMIT. OUTSIDE THIS LIMIT GREAT DISCOMFORT OR PHYSIOLOGICAL HARM IS ENCOUNTERED. OTHER FACTORS NOT SHOWN AND TO BE CONSIDERED ARE: INFRA-RED RADIATION, ULTRA-SONIC VIBRATIONS, NOXIOUS GASES, DUST, POLLEN, CHEMICALS & FUNGI.

compared with the hand. The hand's grip is usually stronger than most of the wrist's movements. Our legs are stronger in pushing and stretching than in bending. For a short period of time, the leg can generate forces exceeding 450 pounds.

In spite of our immense complexity, humans are extremely vulnerable. Perhaps this has motivated our attempts to control our environment through technology. Human beings can survive very well on our planet with few adaptations. We have even developed technical means to survive in space and on other planets. Human operation is affected by the physical variables of the environment. The human body has a tolerance limit. Beyond this limit great discomfort and even physi-

limits of tolerance

cal harm takes place. Specific limits become important in the design of products such as camping equipment, air conditioners and space vehicles.

The Senses

senses

What goes on both outside and inside our body is experienced through our **senses.** In order to adapt to or change the environment, we must gather information about it. Our senses gather information and our brain interprets it. In order to detect something, enough stimulation must be received. Otherwise, we do not pick up the information and it is not passed on to the brain. You usually cannot hear a mosquito in another room. When you are tired, or when you are watching the same movie for the third time, you may miss a message. On the other hand, when too much stimulation occurs, the body acts to protect itself. Sometimes, our senses become overloaded and turn off (fainting, for example). The information must not be too much or too little. We will either fail to notice the information or defend against the stimulation. Tools have been developed to extend people's capabilities and to compensate for human tendencies to distort or delete information.

vision

Vision includes the ability to judge distance, to see both near and far and to see colors. Vision, for some people, diminishes as they get older. In general, people's eyes tend to focus slower as they get older. Some people's ability to see colors decreases with extreme age. Some people are born blind or color blind. Many others become blind due to injury or disease. Studies of vision and the structure of the eye are important in correcting visual problems.

hearing

Hearing is the second major sense. People can hear the frequency or pitch and intensity of a sound and can detect the time interval between sounds. People hear within a narrow range of sounds and continued exposure to a high volume of sound can result in decreased ability to hear.

balance

A sense of **balance** keeps us oriented to the planet. We know the importance of this sense when we try to walk after riding a dizzying carnival ride. This natural sense of balance makes us prefer to live where the floor is level. Interior designers have found that even the color and shape of walls can have an effect on human balance.

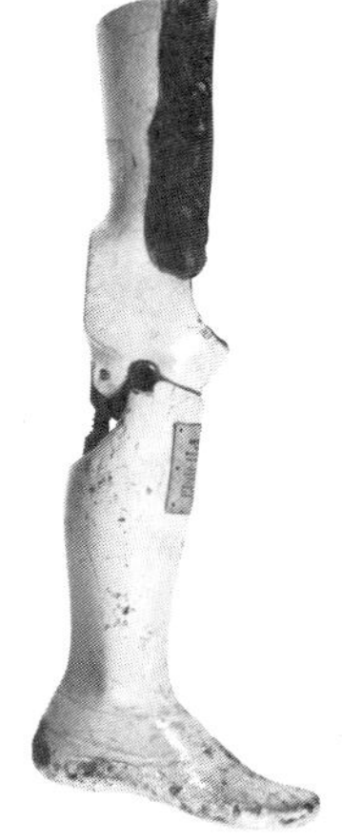

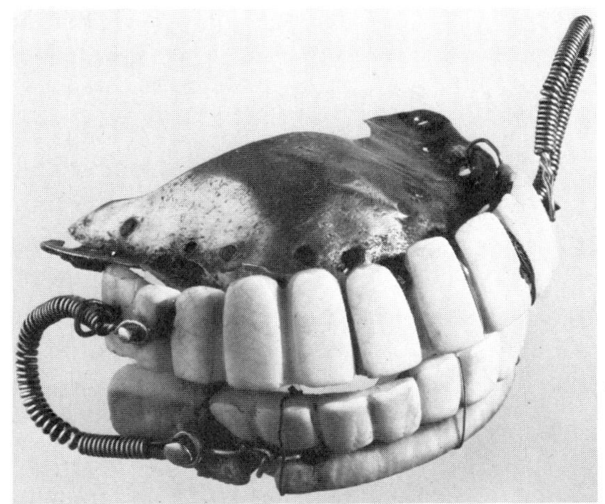

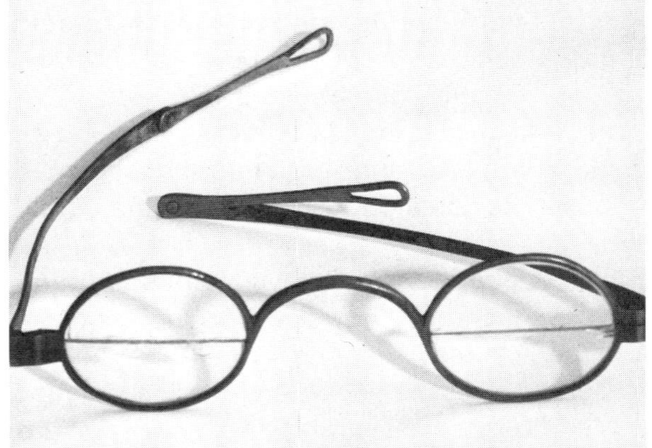

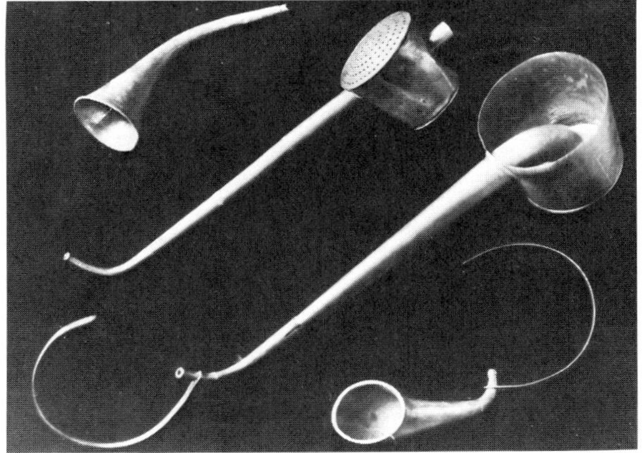

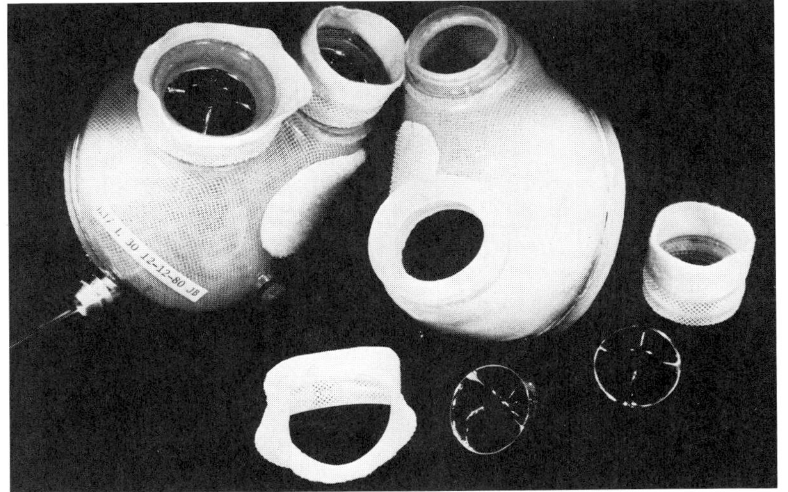

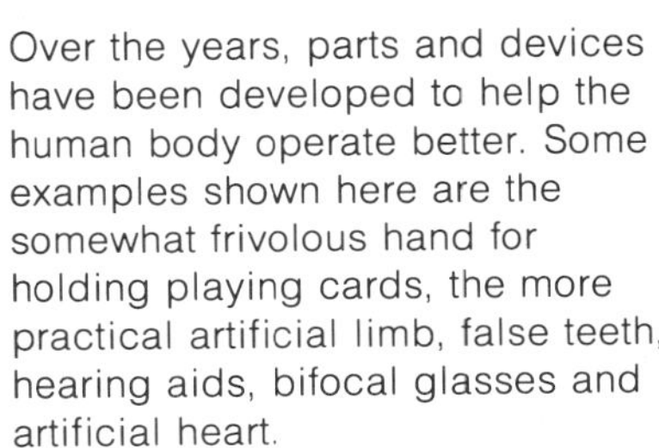

Over the years, parts and devices have been developed to help the human body operate better. Some examples shown here are the somewhat frivolous hand for holding playing cards, the more practical artificial limb, false teeth, hearing aids, bifocal glasses and artificial heart.

taste

The sense of **taste** is closely associated with the sense of **smell.** Food may seem relatively tasteless if your nose is stuffy. The sense of taste is often weakened in persons who smoke. Some change in the sense of taste and smell also occurs as people get older. Our preferences for certain foods may change as our senses change.

touch

General body sensations include the sense of touch, pain, muscle movement and vibration. **Touch** can be an important form of message between people. Pain is a warning that something is wrong inside or outside the body. We are just beginning to understand the nature of pain and to develop methods of controlling it. Humans, in general, try to live their lives in a way that brings as much pleasure, and as little pain, as possible. Can you think of how this has an effect upon the technology we develop?

All the human senses must be put together and interpreted by the brain. Not all sensations are given equal value. A machine handles all data equally, but humans may add to or ignore what is sensed if it does not fit what is expected or wanted. For example, if you are hungry you might hastily read the sign "steam house" as "steak house." A machine would not make that error.

People's perceptions change with age, previous learning and present motives. Machines, however, do not vary as much as people. A person may respond in many ways to a given situation. A person's biases and variations may or may not be beneficial. Machines, however, are predictable and are limited to the functions and actions for which they were designed.

Humans as Material Users

Humans use materials to take care of basic survival needs and to supply their wants. Humans are vulnerable and have relatively narrow limits of tolerance. If we are to survive we must inhale enough oxygen and exhale waste gases. We must stay not too cold and not too hot. We must be able to work and to rest. We need some bacteria to survive, yet others will threaten us with illness and infection. A delicate balance must be maintained within our cells. Required nutrients must be supplied and wastes and poisons removed. Although we walk around as distinct units, we constantly interact with our environment. Our skin is usually considered the boundary line, where the "self" stops and the rest of the world begins. Yet, there is continual exchange of materials between our bodies and the environment around us.

When we think of survival and its relation to materials, we often think of food. **Food** is a form of material that we ingest, burn and store. Unused food is excreted as waste, along with the by-products of the food that is used.

water

Perhaps the most important material for body use is **water.** Many of the chemicals necessary for life are water-soluble. Your body may have sufficient stores to go without food for several days, but a lack of water would threaten your survival first.

nutrients

proteins
carbohydrates
fats
minerals
vitamins

Humans can satisfy their requirements for **nutrients** in many ways. However, in order to survive and stay healthy, certain chemicals are required. These chemicals are found in proteins, carbohydrates, fats, minerals and vitamins. The imbalances caused by illness may require special medicines or nutritional supplements. But, in general, most people in the United States can receive the nutrients they need if they eat regularly from a wide variety of nutritious foods. Artificial vitamins, food additives and supplements, synthetic flavors

Over the centuries, humans have discovered and developed food-plants. Pictured here are some of the many sources of basic nutrients for humans.

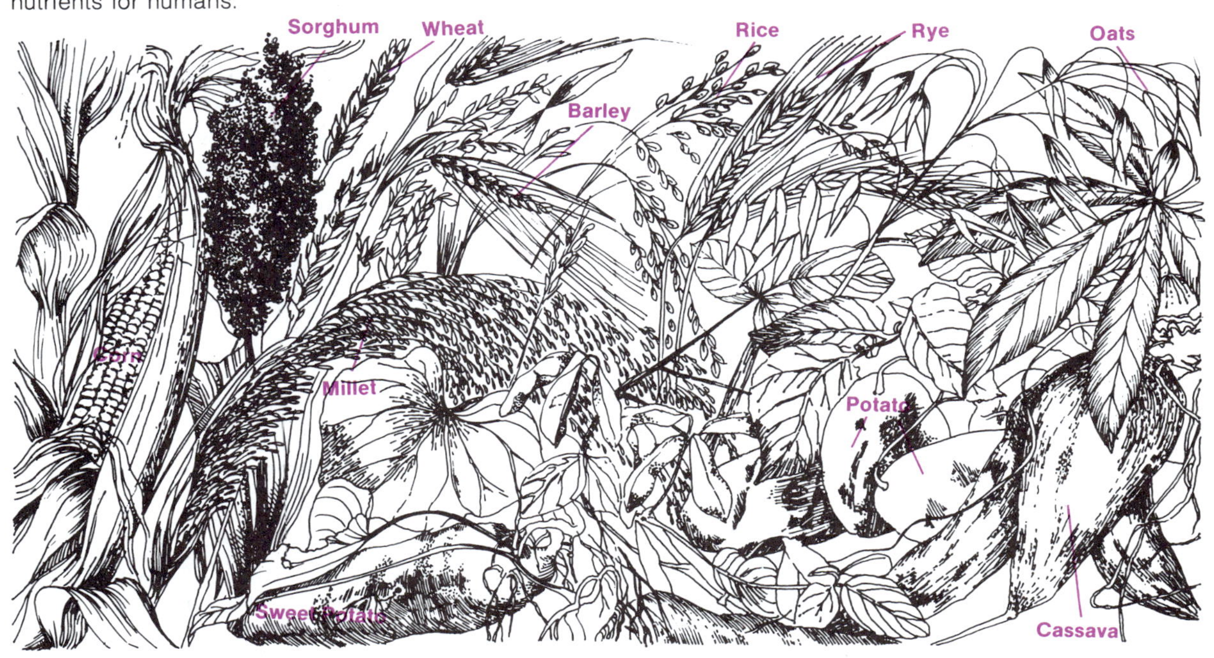

and sweeteners have been developed in recent years. Some of these are needed in special cases and may become more important if the world's population is to be fed. However, we do not yet know the long term consequences to people's health when large amounts of additives or synthetic materials are eaten.

People also use materials to cook food. Some foods cannot be digested well if eaten raw. Cooking also kills dangerous germs. A major activity of many people in developing nations is the production, storage and preparation of food. In technologically advanced societies, less effort may be focused on food. In the United States, for example, only a small percentage of the population's efforts goes into the production of food for the nation. Some developing countries currently face serious energy problems as the supply of firewood becomes scarce.

Materials for keeping warm or cool are vital to humans. These materials may be readily available to people or a great deal of effort may be needed to assure a dependable supply. Early in history, it was a warm fire that led groups of people to gather together. This need to keep warm may have led to other forms of social learning. If you have ever been camping, you may have had a similar experience. You might have gained a sense of belonging to the group by learning camp songs and games.

clothing
ventilation

Materials to maintain temperature are also used in **clothing** and other means of insulating and ventilating the body. People in warm climates wear considerably less clothing than people in colder climates. You probably wear much lighter and briefer clothing in summer than in winter. Clothing expresses many other aspects of the person and the social group. The major purpose of clothing, however, is to protect the body and skin from temperature extremes and potential injury. Clothing may be considered a synthetic skin.

synthetic skin

A tough pair of pants can help protect your legs if you hike through thistles and other weeds. Foot coverings have evolved according to terrain, the covering's use and available materials. Hats and caps can be extremely important in cold weather because as much as 40% of the heat lost by your body escapes from your head.

Bulky materials were used by early humans for protection from the cold. Animal skins or hair, wool or fur trapped air and provided an insulation that kept in body heat. Modern

insulation

The demands of space travel make us aware of humans' need for protective clothing to survive in hostile environments. Although our clothing is often chosen for fashion reasons, early people used clothing strictly for protection.

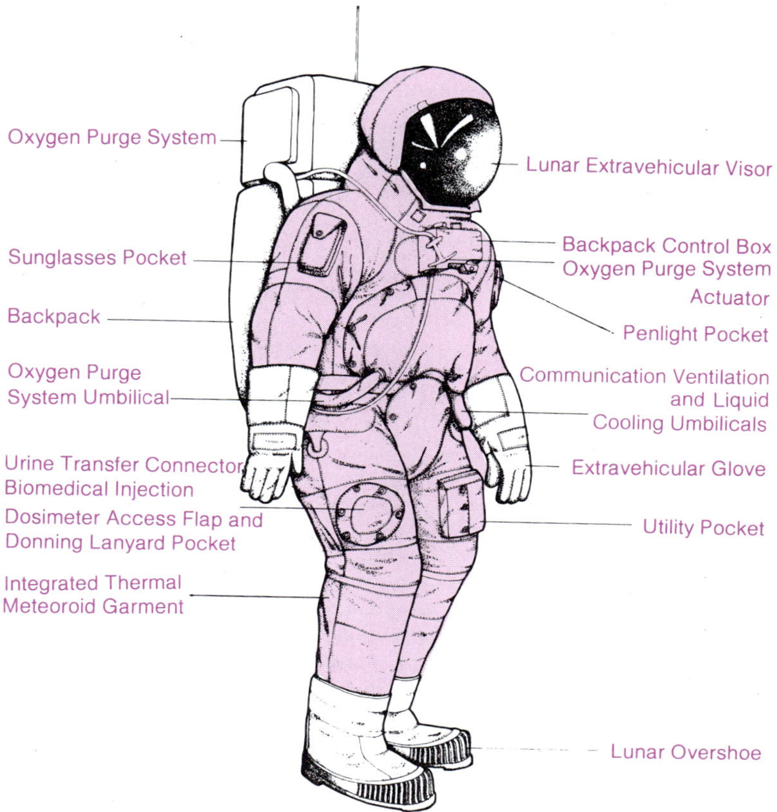

DIAGRAM OF APOLLO LUNAR SPACESUIT

clothing materials are much lighter. Some materials used in the space program are designed to keep the wearer warm or cool, depending on which side of the material is next to the skin.

shelters

Humans have built **shelters** out of many kinds of materials. The purpose of human shelters is to protect us from outside forces, as well as to keep us warm or cool. Early shelters included caves, mud huts and many types of portable structures. They were built with available materials and according to the type of protection needed. It was in these early shelters that humans began to record their ideas and life experiences. Cave drawings are some of the oldest human records we have. Many people today keep sketchbooks or diaries.

Humans need materials for nutrition, protection and maintenance of body temperature. Certainly, we use materials for many other reasons. But these reasons might be called "wants," because we could survive without them. In advanced parts of the world, technology is directed toward satisfaction of human wants, rather than needs. Imagine your home if it only included what you need to stay alive. Record the money that you and your family spend in a week. How much of that money was used to buy materials and products you really need to survive? In many parts of the world, families use most of their resources for food and other survival needs.

Humans and Energy Use

Early technological activities were designed to replace human muscle power with other sources of energy. When early humans used tools, the person supplied the guidance and control, the support and cover and the power transmission. When animals were harnessed to do work, humans had access to more power. Wind and water power helped humans with great amounts of work. Atomic energy may free us even more, but there are many potential dangers involved.

There are many other ways in which energy affects our lives. We produce sound and use it to communicate thoughts and feelings. Energy is needed to make sounds. For example, we control the airflow over our vocal chords. These vocal chords are stretched or relaxed to produce different frequencies (pitch). Control of muscles in the larnyx, mouth and lips form words and other utterances.

Our hearing provides feedback so that our sound production can be corrected. People vary in the degree to which they can accurately reproduce sounds. Deaf people can't hear themselves, so their speech is often flat and off-pitch.

Simulation of speech has been a complicated process attempted by designers of computers. People who have lost their speech capabilities through disease can sometimes be taught to use devices that simulate the original voice box.

Sound is also used by humans to communicate over great distances. Use of telephones, television and radio have extended our knowledge of people in other parts of the world and allowed us to communicate with them. As a consequence, some people feel that the world is "shrinking."

light energy

Light energy is important in vision. The ability to visually sense direction, movement, color and shape is important to human behavior. Imagine what your world would be like without colors. Images are quite different on a black and white and a color television set. Your emotional reaction is probably different.

Some people believe that certain shapes and colors are almost always connected with certain feelings. Red may bring out strong feelings, and blue may bring out more neutral feelings. Whether you believe this or not, you probably do have color preferences.

colors
pigments

Colors come from light or from **chemical pigments.** The combination of colors in the light spectrum are seen on a color TV. These colors may be different from combinations of pigments in paint. Colors can be mixed to varying intensity and hue. Brilliant orange letters on a black background produce a very different effect than light green letters on white. Most of the effect comes from the impact of the light on your eyes. Humans use color for decoration and communication purposes.

There is much concern over the impact of using so much energy and so many materials for human wants rather than needs. Some people think we should return to natural foods and natural energy sources. There was a time when people had no choice—"natural" was all there was. When food became scarce in an area or when they had used all the fuel, people moved.

With technological changes we have increased our average life span, and our world population. Few of us would like

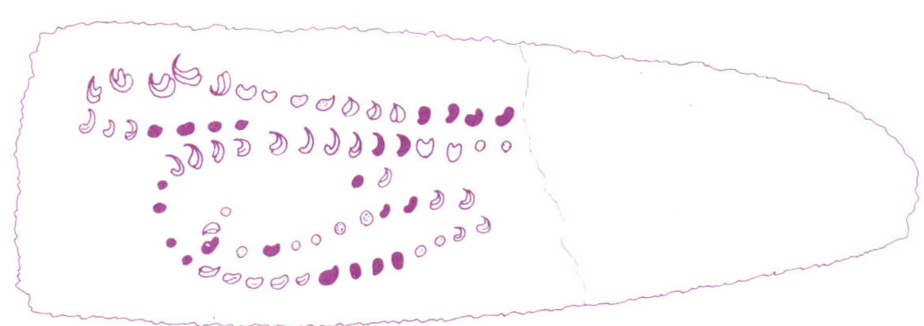

This drawing, made on a piece of bone, is believed to be a record of the phases of the moon. It is one of the oldest known scientific records of human observation.

to return to an average life span of 25 years. Few of us would like to return to a life in which most of our time is spent obtaining what we need to survive. But we can no longer afford to use resources at a faster rate than they are replenished. We cannot afford to use up the resources in one area and move on. Many people are becoming aware of the value of conserving and recycling our resources. Some hard decisions must be made by weighing the costs and benefits of what we ask technology to do. Humans have the capability to make such decisions.

Humans as Information Processors and Convertors

Our use of language and our ability to learn set us apart from other animals. Several other animals have communications systems and many animals are capable of learning. Some chimpanzees can even be taught to use language similar to that of two-year-old humans. However, the infinite complexity of the human's ability to receive, process, store and retrieve information is unmatched by any other animal in the world. The most complex machines cannot compare with the information processing capabilities of even young children.

This machine has gone through a long process of improvements. The early prototype looked like something from outer space (left) and was far less efficient than the compact hand-held viewer. Without the device people suffering from a certain eye problem would be virtually blind at night.

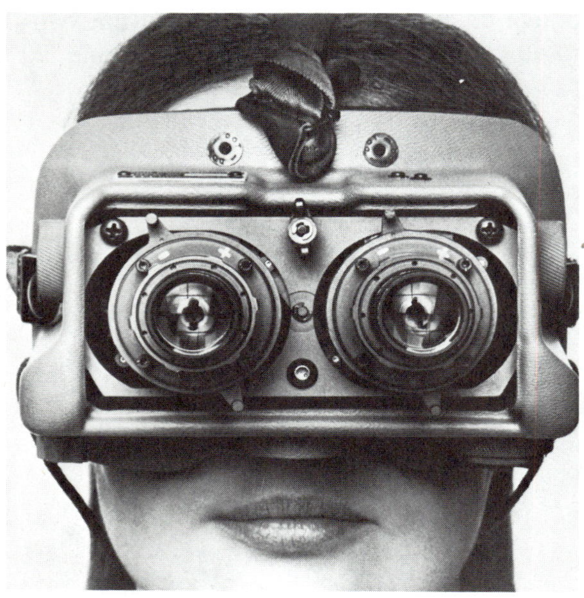

The tools that have been developed to explore human intelligence have helped us build machines that mimic some human processes. Technology can be used to learn more about humans, and the knowledge about ourselves improves our ability to design even more useful technological devices.

The range in which humans receive outside stimulation is limited. Individuals are also limited in their sensory acuity. Our sense organs *select* from the "buzzing confusion" of what is going on around us and send those sensations to the brain. The brain then makes some order out of our experience. **Language** is the major tool for the organization of experiences into thoughts and for memory and communication. Language also has some impact on what our senses select to perceive. For example, a person with a language that has fifty words to describe snow probably sees snow differently than a person who has only one word or none at all.

language

Perhaps, as some researchers believe, our brain records *every* experience. Perhaps, the brain could be made to call back any given experience. Humans usually select, categorize, receive and recall only *parts* of the actual stimulation. You probably have had an argument over the right words to a song on the radio. You heard the words differently than your friend, even though you both heard the song at the same time. Two people may easily perceive the same event in very different ways. People selectively remember and forget.

In general, we remember events connected with strong emotions. We repeat and learn actions that make us feel good, whether the good feeling is from the event or from a reward. We remember the actions that make us feel bad, too. We learn to avoid these actions.

Some painful events are "forgotten." Effort is required to not remember these experiences. Events and experiences that do not produce either good or bad emotions are most easily forgotten and repeated less often. The way an experience makes us feel is essential information for learning and for modifying future behavior.

information processing

Our information processing abilities make us creative, inventive, spontaneous, interesting to other people and able to learn from a wide variety of situations. Yet, these same abilities sometimes distort and delete important bits of information from sensation and memory. They also allow us to add bits of information to our memory that were not in the original experience. We can remember the experience the way we wanted it to be. We may overgeneralize or use inappropriate words to categorize information.

People have invented machines to correct some of these tendencies. A calculator can handle volumes of numbers and processes without making a mistake. A photograph will show all the details that you may want to ignore. These machines, however, only carry out the processes and use the information that human designers and operators dictate.

Humans Use Information for Decision-Making

Since people began to write and draw to communicate, humans have shown a tendency to generate new information. We have information that allows us to study the past, the present and predict the future. We are the only animals that can be aware of our "roots" and are concerned about aging and death. We can almost instantly be aware of events that are happening thousands of miles away.

The vast availability of information has both positive and negative effects. Once, a well-educated person could read every important book in the world. Now, people have trouble keeping up with monthly magazines. It is possible to feel overwhelmed by an information "over-load." We may decide to let

Human-machine systems continue to be developed that intergrate the six elements of technology. These systems extend our abilities to allow us to perform difficult tasks. Such systems, because of their potential impact, also extend our responsibilities.

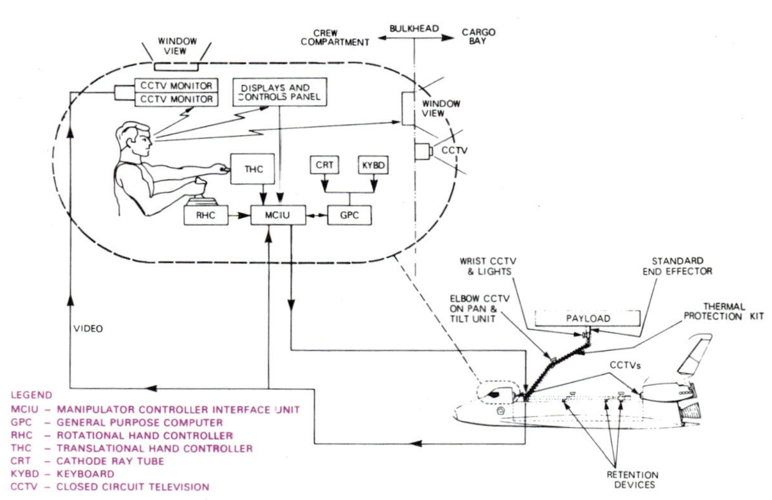

REMOTE MANIPULATOR SYSTEM

others stay informed and make the decisions. Or, we may make the decisions without even considering the available information.

The processes of communicating, retrieving and sorting information related to a problem have been made easier by the machines, processes and materials that people have made. It is still a human function, however, to read and interpret the information, decide what you want and reach conclusions. Technology for the generation and processing of information has made decision making more complex, difficult and critical. At the same time, this technology provides us with the means to deal effectively with these complex decisions.

Summary

There are many ways in which humans interact with and use machines, energy, materials and information. Humans have developed tools suited to human dimensions and abilities. We have also developed tools to extend our senses and increase our available power. Materials are used for survival and comfort. In general, people have tried to design and use technology to make their lives contain more pleasure and less pain. We attempt to satisfy our curiosity and learn more about ourselves and the environment through the use of machines and processes that create information. Information processing becomes vital to developing useful technology.

Key Concepts and Terms

abrasive
balance
biceps
carbohydrates
cartilage
clothing
colors
conscious
extensor muscles
fats
hearing
human engineering
information processing
insulation
joints
language
light energy
limits of tolerance
lubrication
minerals
motivation
muscles
nerves
nutrients
pigments
proteins
self-propelled
senses
shelters
skeleton

skin
strength
supplements
synthetic skin
taste
tendons
touch
ventilation
vision
vitamins
water

Summary—Unit One

A Model of Technology

Tools, materials, processes, energy, information and humans are the elements of any technological event. We can use these six categories to take apart a technological event and study its pieces. But how do these pieces go together?

Technology has its own vocabulary. One commonly used term is "model." A model is a simplified view that helps us look at and understand real events. We are familiar with mod-

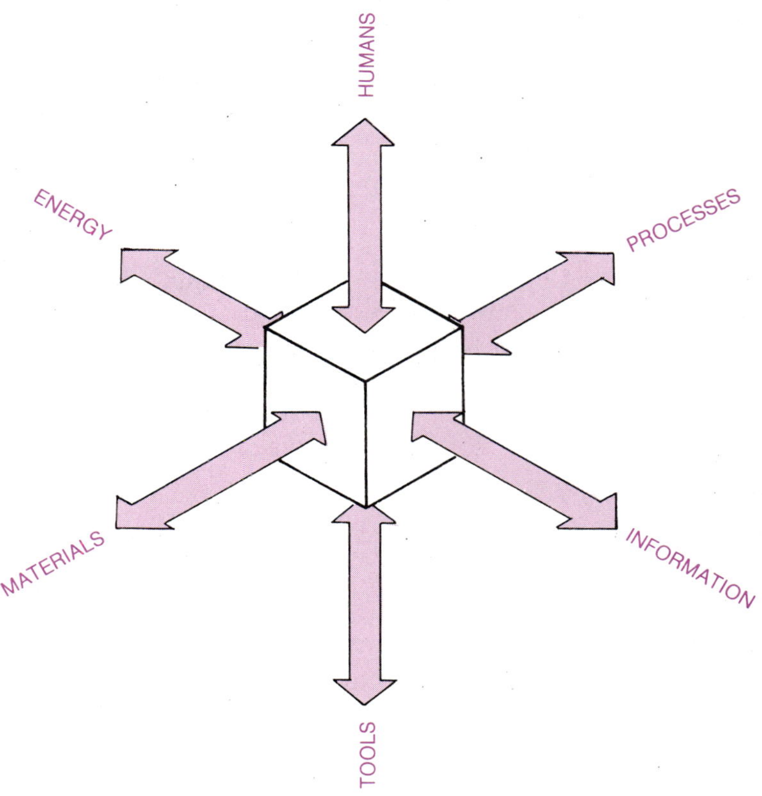

The six elements of technology and how they relate provide us with a model of looking at new and unfamiliar instances of technology.

els of airplanes, models of cities and models of our solar system. These smaller versions help us understand the real thing more easily.

Our Model of Technology suggests how the six elements relate to each other. The following set of statements explains how the pieces go together. The statements describe only some of the ways that the elements relate to one another.

1. Humans combine materials in specific ways to design and develop machines.
2. Humans use selected materials to provide energy for themselves and for their machines.
3. Humans use tools and machines to provide processes for changing materials.
4. Machines provide processes for changing the form of materials, energy and information.
5. Information results from the conversion of a pattern of material or energy into a signal or symbol.
6. Information between machines flows as signals and must be converted to symbols for use by humans.
7. Information between humans flows as symbols and must be converted to signals for use by machines.
8. Machines are controlled (by humans and other machines) by signals to their guidance and control systems.

UNIT II: The Activities of Technology

Introduction

The six elements of technology work as a system, interacting with each other to produce a technological event. The record player used to illustrate an instance of technology in Unit I was a simple system that is part of a larger set of systems. The larger systems of constructing, transporting, communicating and producing are called **activities of technology.** It is through the activities of technology that products such as record players are made available to you.

Records and record players are made within a production system. The materials and energy for making a record player have to be transported to a common site. The finished product then has to be transported to you. The building in which the player is produced, the furniture on which it rests, the store in which you purchase it and the roads and vehicles are all constructed. A communication system is used to record the music and to advertise the product.

Four activities of technology are described in this unit. The elements interact to produce products and services that can make our lives more comfortable and enjoyable.

Structures are the products of constructing. They have become more and more complex in their evolution. The examples shown here include the mammoth pyramids of ancient Egypt, the longest cantilever bridge over the Firth of Forth in Scotland, the Great Eastern—the largest moving structure of its time, the geodesic dome at Epcot Center in Florida and the multi-stage moon rocket.

Chapter 8 Constructing

Where will you live twenty years from now? Perhaps your home will be suspended from a mountain or buried beneath the ground. You might choose to live in environments such as the sea or in space. Wherever you live, you will need some form of shelter to protect you and make you comfortable. You will want containers and structures for storing and processing the materials you need. You will also need structures to get those goods to you and to transport yourself.

constructing
structures

The technological activity of building these structures is called **constructing.** In this chapter, we will explore the types of structures people build. We will also describe the parts of a structure. We will study what makes structures hold together and serve the function for which they are designed. When you have studied this chapter, you will be able to look around you and recognize how structures are built, how they are alike and how they are different. Perhaps you will design some new structures that are more effective than those that have been built before.

Up, Out, In and Away

Suppose you are one of the first humans to land on a far-off planet. You will need to build some type of shelter. Your spacecraft might be adapted for use as a temporary home. If you go exploring far from the base, you will probably carry along the materials to construct a shelter for sleeping.

supporting

People build structures to serve a **supporting** function. If you wanted to raise a signal flag outside your space colony, a flagpole could be constructed. *Supporting structures* hold themselves and other things up. Bridges, lighthouses, antenna, chairs, tables, shelves and stairs are examples of supporting structures. It could be vital to your colony to get your communication equipment up high enough to clearly transmit to home base.

shelter

Another reason that people build structures is to **shelter** themselves and their tools, animals and goods. How the structure is built and the materials that are used will depend upon the environment. The planet may be too hot or cold, too windy or dry, prone to landslides or ground cave-ins or surrounded by poisonous gases. The shelter will be designed to help you survive and be physically comfortable in this new environment. You may seek to build strength into your shelter. Its color and shape and other aspects of its design may be chosen because of personal preferences. Safety factors may be of primary importance in your new setting. You may need a lookout tower or protection from animals. You might build a shelter much like the early castles and forts on Earth.

You may provide private living areas within your colony. The need for privacy will differ considerably from one person to the next. You may also build sheds and barns and other buildings for your animals and machines.

sheltering structures

All of these structures are sheltering structures. **Sheltering structures** hold themselves *up* and something else *out.*

After your safety is secured, you might build other kinds of structures. Walls could be built to keep tame animals in. Containers of various types could be constructed to store food, fuel and other materials. Cabinets, jars and closets may be used as containers inside your sheltering structures. **Containing structures** hold themselves *up* and something else *in.*

containing structures

You may also need to build structures for moving materials and people. **Directing structures** control the flow of materials. Liquids and gases can be moved in pipelines. Even some light solids such as grains, powders and plastic granules can be moved in special directing structures.

directing structures

Vehicles that hover in the air, fly, roll along on wheels or float are transport structures. Structures might be needed to move materials from mines. Heavy weapons for defense are moved more easily on constructed pathways.

transporting structures

Transporting structures hold themselves *up* and *move* themselves or something else. Sometimes, they also hold something else *away*. Boats, for example, must hold the water away from the cargo.

Most structures can be classified into these five basic types: supporting, sheltering, containing, directing and transporting. Many structures are combinations of these types. An oil tanker, for example, is both a containing and a transporting structure. The function of the structure is important in deciding where the strength is needed and how the parts are put together. All of these five types of structures have common problems in construction.

Keeping it Together and Keeping it Up

foundation

A structure may have walls, a roof, a foundation, a floor, or a combination of these. Walls can be built as fences and have no roof or floor. Roofs may be erected with just a few supports to hold them up. A road can be thought of as a floor that is lying on the ground. A wharf is an example of a support structure that only has a floor. It is held out of the water by piers.

Did you ever build card houses? Stacking the cards to make them stand was a measure of your skill. These houses did not last long because the parts were not joined together and so were not strong or stable. If you taped the edges of the cards together, the house was more stable. But a card structure with taped joints is not very strong when turned up on one side. A small amount of force will cause it to collapse. The structure collapses because its joints act like hinges and allow the structure to fold.

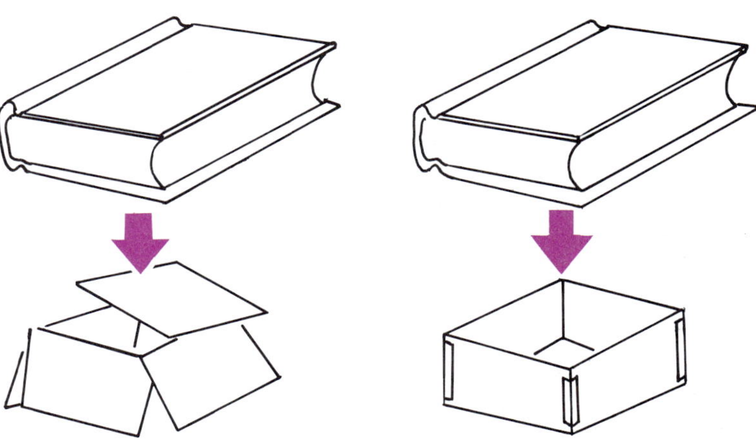

Card houses become stronger when the cards are taped together. The taped joints act as hinges when a force is applied to them.

A hinged frame may be useful if you want to collapse a structure. You could then store it or transport it in its flat form. Most of the time, however, you want a structure to remain in place. There are a few simple and direct ways to make a structure strong and stable.

arch

Many stable structures have triangles in them. An arch is not quite a triangle, but supports the stresses and strains on the structure in a similar way. Folding chairs, tables, kites, tents and the frame of bridges often have a triangular shape for strength and stability. The triangle provides stability because of the two basic principles of compression and tension.

stability

Fasteners such as nails, rivets and bolts are used to hold a building together. Some fasteners provide the strength of the triangle from their wedge shape. Often, however, you will see that a brace or truss has been used to stabilize the structure.

brace

Structures can be made more stable by using one or more of these techniques:
(A) Add a solid brace.
(B) Add two internal flexible braces.
(C) Use a minimum of two external braces.
(D) Use a solid joint to provide a rigid frame.
(E) Use the arch, a more stable form.
(F) The triangle.

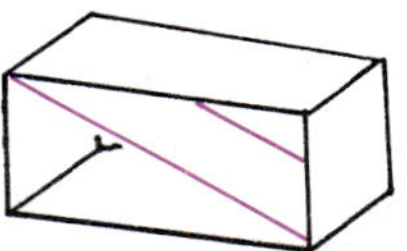

(A) Add solid brace. (Only one is needed.)

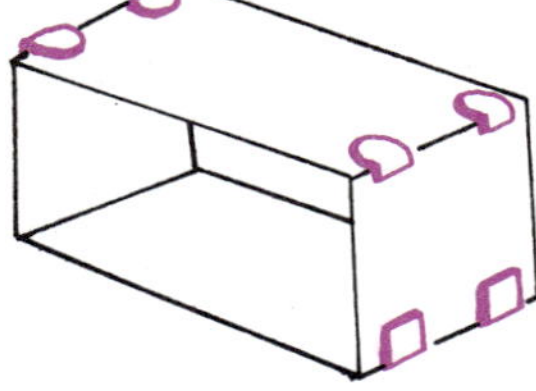

(D) Use a solid joint to provide a rigid frame.

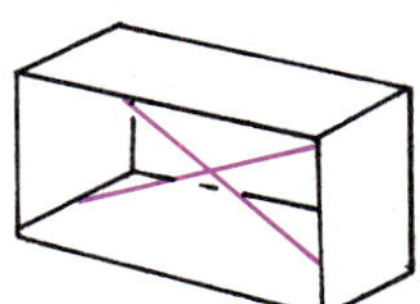

(B) Add internal flexible braces. (Two are needed.)

(E) Redesign and use a more stable form—the arch

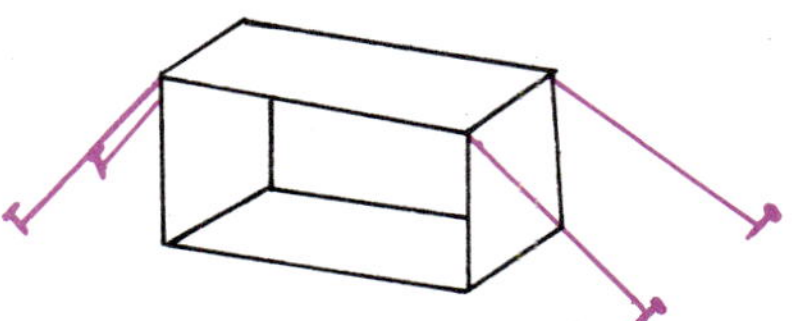

(C) Use external braces. (A minimum of two are needed.)

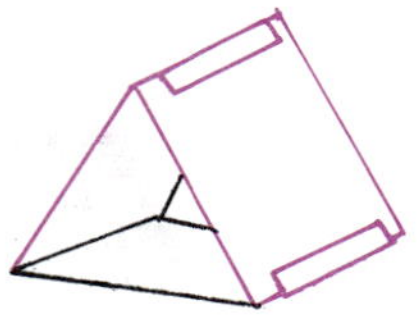

(F) Redesign and use a more stable form—the triangle.

Push and Pull

Remember, any structure must hold up its own weight and the weight of the materials it holds in, holds out or carries away. Some of these loads act on the structure with a force called **compression.** This force causes the molecules of the material to squeeze together. Other parts of the load act with a force called **tension.** These stresses pull apart the molecules of the materials in the structure. Most of the time, the structure handles both types of forces at once. The structure usually is designed to balance these forces.

compression (strength)

tension

Suppose you use a thick sheet of styrofoam as a mattress. It will support you as you lie down. You might use the sheet of foam as a supportive structure to help you float in water. Your weight is a compression force. You can see the effects of compression as the foam becomes thinner in places where your weight compresses it. The structure may be able to withstand a lot of compressive force.

Suppose this same sheet of foam is used to build a bridge. The sheet could be laid across a small stream of water with each end resting on a bank. As you attempt to walk across, the sheet is subjected to compression and tension. The force exerted by your weight pushes the material together on one side and pulls the material apart on the other. If a cut is made in the top and bottom sides of the foam, the effects of the compression and tension forces will be obvious. A load on the top of the sheet will push the cut closer together while the cut on the bottom will separate.

A Styrofoam bridge is not strong enough to support your weight and the resultant compression and tension forces. It will sag or break. To counteract this, you can combine the

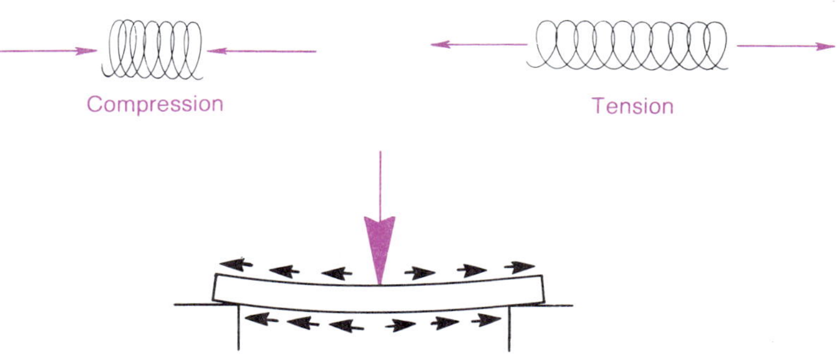

The forces that act on a structure cause stresses on its parts. The two principle types of stress are tension and compression. Tension tries to stretch or pull apart. Compression acts to squeeze or force things together.

A simplified plastic foam model of a bridge illustrates compression and tension forces.

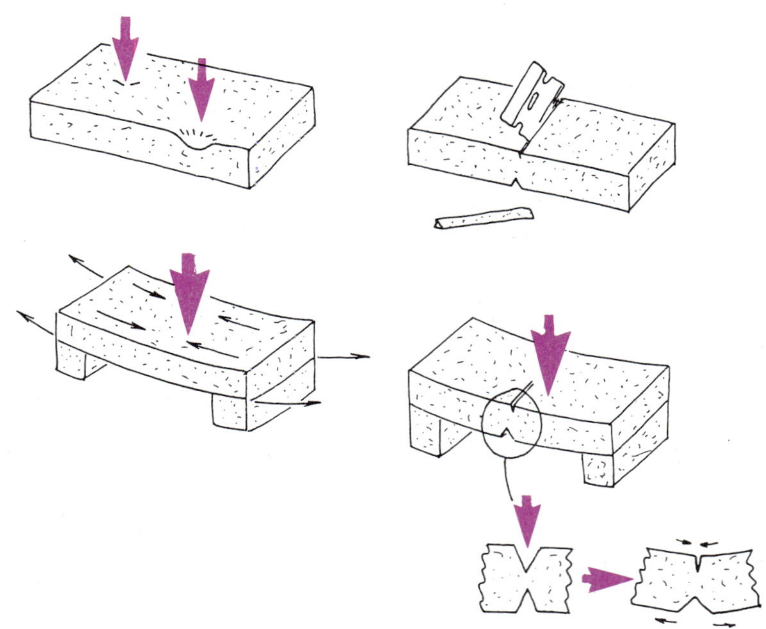

tensile strength

compression strength and the tensile (tension) strength of the sheet of foam. Brace the ends of the sheet against each bank to form an arch. This will increase the overall strength considerably. Tension and compression forces are then brought more into balance.

trusses

A triangular shape used to brace a building also helps balance compression and tension forces. The **truss** is a special kind of triangular brace often used in structures. The triangle provides the truss with a stable geometric form that can be applied in numerous ways.

All structures must be built to withstand forces of compression and tension. Structures must withstand several kinds of weight. If the material itself cannot support the load, a triangle is often added.

Supporting Loads

Structures are designed to support, shelter, contain, direct or transport something. They must resist the pull of gravity on the structure. Bridges, for example, are designed to support the weight of the vehicles and people who travel over them, as well as their own weight.

This truss structure for an oil-drilling rig uses compression and tension forces in the tower and tension forces in the guy lines. The structure has a unique way of dealing with the forces of storms at sea. The base works like a swivel hinge to reduce the strain on the tower. The heavy weights actually lift off the ocean floor to protect the guy lines.

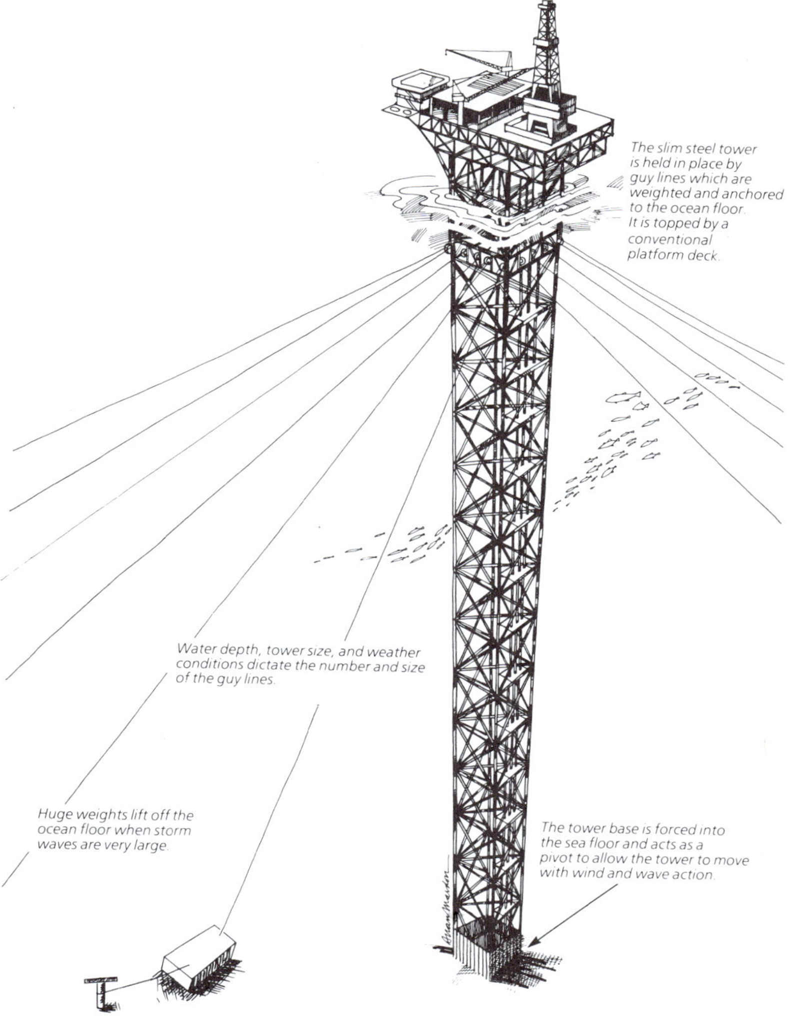

There are two types of load that any structure must support. The materials and design of the structure must work with both types.

static load

The first type of load is called **static load.** This is weight that does not move. Static load includes the weight of the structure itself. It also includes the weight of the objects or materials that the structure must support. Your body on a mattress and snow on the roof are examples of static load. The weight of a sidewalk may not be very important since it lies on the ground, but a walkway over a road will need to support its own weight, along with the weight of traffic. A picnic table may be able to stand by itself, but it may collapse under the weight of food.

dynamic load

The second type of load is **dynamic load.** Dynamic load is weight that moves around. You have probably been warned not to jump on a bed. Your weight while sleeping exerts much less force than when you jump around. In fact, if you jump on the bed, your dynamic load may be double or triple your regular weight. Obviously, such an increase could cause the bed to collapse. Walking or running on a footbridge exerts a different load than just standing on the bridge. This is called a **dynamic load.**

Many structures are subject to strong winds or unequal amounts of air pressure in the form of eddy currents. Strong winds have caused some bridges to collapse. Earth tremors and quakes are important causes of dynamic loads. In regions of the earth where such forces are common, structures must be designed to withstand these additional dynamic loads.

Designing for the Load

In the past people placed a piece of material on top of others in their building attempts. This picture of a Greek building shows a structure that required transporting, lifting and plac-

The ancient Greeks took advantage of the compressive strength of stone. They often built structures that used the post and lintel method. The posts were placed close together because the stones (lintels) that stretched across the top of the posts cracked if the tension stresses got too high.

The Romans developed the arch as a better way to support heavy loads over larger expanses. Since the arch relies almost entirely on compression, it is a structure well suited to stone.

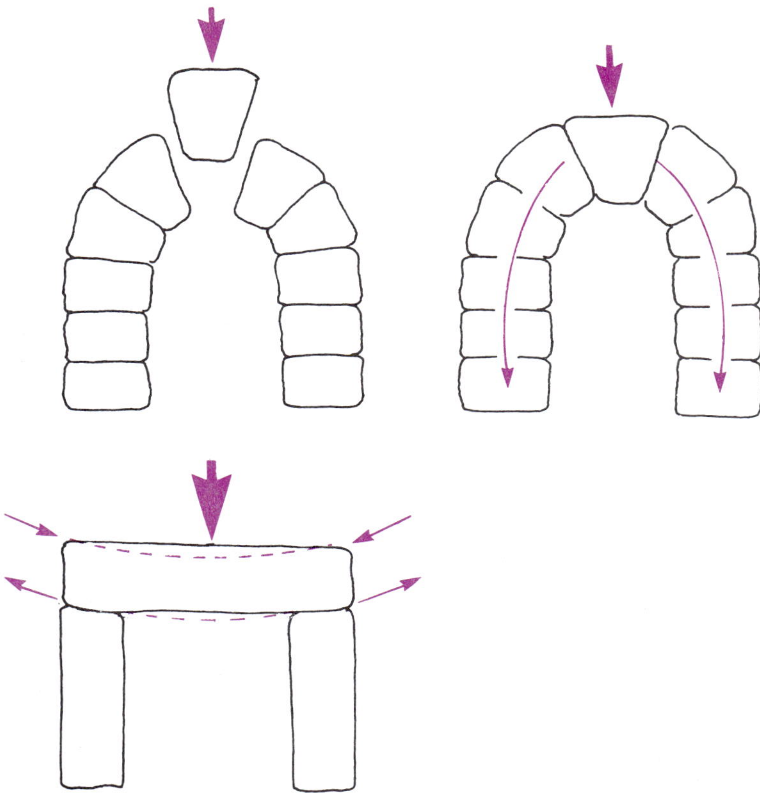

post and lintel structure

ing huge stones. This kind of structure is called the **post and lintel structure.** The support columns are called posts and the piece across the top is called a lintel. This structure is still used in many buildings today.

Later, the arch was introduced as a design feature of support structures. The arch supported loads far greater than could be supported by the same materials in a post and lintel structure. Can you identify modern buildings that use arches?

Early structures were made of stone, wood and other natural materials. As stronger materials were developed, more ambitious structures could be built. Some arch bridges are built of steel and concrete.

flying buttress

The arch also served as the forerunner of the flying buttress. The **flying buttress** is an external support to a building—much like an extended arch—that was used to build truly inspiring cathedrals.

transparent structures

Bridges are good illustrations of construction because

The arch works well with concrete structures, too. This modern looking bridge in Switzerland was designed in the late 1800's by Robert Mallert, the first engineer to use concrete to advantage.

The builders of the Middle Ages controlled the compression forces to construct cathedrals of truly magnificient size and beauty.

they are **transparent structures.** They seldom have walls or roofs to cover up the frame. You can clearly see the structural design of the bridge. The same designs apply to skyscrapers, water towers and railroad tracks. All of these are examples of **rigid frame structures.**

Choosing and Making Materials to Withstand Loads

Early structures were designed to distribute the load. This approach is still used. However, new materials and knowledge have allowed new structures to be built.

Stones in an arch pack together under load. They have high compression strength. A rope tends to fold over itself when pushed together. Ropes have low compression strength. However, they are very strong under tension. Early ship builders took advantage of this in their choice of materials.

Shipbuilders matched materials with the kind of force that each part of the structure was to withstand. The intent was to

The arch can be used effectively as part of truss structures. This bridge in the mountains of West Virginia is the world's longest arch bridge.

tensegrity

reduce the weight of the structure. Sails with ropes attached for control were combined with heavy, wooden masts. This approach is called **tensegrity.** Solid, heavier materials are used where compression is expected. Light, flexible materials that are very strong when pulled are used for parts that must withstand tension. Light line, strong wire, cables and cords are examples of materials used for tension structures. Masts for ships, circus tents, suspended bridges, hang gliders and tents use the tensegrity approach.

Sometimes materials are changed or developed to achieve the design purpose. New synthetic materials are light, but very strong. Even some of the conventional construction materials have been improved.

The first iron bridge used the same principles developed by the cathedral builders. The use of cast iron materials was so new that Abraham Darby, the bridge's designer, used many techniques developed for use in wooden structures.

The sails used to propell ships are relatively light weight structures of tremendous tensile strength. The sails place huge tension and compression forces on the masts of a ship.

Cement was known to the Romans, but was probably discovered by accident. The Romans apparently mined the crude cement they used as mortar in their construction. The "recipe" for cement and concrete was not developed until much later. The Romans would be amazed at today's uses of concrete. Concrete, like stone, compresses well but does not withstand tension. Techniques have been developed to overcome this weakness. Wires are pulled so that they are under high stress or tension. These wires are then covered with wet cement. The cement hardens around the prestressed wires. When the tension on the wires is released, the concrete is packed even more tightly. The prestressed concrete can withstand heavy tension and compression loads.

Plastic and other materials can be reinforced in production to provide strength. One piece of lumber is sometimes glued on top of another to produce strong, laminated beams.

As cathedrals were built taller and taller, builders developed the flying buttress. The flying buttress transferred the stress from the roof of the cathedral to an outside support and then to the ground.

Holding Things Up—Support Structures

support structure
footings
piles
piers

Support structures are usually stable and firmly anchored in one place. They are often built to provide a flat, usable surface, such as a floor. Examples are roads and foundations of other structures.

Most roads, other than those made of dirt, are composed of materials that make a strong, smooth surface to support vehicle traffic. A concrete road is very similar to a concrete wall laid on its side. Over two hundred years ago, a new type of road construction was introduced. The cobblestone approach was quite similar to stone wall construction of that time.

A short section of road may look very similar to a foundation slab for a building. But a foundation also has to hold up another structure. Additional support may be needed in the form of **footings, piles** or **piers. Footings** are solid pieces of material at the bottom of a structure. Footings support a structure just as the edges of the bank support a bridge. **Piles** are solid pieces of strong material that are driven into the earth to hold the structure in a more stable position. **Piers** are supports that are constructed of reinforced concrete, wood or other strong materials. In some cases, only one type of

This modern tent employs the "tensegrity" approach. It uses the minimum amount of materials (in compression) to support the tent shell.

foundation will be used. In bridge construction, it is not unusual to see all three used together on a single structure.

poured footings

Poured footings, usually of concrete and metal, are used when the earth can support the weight of that building fairly easily. If a structure must be erected on earth that cannot support the structure, such as clay or sand, piles are driven through the soil until they reach bedrock. **Bedrock** is a layer of rock under the topsoil that does not shift as soil does. A foundation is poured on top of the piles. Sometimes, bedrock or suitable load-bearing soil is too far below the ground surface. Holes must be dug deep in the ground and concrete poured into them. These concrete piers then provide support for the building.

bedrock

Support structures may also hold materials high in the air. Towers are often truss structures that are built for height. They are similar to bridges except that they are designed to support their own weight plus pressure from the wind. Some towers are used for communication, such as TV broadcasting. Towers may be strengthened by the use of **guy wires** or cables. These wires provide vertical stability and add little to the weight of the structure.

guy wires

In 1764, Tresaquet developed a new form of road construction that used a base of vertical stones, like a wall laid on its side.

In some cases, it is necessary to sink supports into the ground to provide a good foundation for a structure. A builder may choose to use piers of poured concrete or to drive piles of wood or metal deep into the ground.

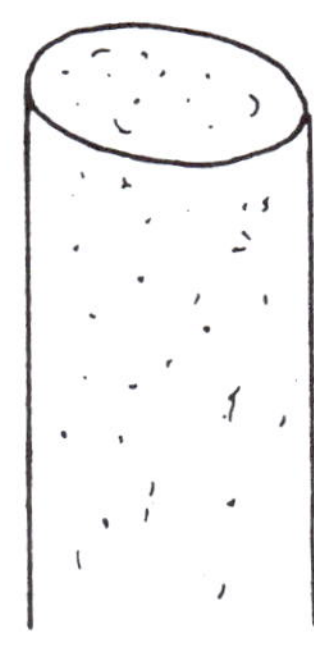

Piers

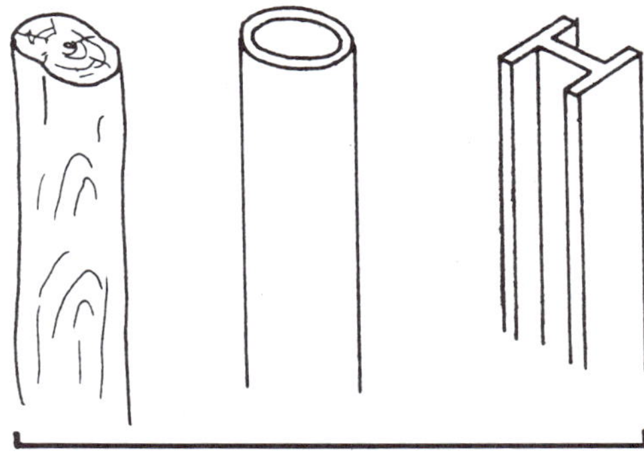

Piles

This example of a foundation for a modern building uses a poured concrete footing with a keyway to lock the wall to the footer. The reinforcing bars add tensile strength to the structure.

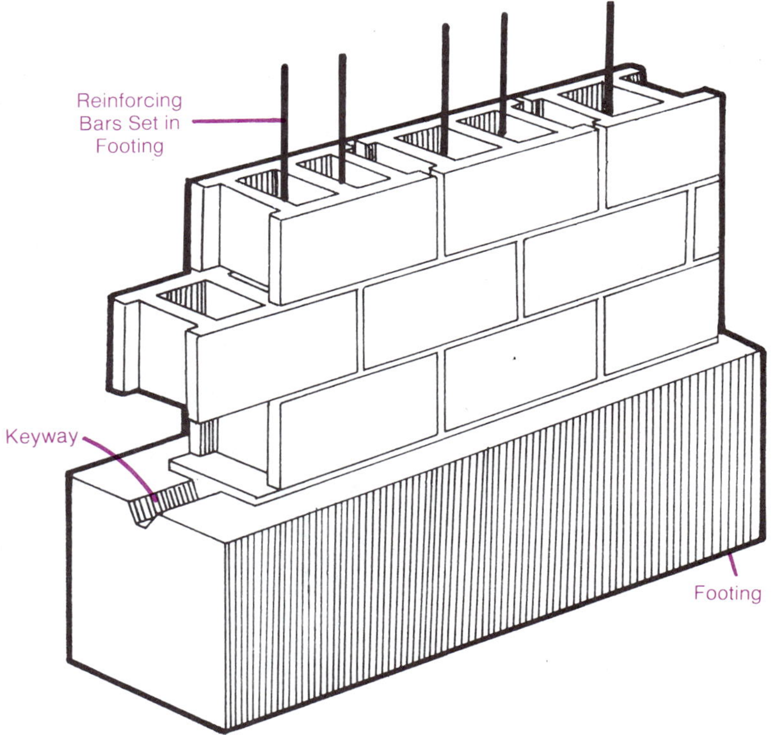

This stone wall was constructed without mortar. Each row of the wall is sloped slightly to make the top smaller than the bottom. In this way the wall presses in on itself.

Some support structures must withstand the moving load of the wind. Some must carry heavy moving loads far above the earth or water. Others support the structures that move over them (such as trains). Still others stand up under loads of snow, earth or water.

Holding Things Out—Sheltering Structures

Sheltering structures often need little more strength than that needed to hold themselves up. Most sheltering structures have roofs. Some have walls, and others have floors and foundations. This all depends on what the material inside is being sheltered from.

Early wooden bridges were covered to protect the bridges from the effects of the weather. The covered bridge is a simple example of other sheltering structures. Most shelters have a roof to protect the structure itself. There are many different kinds of roofs, with many shapes and styles. All roofs, however, are designed to provide a watertight covering, primarily to hold out the weather.

Support structures for roofs are quite similar to those of bridges. Roofs can be designed to use the same principles found in suspension bridges. This approach is not as widely used as the truss, arch and post and lintel. The suspended roof, however, provides more freedom in design.

Railroad ties form the support structures for railroad tracks. The tracks serve as both support and directing structures for railroad trains. The replacement of railroad ties and the maintenance of the track is a continuous effort to insure that the railway—the foundation for the trains and cars—is strong and safe.

Tension structures, such as tents are usually not considered for large buildings. These structures, used in an airport in Saudia Arabia, illustrate the potential of the tent. You can get an idea of the size of the structures by comparing them to the nine trucks parked in front.

In geographic areas where there is a great deal of snowfall, roofs are constructed to withstand the extra weight. Likewise, roofs on shelters where severe storms and strong winds often occur must be built to withstand those forces.

Walls are not always required. Some shelters need only be strong enough to support the roof. In those cases, posts may be used to hold up the roof. The sides of the building remain open.

Floors and foundations may or may not be required. Portable shelters, such as tents, need little foundation for stability when erected. The floor may be the earth itself. Mats may be added for comfort, or a waterproof floor may be desired.

A computer analyses of the proposed structures helps designers and builders maximize strength of the support using a minimum of materials.

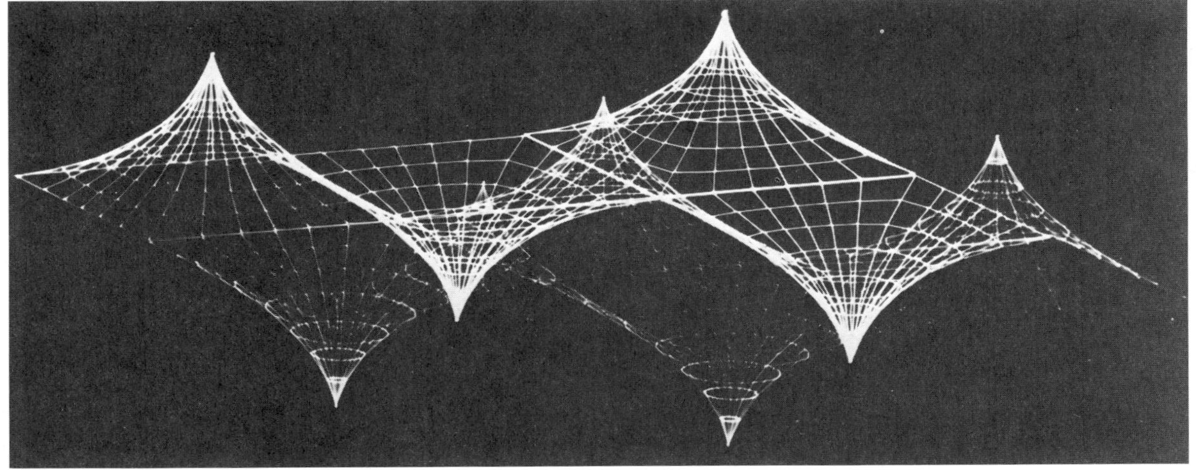

Another type of tension structure can be inflated like a balloon.

tents

Tents are sheltering structures that integrate the roof and walls into one unit.

inflatable structures

Inflatable structures are essentially large balloons of durable materials held up by air pressure. The inflatable building is relatively inexpensive and requires almost no support. The structure is supported by the difference between the air pressure inside and the pressure outside the balloon. Inflatable structures are often used to cover large athletic fields because they require no posts or columns.

shell structures

Quonset hut

One of the early **shell structures** was the Quonset hut used during World War II. The **Quonset hut** is like a large barrel cut in half lengthwise and erected on a concrete or wood founda-

Inflatable structures used to cover football stadiums are giant balloons.

The geodesic dome combines tension and compression forces in a unique tensegrity structure. It is easy to spot the triangular form similar to the truss.

tion. The walls and roofs of Quonset huts were combined in the structure to minimize cost. **Space frames** are like shell structures, but use less material. The cost of such buildings is lower than older approaches. The geodesic dome is one type of space frame.

space frames

geodesic dome

Holding Things In—Containing Structures

Many structures hold or store things. Containing structures serve not only to store materials, but also to restrain and direct materials. Containing structures help control the environment in which we live. Usually, the walls are the most important part of containing structures.

There are many different kinds of storage structures. Water towers store water for use in homes and businesses. Oil and gas tanks store fuels. Grain elevators and silos store feed for animals.

This earth-dam uses a water-proof core to contain a large volume of water.

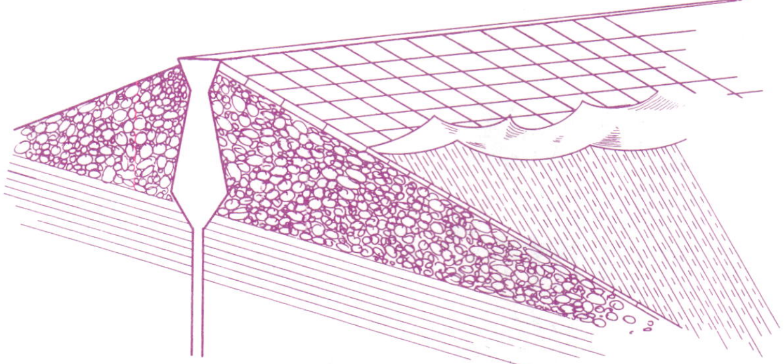

restraining structures

Restraining structures are used to keep something in place. Jails and animal pens are restraining structures. Levees are used to keep river water within a passageway. Ocean and river waters can be restrained with a dam. The most extensive and famous system of dams and levees is found in Holland. Two-thirds of the people there live on land that has been reclaimed from the sea.

Structures That Direct

pipelines

Directing structures provide passageways by which vehicles, people and materials are transported. The uses of these "passageways" are important in the technological activity of transporting. They are a unique kind of structure. Directing structures contain materials, usually liquids, and control the movement of those materials. Pipelines have been built in many parts of the world. Pipelines are designed to carry gases, liquids and some solids. Solids, such as powdered coal, can be mixed in water or oil to make a slurry. The slurry, something like runny pancake batter, can be pumped through a pipeline. The shape of a pipeline and the path it takes directs and controls the movement of the materials to the desired destination.

aqueduct

The Alaskan pipeline is a sophisticated offspring of a similar structure nearly two thousand years older. The Romans built directing structures to move water to their cities. They used gravity to move materials, a principle that still makes sense today. All directing structures are much more efficient when they use gravity. Much can still be learned from the work done by early aqueduct and canal builders.

The Romans made extensive use of the arch as illustrated by their impressive systems of aqueducts for directing and moving water.

Moving Things—Transport Structures

Transport structures are portable containing structures. They usually have a floor and walls. They may also have a roof. Flat railway cars can carry more cargo if they have containing walls. A roof may be needed to shelter the cargo from the environment. Transporting structures for gases need a floor, walls and a roof to keep the vapors from escaping. Often the structure is built in such a way that the roof and walls are in one piece, like a bottle.

vehicle

Transporting structures provide the vehicles that are essential to the activities of transportation. The work of early builders of transport structures has given us the knowledge necessary to build modern structures. For example, the concept of "rigid frame" is the basis of any skyscraper. This concept came from early barnbuilding techniques. Barnbuilding methods developed from earlier shipbuilding techniques.

rigid frame structures

There are three basic types of transporting structures: **floating, rolling** and **flying. Floating structures** can be awe-

floating structures

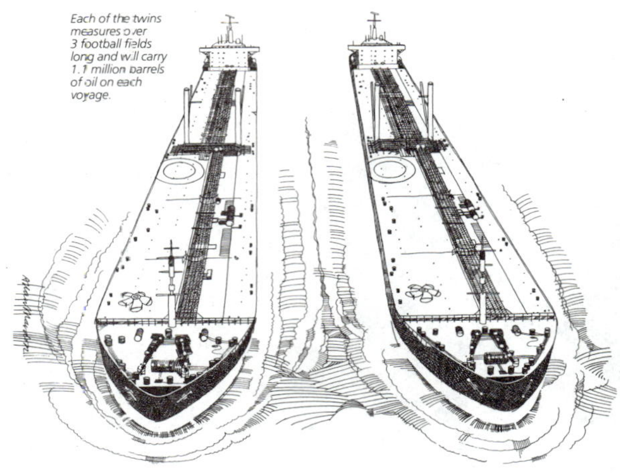

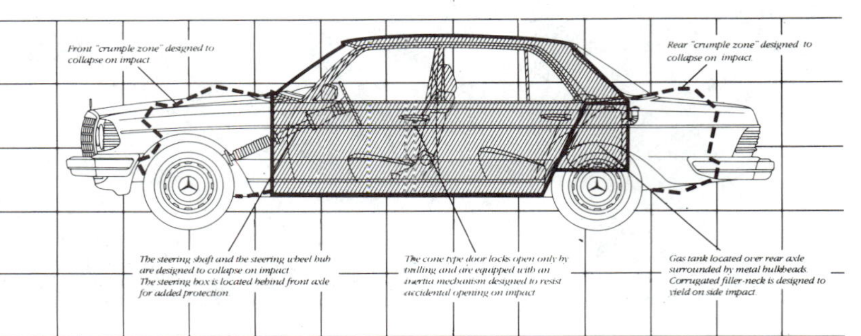

These moving structures include oil tankers longer than three football fields, a one-person kayak, an automobile designed to protect its passengers, a medium sized commuter jet and a railroad car from the Civil War era. Can you identify different design and construction problems these structures present?

Canals are viaducts that use some of the same principles of construction as the Roman aqueducts. Viaducts use gates or "locks" to allow a boat to be raised or lowered to different levels.

some. From the stern or back end of a modern supertanker, the front of the ship is nearly a quarter of a mile away. On the other hand, a kayak is so small and light that it can be lifted with little difficulty. Despite its size, the kayak can be used in rough water by those who are skilled and brave enough to venture out on the ocean or on white water rivers.

There is much diversity in the types of floating structures. Many different materials were available to the early shipbuilders. In some places, reed and papyrus were the only materials that were plentiful enough to use for boat building. These boats had limitations, due to the delicate nature of the reed and papyrus. All materials have limitations, however. Reed, hide, wood, iron, concrete and steel all have distinctive characteristics that influence the design of floating structures.

All structures are influenced directly by the technology available to the builders. Boats built today are very different from the boats built 500 years ago. Compare the tools and materials of today with those that were available to the early builders. The available technology is very different.

Boats and ships were the first sophisticated moving structures. The building of this ship in Japan, about 1766, used techniques that were developed and improved over centuries.

Rolling Structures

rolling structures

impact structures

The form of a **rolling structure** is significantly different from a floating structure. Floating structures have large surface areas over which the weight is distributed. Rolling structures focus their load on very small points of contact between the wheels and the frame supporting the vehicle. The frame of a rolling structure, such as an automobile, must be able to support the total weight of the vehicle and its load. Compare a car body to a bridge. The foundation of the bridge represents the

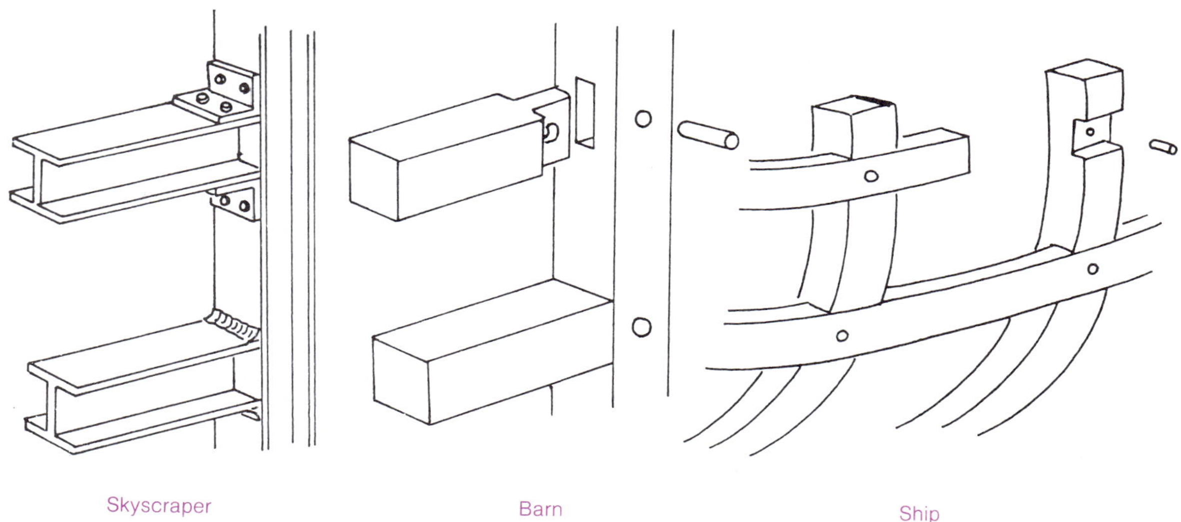

The techniques for support of a skyscraper evolved from techniques used in building large barns and wooden ships.

The wheel is a rigid frame structure of tremendous technical complexity and historical importance.

wheels of the car. The major difference between the car and the bridge is that the load in the car does not move, whereas the load on the bridge does. Another difference is that the car itself moves over its foundation, while the bridge does not.

Rolling vehicles often become **impact structures.** A car must be designed as a high-impact, collapsible structure. The materials and design of the vehicle allow for the loss of mechanical energy if it should collide with another vehicle or non-moving object.

The shape of a rolling structure is a rigid frame for the passenger compartment that sits on a main frame. The main frame transmits the weight of the vehicle to the wheels. The rest of the structure supports the necessary parts of the vehicle, such as the engine, and serves as a buffer for the passengers, in case of impact. Cars, trucks and other transport structures have not always been designed as efficiently as possible. Only recently have trucks been equipped with devices that serve to cut down wind drag. Very few trucks have been designed with the consideration that they are vehicles that must pass through an ocean of air. Similarly, the bicycle has been improved primarily by the reduction of its weight, not by adapting its shape to the environment in which it operates.

Flying Structures

flying structures

The development of **flying structures** has been considerably different from floating and rolling structures. A different kind of knowledge was needed to get a structure to fly. Boats could be made to float and carts to roll with little need to understand why they worked. Airships and airplanes were more difficult to design, however.

Airships are *lighter* than air. They are similar to boats in that they must displace enough of the material in which they travel to equal their weight. A blimp is an example of an airship.

lift

Airplanes do not work as airships do. Airplanes are lifted by the flow of air over their wings. This flow creates a vacuum that pulls or holds the wing and the airplane up. The wings of an airplane are very strong beams that are specifically designed to hold up the weight of the plane during flight. The airplane structure must be able to withstand the static and dynamic loads experienced in flight. When an airplane is flying straight and level, the weight that the structure must support is very predictable.

G-load

An airplane's dynamic load and stress increase a lot as the plane moves through turns, dives and pullouts. These movements must be made against the pull of gravity. These maneuvers increase the "G-load" of the plane. The **G-load** is the weight that must be pulled or lifted against Earth's gravity. A G-load of *one* is equal to the plane's weight at rest. A plane pulling out of a dive has a G-load much greater than one.

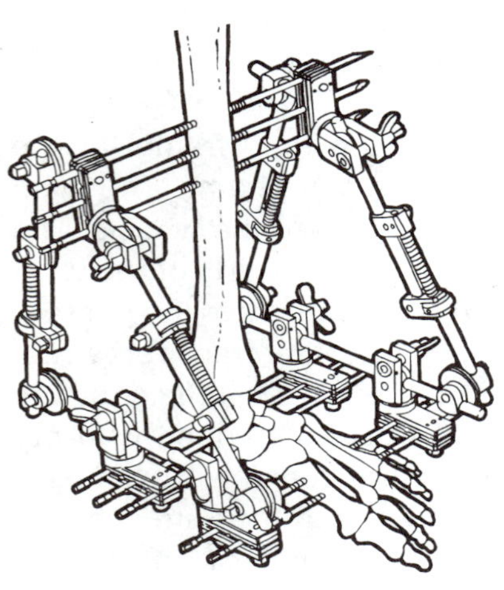

The stability of this rigid frame structure is designed to keep damaged bones immobile as they heal.

There are immense structural loads on an airplane pulling out of a steep dive. Lifting force must be developed to keep the plane from continuing the dive and crashing. The lift must be developed by the wings and may have to exceed two, three or four times the plane's weight. This is equivalent to 2 G's, 3 G's or 4 G's of load. Only acrobatic airplanes are built to handle loads of above 4 G's. Steep dives at high speeds can snap the wings off a standard airplane.

Spacecraft also present many construction problems. They must be built to withstand the large G-loading required to get them out of the earth's gravitational force. They must withstand that pull again during landing. Yet, the demands on a spacecraft are not as severe as one might think. The major problem is to contain the required environment so that equipment can operate properly and humans can live in safety and reasonable comfort.

Summary

Structures play an important role in the technological activity of constructing. All structures must be designed to hold themselves up under moving and stable loads. This is accomplished by the choice of materials and the structural design. There are supporting structures that hold something up; sheltering structures that hold something out; containing structures that hold something in; directing structures that move something away; and transporting structures that move themselves and their loads away.

Constructing provides necessary environments for shelter, movement and work. The relationship of constructing to transporting is complex and varied. There are directing structures, rolling structures and flying structures. Each has unique construction problems.

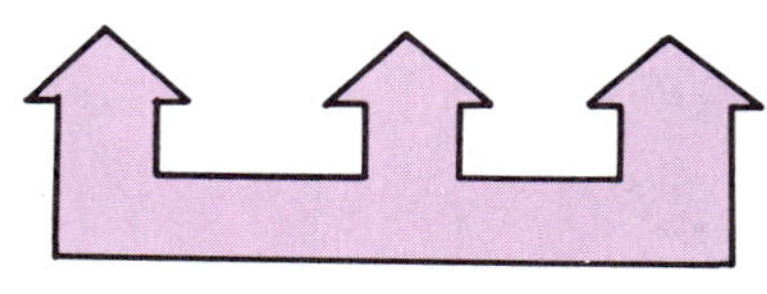
Supporting

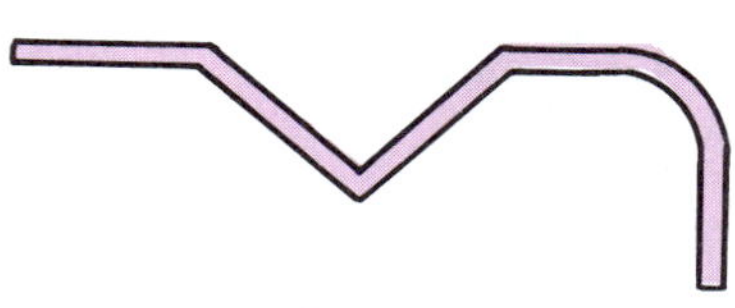
Sheltering

Containing

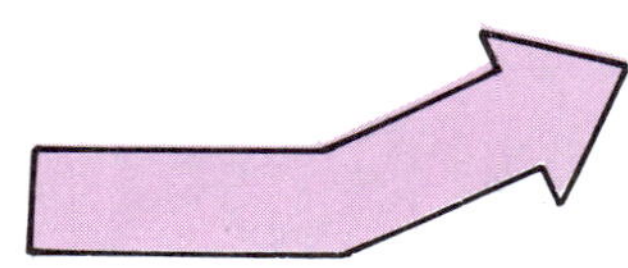
Directing

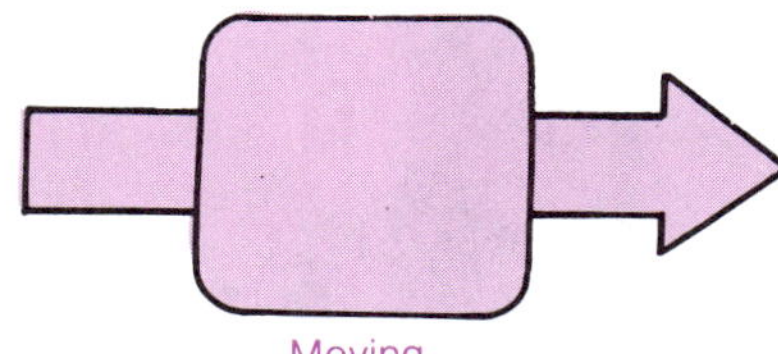
Moving

Through constructing as a technological activity, we can build an almost countless variation of structures. All these structures, however, take one or more of five basic forms to serve the function of supporting, sheltering, containing, directing or moving.

Key Concepts and Terms

aqueduct
arch
bedrock
brace
compression (strength)
constructing
containing structures
directing structures
dynamic load
floating structures
flying buttress
flying structures
footings
foundation
G-load
geodesic dome
guy wires
impact structures
inflatable structures
lift
piers
piles
pipelines
post and lintel structure
poured footings
prestressed concrete
Quonset hut
restraining structures
rigid frame structures
rolling structures
shell structures
shelter
sheltering structures
space frames
stability
static load
structures
support structure
supporting
tensegrity
tensile strength
tension
tents
transparent structures
transporting structures
trusses
vehicle

Chapter 9 Transporting

subterrene
VHST

Imagine going from California to New York by subway in twenty-one minutes. Such a subway would only use wheels for slower speeds. The cars would be suspended by a magnetic field and would run in tunnels melted by a nuclear powered machine called a **Subterrene.** The Subterrene would melt rock to form a glass lined tunnel in which the **VHST** (Very High Speed Transit) vehicles would run or, more accurately, fly. The air in the tunnels would be evacuated so that the projectile-shaped tubecraft could achieve speeds approaching 14,000 miles per hour.

transport system

According to researchers of the Rand Corporation, this system of transportation is possible even now. The system would be very expensive if implemented. Supporters of the proposed system claim that it would be considerably cleaner and more efficient to operate than many of the existing forms of transportation.

A nuclear-powered subterrene that could melt underground tunnels in rock has been designed. Tube-shaped vehicles would carry passengers through these high-speed subway tunnels faster than the speed of sound.

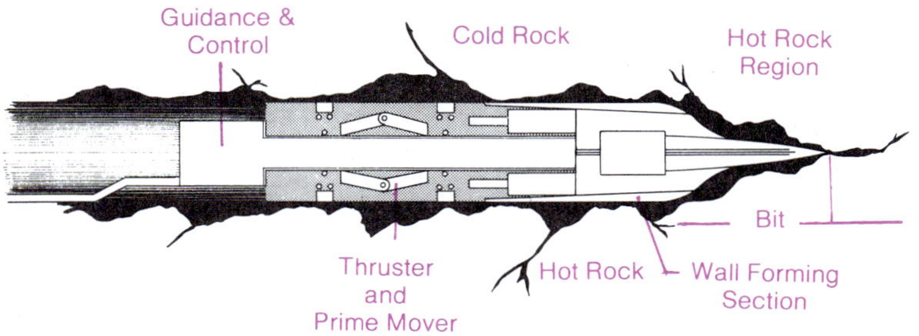

The last 75 years has seen huge advances in transporting—from a horse and buggy ride on unpaved roads early in this century to a buggy ride on the plains of the moon only a few years ago.

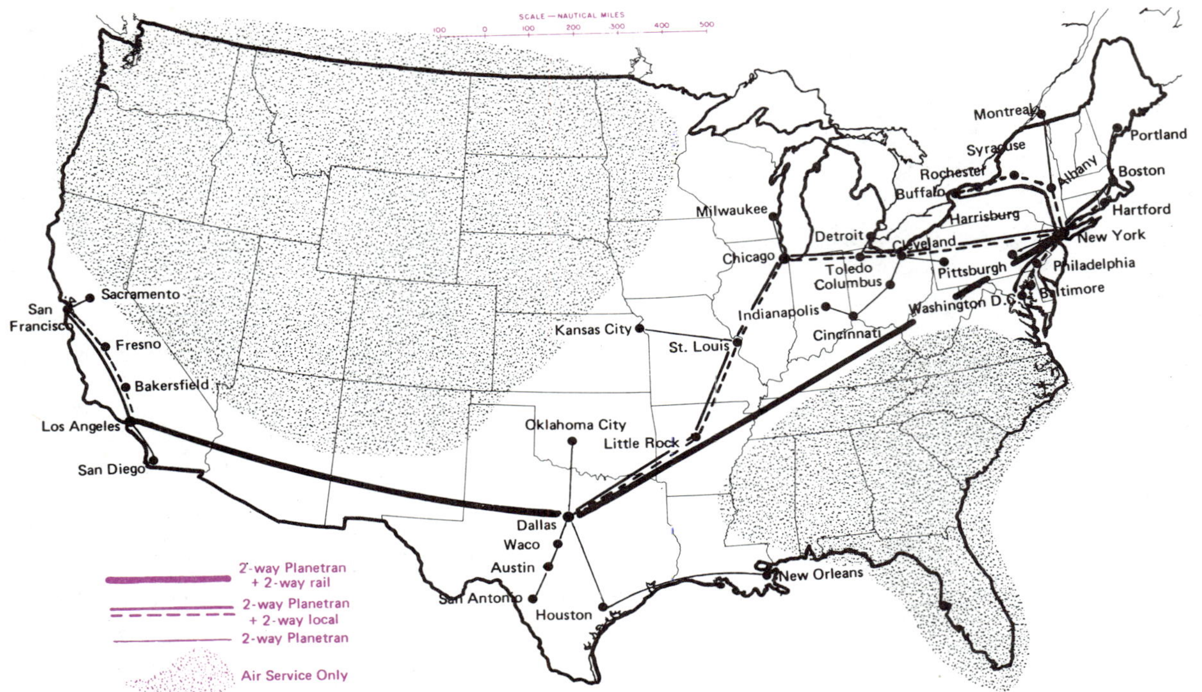

Rand Corporation has proposed a system of airless tunnels linking major cities from coast to coast.

one-directional transport

The VHST system may seem very unusual. But it is merely a sophisticated example of a transport system confined to two directions. It can only go back and forth within its underground tunnel. An example of **one-directional transport** is piping water through a garden hose. The water moves in only one direction, away from the faucet and out of the end of the hose. In this chapter, we will look at other examples of transport that move in one or more directions.

Moving

use of transport

Transporting is the technological activity in which people, machines, materials or other substances are *moved* from one place to another. Transporting involves all six elements of technology. For example, transport machines use materials for energy conversion processes, which are controlled by humans through the symbols and signals of information.

Vehicles, Pathways and Purposes

vehicles
pathways

All transport systems include some form of *vehicle.* This vehicle travels along a *pathway.* The travel is for some definite

Electrical power is transported (distributed) through these high tension lines.

purpose. The key ideas in transporting are **vehicles, pathways, loads** and **purposes.**

loads
purposes

The simplest transport systems have the most controlled pathways. The vehicle travels in only one direction. Some transport paths are more restricted than others. For example, compare the VHST tunnel with the airways used by standard airplanes.

Paths: One-way, Two-way, All-ways

ways
directional
two-directional transport

The concept of "ways" is a key to understanding the differences between types of transportation. **Ways** are the pathways from one place to another. Ways include roadways, highways, seaways and airways, as well as tracks and canals. The type of pathway used depends on the number of directions in which the vehicle is to move. Transport in one direction usually requires the most expensive forms of ways. The vehicle must be more confined and controlled in its movement. Examples are railroad tracks, high speed tunnels, subways, elevated rails, conveyors, and long pipelines. They all cost a great deal of money to build and maintain. Movement in one direction also needs a separate path built for flow back to the source. The pathways used in **two-directional transport** also cost a lot for construction and maintenance. In comparison, seaways and airways are relatively inexpensive. However, they provide relatively no control over the vehicle. Each type of path has its advantages and disadvantages.

One-way transport allows movement in only one direction. Ski tows, escalators and firefighter's poles support travel in only one direction. Materials moved by such devices as

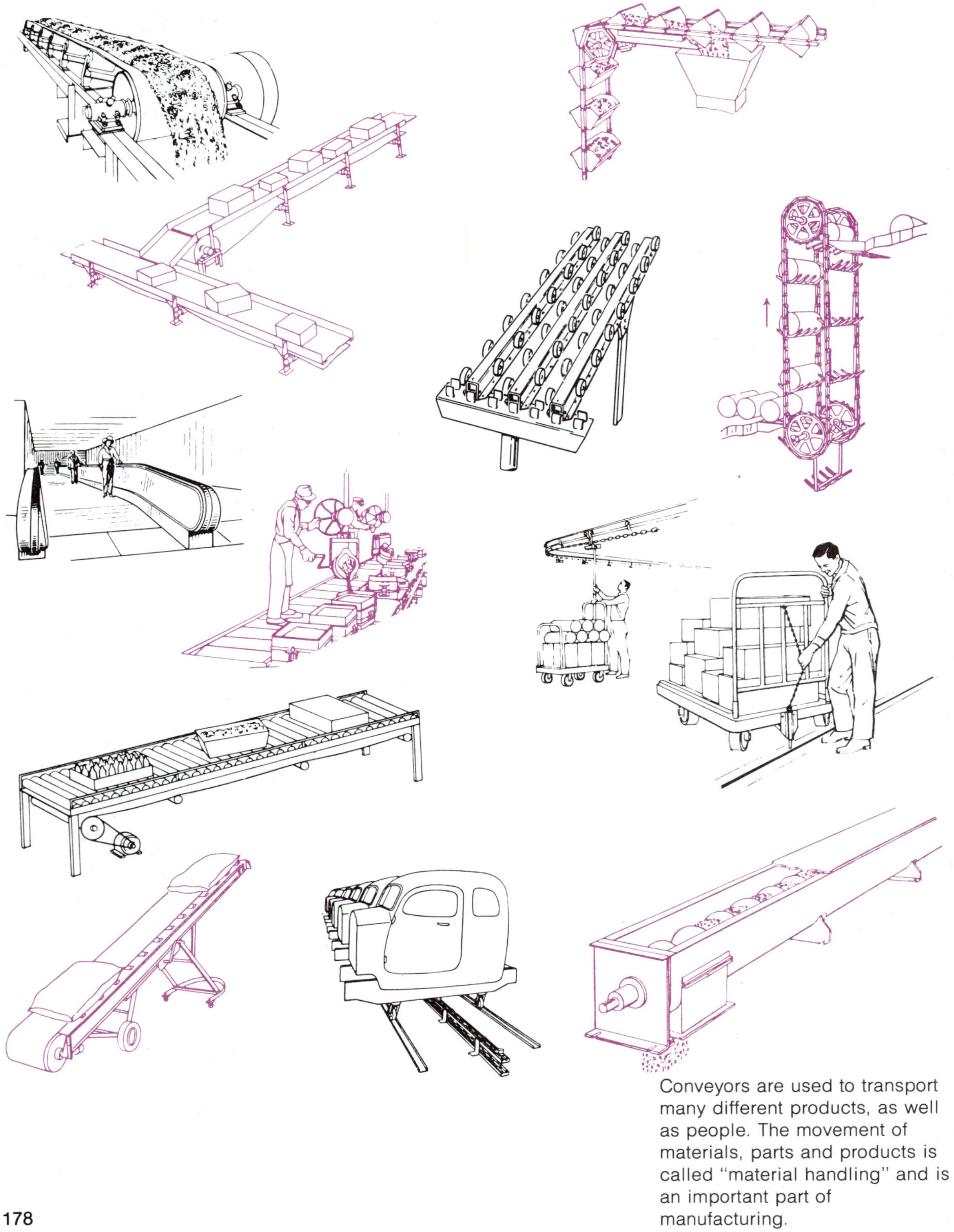

Conveyors are used to transport many different products, as well as people. The movement of materials, parts and products is called "material handling" and is an important part of manufacturing.

When it is difficult to build a track on the ground, one solution is to hang the train from an elevated track. This modern looking system—built about 75 years ago in Wuppertal, West Germany—may be considered the world's safest railroad. The built-in safety factor of "two degrees" of freedom, a sophisticated guidance and control system on the trains and the overall systems approach to the design and operation of the train have prevented anyone from being killed since it was opened.

pipelines, aqueducts and conveyors are also restricted to one direction of travel.

degrees of freedom

Two-way transport allows travel in two directions. Examples are railroad tracks and office elevators. The travel of a railroad train is restricted. It can only move forward and backward on the tracks. The train has no freedom to move anywhere but along those tracks. Travel in two directions is not limited to trains, however. Some ferry boats move along a fixed rope or chain as their guide track. An early bridge used a supported rope as its track. Some people use clotheslines

This personnel carrier moves automatically along a concrete track. Since it can move forward and backward, it is considered to have two degrees of freedom.

that work in the same way. Self-propelled people movers also operate on this principle.

three-way transport

Three-way transport allows movement forward and backward and to one side. An example of this type of transport is a sailboat that uses square sails. It can sail in all directions except directly into the wind. It was not until the invention of the lateen sail and the stern rudder that sailing ships could tack or zig-zag into the wind. Earlier boats with square sails and side and steering oars were, to a great extent, blown with the wind. In the South Pacific, a different approach was used to tack into the wind. Rudderless boats that could sail either frontward or backward were used. These boats had a sail that could capture the wind at a sharp angle, and move the boat into the wind.

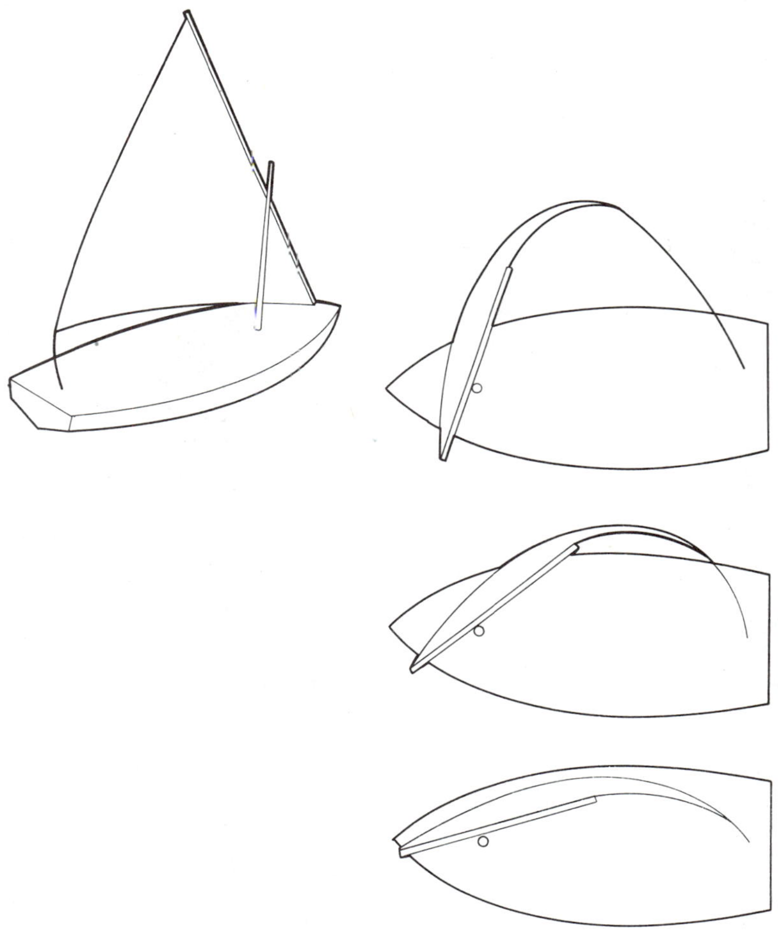

The development of the lanteen sail allowed more precise control in the piloting of sailing craft. The sail could be turned to catch even light breezes, making the sail more efficient.

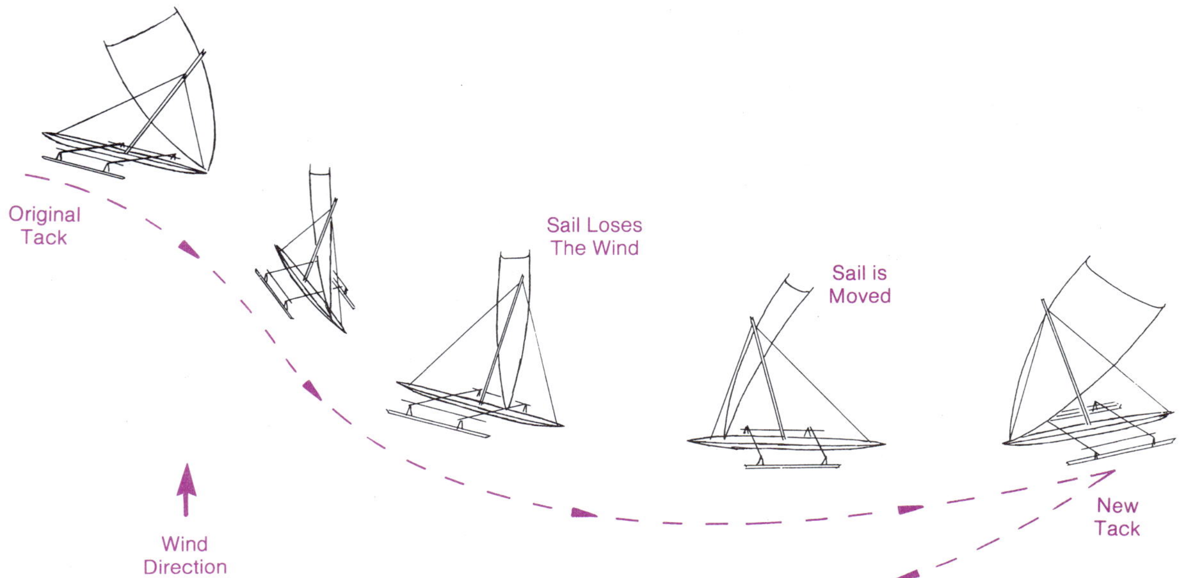

The type of sailing craft with outriggers is still used in the South Pacific. The craft has no front or back and can sail in either direction. To change directions, the sail and its support structure must be moved to the other end of the boat.

four-way transport

Boats that float on the current of a river are also examples of three-directional travel. They can move forward, right and left but cannot move backward against the current.

Four-way transport is the most widely used of any of the forms. Travel by foot allows four directions of movement. The early transport of goods and materials was done by people carrying the burden on their backs or heads. Pack animals made the transport easier.

The introduction of conveyances for dragging, such as the sledge and the Indian travois, also influenced four-way transport. They were used on well-worn pathways or trails. This approach became common in land transport and the beginnings of roads were seen. When the wheel was used as part of a wagon, the need for roads was already established. Roads, however, were slow in coming. The entire Greek empire had only one paved road. During the time of the Roman empire, roads were improved. But in most places, roads were primitive pathways until the Industrial Revolution. The lack of good roads, however, did not interfere with the significance of the wheel in the growth of transportation.

The relationship of the road and the wheel in moving heavy loads provided the basis of a land transport system that continues today. Roads restrict four-way movement and change it to two-way movement. With heavy loads, the wheel requires

The first automobile was actually a tractor to pull artillery pieces. It was steam driven and developed by the frenchman Cugnot although slower than a person on foot, it was the first self-propelled vehicle that had a full four degrees of freedom.

Air effects vehicles such as this one can carry passengers and cargo with four degrees of freedom over land and water.

the smooth surface of a road. The vehicle then has to follow the direction set by the people who built the road.

For years, powered water travel, especially on the seas and oceans, represented the extreme in four-directional travel. There were no hills to climb or go around. There was only a flat surface on which to sail. Even the earliest of water travelers had this range of freedom. The boats they paddled or poled could move in four directions.

Many early people had some form of large water craft. These vessels were usually limited to drifting or, at best, sailing. They were limited to two or three directions. The introduction of steam and, later, engines powered with fossil fuels changed this. Today, four-directional travel on the seas with very large and fast ships is possible.

five-way transport

Five-way transport or travel in five directions can be done in the air and beneath the surface of the ocean. Fixed-wing aircraft can move in five directions, every way but in reverse. Some submarines travel in five directions. Airplanes that go up will come down; and hopefully, submarines that go down will come up. In both cases, however, there are set limits in which the travel can be accomplished.

The space shuttle, like most other airplanes, has five degrees of freedom. Space craft and airplanes than can fly backward have six degrees of freedom.

lighter-than-air vehicles

Lighter-than-air vehicles and submarines both float in their own oceans. Lighter-than-air craft, such as hot air balloons and blimps, depend on the air to hold them up as they move. A submarine moves under the surface of the water. The movement of a balloon up and down in the air and a submarine up and down in the water is accomplished by changing their buoyancy, or ability to float. A hot air balloon can be made lighter by dropping weight out of its carrier basket or by heating the air in the balloon. It becomes heavier, in a manner of speaking, by letting off small amounts of its lifting gas or reducing its hot air source. A submarine is made heavier by taking on more water; lighter by pumping out the water.

heavier-than-air vehicles

Heavier-than-air vehicles operate in a similar manner. They move in an ocean of air. A great deal of lifting energy is required to get a heavy airplane or helicopter off the ground and into the air. The aircraft must continue to move through the air rapidly enough to create a low pressure area above its wings. This low pressure zone creates a vacuum, a form of suction, that holds the craft up.

propulsion
navigation
control

Vehicles capable of traveling in five directions must have its methods of propulsion, navigation and control. All must maintain an established relationship with the force of gravity. The propulsion, navigation and control of these craft are accomplished in relation to the closest horizontal plane of the earth's surface.

six-way transport

Six-way transport is possible by helicopters, submarines and space craft. True sixth direction of movement removes

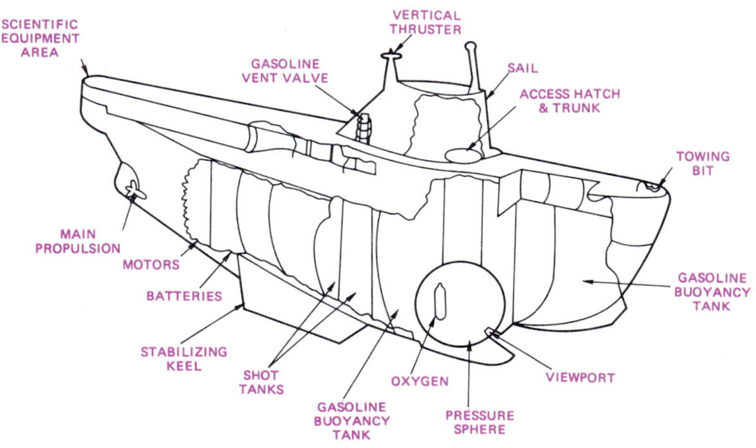

This submarine is especially designed for deep water exploration. Since they can move backward submarines have six degrees of freedom, limited only by the depth they can reach and the ocean surface.

the restriction of gravity. Consequently, the spatial reference that gravity provides is also removed. Right and left, up and down and back and forth can no longer be determined by referring to the earth. This freedom of movement is possible until another planet catches the spacecraft in its gravitational field. Then, a new reference plane is established. "Up" and "down" are in relation to the surface of that planet. The freedom of movement possible in outer space can pose some problems for space travelers. Without reference points, it is difficult to keep track of where you are going. It is also difficult to keep a sense of time.

movement

All spacecraft have definite limitations in the amount of energy they carry. Because the planets themselves move, the possibility of reaching other planets from Earth is considerably better when they are closer to us. It would be entirely possible for a craft to use up its energy before reaching the target. Time, speed and available energy are critical. There are dangers in having too many directions of movement possible in transportation devices.

More Freedom—Less Control

The VHST tubecraft discussed at the beginning of this chapter is an advanced form of two-directional transport. The uniqueness of this system is that part of the controls are built into the path. The flow of vehicles or materials is controlled by the pathway itself. Obviously, many more controls must be built into the tracks of a train and the high-speed tubecraft than into a slower pipeline or canal. The potential of one- and two-directional movement is high because the control can be built in. To hold moving materials within a pipeline or move a

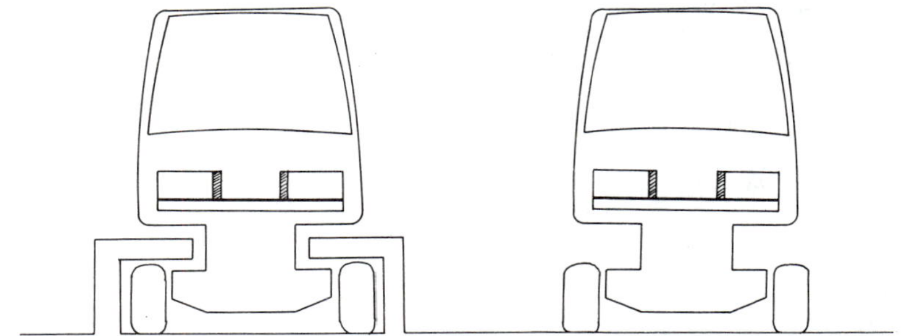

This vehicle is designed to run either in special tracks or on regular roads. Such an approach would provide the safety that can be built into a tracked vehicle and still allow the flexibility of a regular car with four degrees of freedom.

vehicle on tracks enables the automatic control of that movement. Travel at thousands of miles per hour in a tunnel below the surface of the earth can be made safe. It can be safer than travel at the same speed in aircraft flying above the surface of the earth.

Freedom of movement has both advantages and disadvantages. More freedom means less control. Safety is not the only factor to consider. It might be desirable to turn the navigation and guidance of a vehicle over to an automatic system. It becomes more difficult to attain this automatic guidance and control as the number of potential directions increases.

Containing and Propelling

When people think of transportation, they often think first of trucks, cars and planes. But water is transported in pipeways, as is oil, natural gas and other materials. Some materials are transported in pathways that also provide the method of containing the material. There may be devices built into the pathway to keep the flow distributed evenly. Other devices may measure the amount of flow.

Railroad tank cars, especially those that carry dangerous liquids, present unique design problems for their builders.

Some types of transportation require separate structures for containing the goods, materials or people. These containers are often called the **transport unit.** Examples are cars in a train and modular containers to load on a truck bed.

transport unit

Some of these transportation units have their own power or propulsion. Some must be attached to a power vehicle. Others are propelled by devices within the pathway. Tractors, trailer trucks, railroad engines and tugboats are examples of vehicles that provide the basic power for moving the transport units.

Vehicles that combine a propulsion unit with a container unit are relatively common. Such a vehicle is the automobile. The automobile unites the engine and the cargo space, or passenger compartment in this case, into one combined unit.

Load of Feathers versus a Load of Bricks

Any vehicle will have a limit to the amount of cargo it can transport. In some cases, such as the hauling of rolls of fiberglass insulation, a vehicle may be filled in terms of space but carry only a fraction of its weight potential. In other instances, such as hauling bricks, there may be plenty of space left in the vehicle but the maximum weight has been reached. At this point, two related ideas enter the picture: **useful load** and **operational load.** The useful load, in the brick hauling example, is the number of bricks that we are able to haul. The operational load is the weight of the truck, driver, driver's helper, fuel, tie-downs, pallets, spare tire and the like. All vehicles have load limits in terms of their strength and power. It is possible to reduce the operational load so that the useful load can be increased. For example, by reducing the amount of fuel in the tanks, removing the driver's helper, and reducing the number of tie-downs and pallets, the total useful load could be increased. Lighter materials in the structure of the vehicle can also increase its efficiency, as long as the strength of the vehicle is not affected.

useful load

operational load

Specialized Vehicles

specialized vehicles & containers

Transportation has long used **specialized vehicles and containers.** The first trains in the early 1800s used horseless car-

Railroad cars are built in many different shapes and sizes to carry different loads more effectively.

The piggy-back train combines the advantages of rail transport for long distance hauling and the flexibility of the trailer truck for on-site delivery.

riages to contain people. In 1865, flatcars were converted to hold barrels of oil. These containers were used for only one kind of material.

standardized containers

There is now a rapidly growing trend toward the use of **standardized containers** that can be used in more than one type of transport. For example, there are containers that can be carried in a trailer truck for land transport and then loaded onto a ship for sea transport. At the next port, the container is again loaded onto a trailer truck for delivery by land to its final destination. Use of standardized containers requires agreement and cooperation among transport companies. Trucks, trains and oceanliners must all have similar spacing attachments and loading equipment.

hybrid aircraft

An entirely different type of specialized vehicle is the **hybrid aircraft.** One such machine combines features of a large blimp to provide lifting capability with those of a helicopter to provide control.

Why Move?

The reason for transporting a cargo of people or material is a major element of transportation. The transportation goal is to

New special purpose machines that combine the advantages of other machines are becoming more common. This machine combines the lifting power of the blimp and the manouverability of helicopters. This system could lift more than five times the load of the largest helicopter.

move a cargo of people, goods or materials in the shortest possible time at the lowest possible cost. The speed of travel may be of primary importance. Transporting people to work places an emphasis on speed. Time lost in travel is time not available to people for work, play or relaxation. Transporting strawberries to market requires a slightly different interpretation of time. The important thing is not how fast the fruit is moved but whether it reaches the market at the point of ripeness. It should also be there when the demand for strawberries is high. Speed is important, but timing is even more important.

Sometimes, the high cost of fast transport may outweigh the advantages of speed. Some people prefer a leisurely oceanliner over a high-speed jet transport. Interstate buses continue to increase their number of passengers even though they are often slower than automobiles.

Summary

Transporting is basically an activity of moving materials or people from one place to another. It requires the use of paths

The major aspects of a transporting system are the vehicle, the ways, the load and the degrees of freedom. An additional aspect of such a system is the purpose or goal, of the transporting activity.

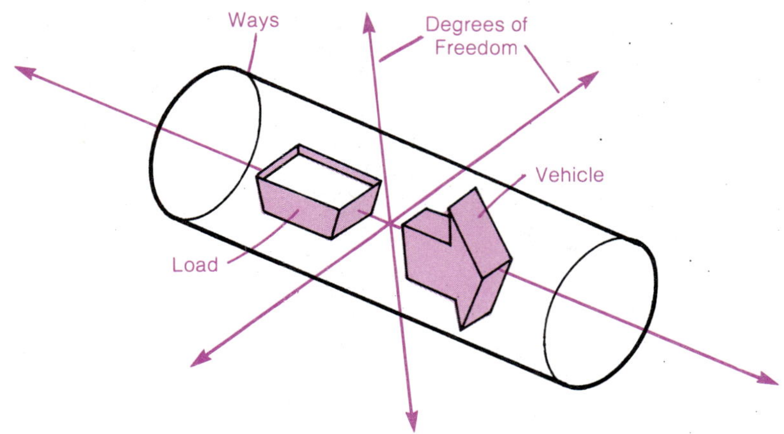

energy conversion

or ways, and some mode or propulsion. The energy may be provided by gravity, but is most often generated by stationary machines and mobile vehicles. The paths used in transporting vary from the high-control/low-freedom of one-directional movement to the low-control/high-freedom of six-directional movement. New hybrid systems that integrate different levels of freedom and control are being developed.

The close relationship of transporting and constructing activities is apparent in the pathways required for floating, flying and rolling vehicles. Those pathways with a high degree of control are provided through special constructing efforts. Transporting by pipelines, highways, railways and the like would be impossible without construction systems.

Transporting and communicating are also closely related. Transporting deals with the moving of physical loads from one point to another. Communicating deals with the moving of messages from one place to another. In most cases, communicating involves the transmitting of signals as energy rather than the moving of physical objects.

Key Concepts and Terms

control
degrees of freedom
directional
energy conversion
five-way transport
four-way transport
heavier-than-air vehicles
hybrid aircraft
lighter-than-air vehicles
loads
movement
navigation
one-directional transport
operational load

pathways
propulsion
purposes
six-way transport
specialized vehicles & containers
standardized containers
subterrene
three-way transport
transport system
transport unit
two-directional transport
use of transport
useful load
vehicles
VHST
ways

[Entered at the Post Office of New York, N. Y., as Second Class Matter. Copyrighted, 1890, by Munn & Co.]

A WEEKLY JOURNAL OF PRACTICAL INFORMATION, ART, SCIENCE, MECHANICS, CHEMISTRY, AND MANUFACTURES.

Vol. LXIII.—No. 9. Established 1845. | NEW YORK, AUGUST 30, 1890. | $3.00 A YEAR. Weekly.

THE NEW CENSUS OF THE UNITED STATES—THE ELECTRICAL ENUMERATING MECHANISM.—[See page 132.]

The U.S. census of 1890 used punched card for recording, processing and reporting data. The system resulted in new ways for communication between humans and machines.

Chapter 10 Communicating

Today, the earth is ringed with communication satellites that bring us in direct contact with other people and events through telephone, radio and television. We are able to communicate with ships at sea and spacecraft through the use of satellites as relay stations. We have also learned to communicate with machines and even with some of the higher level animals, such as dolphins and chimpanzees.

The use of machines to make communicating easier, faster, and cheaper is a relatively recent development. Until recently people did not know much about what was happening in the rest of the world. When they did receive information it was often well after the event had occurred. The importance of information processing machines was made evident in the National Census of 1880. Without the new punchcard system developed by Herman Hollerith, the analysis of the 1880 census would still have been in progress ten years later when the next census began.

information

Communicating is obviously a major activity of humans. This chapter will give only passing attention to communications as a social, psychological and anthropological activity. This chapter looks primarily at communication in a technological context. A model of the communication process can help you understand communication among animals, humans and machines. The discussion here addresses communication involving humans and machines.

Senders and Receivers

sender
receiver

Communication involves a sender, someone who wants to communicate, and a receiver, someone to listen. A **sender** and **receiver** pass information between them. Humans can communicate with other humans and with machines. Machines can also communicate among themselves.

As participants attempt to communicate, one acts as the sender and the other acts as a receiver. The sender has some information to be transmitted. However, the information must first be changed into a code so that it can be understood. With human-to-human communication, words are often used as code. On the receiving end, you hear the sound, decode it into words and puzzle out the meaning. When the meaning of the message finally reaches your brain, the transmission process is complete.

Compatibility Between Sender and Receiver

compatibility
code
signals
symbols

There is no compatibility between sender and receiver unless they use a common code that has agreed upon signals and symbols. If the message is to be understood, the code must

Words, body movement, tone, intensity and expressions are all part of the message when people communicate directly. This enrichment of the spoken word becomes especially important when the participants are not fluent in the same language.

Communication between humans and machines if common today. It requires humans to know how to send the correct signals to the machine and that the machine be designed to provide symbols that the human can understand.

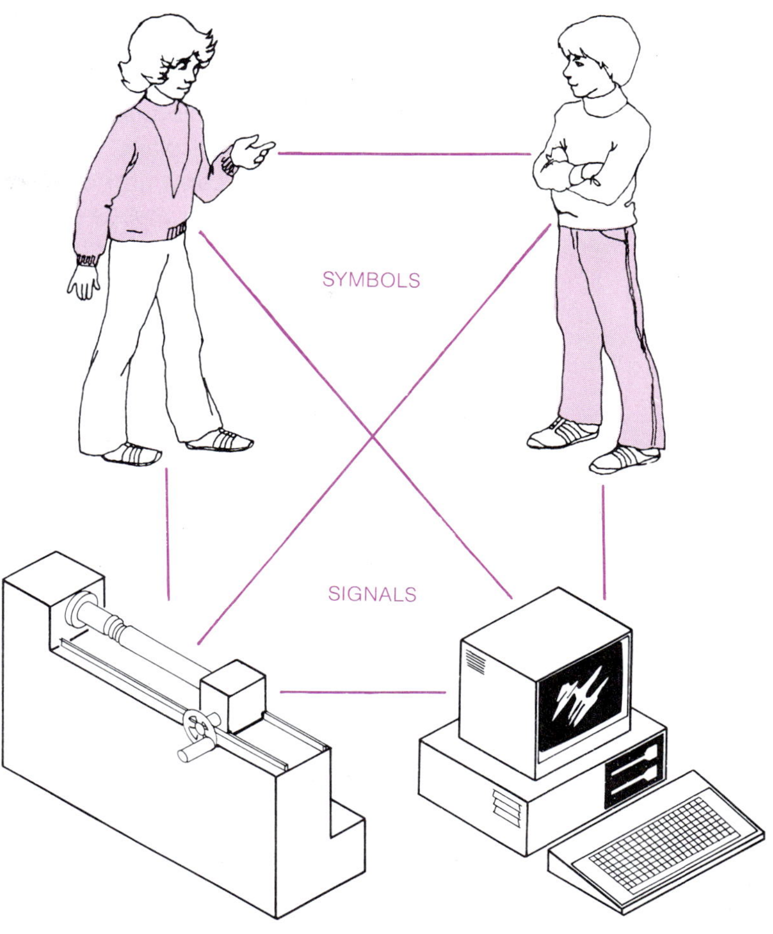

language systems

be understood by both the sender and the receiver. For example, if the message sender speaks English but the receiver understands nothing but Spanish, then the message will not be understood. The languages or codes are not compatible. What other types of symbols or signals could be used to allow communication to occur? How about sign language? Or pictures? Would it work to find someone who speaks both languages?

When humans communicate with humans, a common background is necessary if the symbols are to be understood. Take, for example, "The square of the hypotenuse is equal to the sum of the square on both sides." This sentence is in English and will probably be readily understood by some people in your school but not by others. The difference is in the receiver's knowledge of mathematics. In the same way, most people can understand the mood and cadence of a given piece of music. But only a few can read and reproduce a complicated orchestra score. When humans communicate with other humans, they must use some form of symbol.

At Bell Laboratories in New Jersey, the first machine was "taught" to translate signals into simulated speech that humans can understand. This was the first real "talking machine." Advances since then have led to machines that can translate spoken words into the signals that machines can understand. Now machines are able to receive as well as send spoken words. This represents a giant step in communication between humans and machines. These machines were called **synthesizers.** You may have heard music that simulates the sound of many different musical instruments through one synthesizer.

synthesizers

It is very exciting to have computers or robots speak to people and carry out spoken orders. It is easy to forget that even though the robot seems to understand, it is actually responding to the individual sound waves of the words. These are the signals that trigger specifically programmed events inside the machine.

Recent developments in communication between machines and humans allow many people to use computers. Several language systems have been developed so that peo-

In order to communicate with machines, particularly computers, humans must learn one or more new languages. Many newer generations of computers understand and use languages of people.

ple with little knowledge of the computer itself can communicate with it using familiar symbols. Only a programmer or a computer technician needs to know the highly complex set of signals triggered by the symbols on the keyboard. In other words, when you type "RUN" into a keyboard, the computer does not respond to the inferred meaning, "go ahead and do your work." Instead, the computer has been programmed to turn on specific sets of circuits when the "R" is pressed. Another set of switches is turned on when the "U" is entered into the machine. Still another set operates when the "N" is entered; and still another set of signals is received when the "RUN" appears in combination.

As computers become more and more sophisticated, they are able to interpret complex messages that require very little input from the human. It is easy to forget the "invisible program" of signals that the machine is using. To program a computer to do a simple game hundreds of symbols may be typed. These steps represent the logical breakdown of the sequence of actions in the game. They are in symbolic form so that you can understand them. However, the computer must be built to respond to that particular system.

The computer reads the input as a vast array of signals to its thousands of circuits. Computers made by different manufacturers interpret only the languages compatible with them.

Some machines have become so complex, they can overload their human operators with information. One solution is to have the machine identify crucial information and when it is needed to insure safe operation. At the appropriate time, the computer would generate the necessary sounds and talk to the operator.

The human must use the proper system of symbols and signals to be able to communicate with that particular brand of computer. The processes of communication are the same, however. The human may use keyboards, tape, cards and the like. It is the arrangement of the symbols that differs. There must be compatibility between what is sent and what the receiver can "read." Even very complex information machines such as the synthesizer and the computer are still only machines. They can respond only to signals.

Transmitting Through Media

communication vehicles

transmitter

laser

Information in the form of a code or message is transmitted from the sender to the receiver by means of a vehicle. The vehicle is the carrier of the information. It is the link between sender and receiver through or on which a message can travel. Suppose you write a letter to a friend. You need paper to carry the words and ideas you want to send. If you decide to call on the telephone, you will use the wire between the telephones as the carrier. If your friend comes to visit, you can talk to each other directly. The carrier will be the air between you. Examples of other communication vehicles include radio waves, newspapers, tapes, cables and laser beams.

On paper, a message is sent by the graphic images. Over wire, a message is sent by electrical pulses. When talking, a message is sent by sound vibrations in the air. A communication medium is the result of a message paired with an appropriate vehicle.

A common form of one-way communication is shown here. The sender (transmitter) writes a letter. The paper on which the letter is written is the medium for communicating. The letter is transported by truck, rail, or plane. The one-way transmission is complete after the receiver has read the letter. This is only possible if the sender and the receiver use a common code and common symbols.

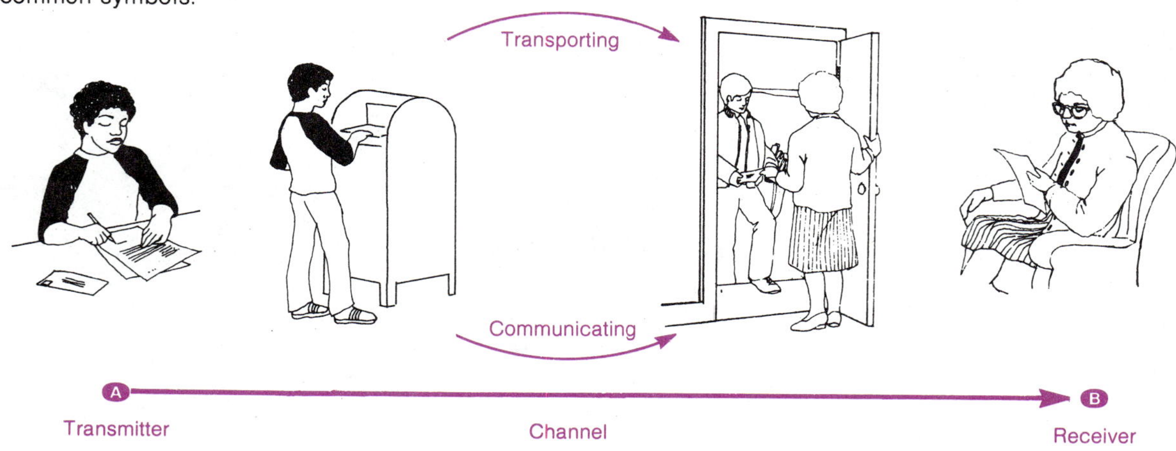

Medium	Message Symbols/Signals	Carrier Vehicle
Print	Words	Paper
Television	Electronic signals	Radio waves
Telephone	Electrical signals	Wire
Voice	Auditory symbols	Sound waves
Recordings	Magnetic signals	Tape
Data processing	Holes making signal patterns or magnetic signals	Cards, tape or disc
Painting	Figures in symbolic patterns	Canvas and paint
Photograph	Figures in symbolic patterns	Paper and silver

carrier

It can be quite important to match the carrier and the form of the message. Each medium has advantages and disadvantages regarding the amount of information that is communicated. Talking directly to a person allows you to see, hear, smell, touch and move as you communicate. The messages returned to you can be received through several senses. You can ask questions or hear each other's tone of voice. Communication by way of two-way television contact is different, however. Communication also changes with the use of a CB radio, tape recording or newspaper advertisement. The effect of the total message and the amount of information that can be sent is different in each case. Some media are static or frozen. They can be stored and replayed. More dynamic media lose the message once it is sent. Some media allow the message to be transmitted extremely rapidly. The receiver can either receive the message rapidly or store it. If stored, the message can be replayed later at a pace appropriate to the receiver.

media

dynamic media

replication

Some media allow easy replication. Millions of copies of a newspaper can be made and transported. This process is impossible for some media. An example is a famous painting. There can only be one original. The image may be reproduced or transmitted by television, but much of the impact of the original, symbolic message may be lost.

multiple messages

The differences in media and carriers are also important when multiple messages are transmitted. Many telephone calls are sent at the same time on the same wire. The telephone company tries to make sure that the call at both ends is tuned in, like a station on a radio. You may sometimes hear

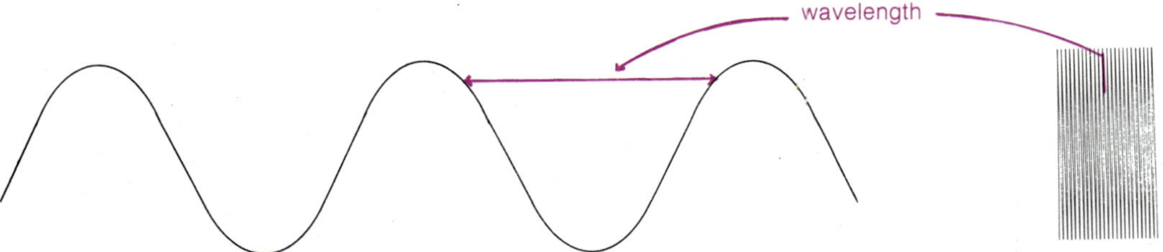

This microwave has a wavelength of about one inch and a frequency of about 10 billion cycles per second.

If this wavelength was reduced about a 1000 times you would have moved up the energy spectrum to the red band of visible light. The frequency of the red light would be about 430 trillion cycles per second.

Information is transmitted on radiated waves. The more waves that are available per second (frequency), the more information that can be transmitted.

parts of other messages when you are on the phone. Even more messages can be sent at one time when radio waves are used. More can be sent by microwave communication. Carriers have different capacities for transmitting signals clearly and accurately. They also differ in their ability to maintain separate channels for each message.

As more and more information is gained, different carriers are needed to provide more ways of communication. More information needs to be stored and decoded later. This needs to be done efficiently and with as little loss of information as possible. Messages sent on solid carriers of paper, canvas and the like must be transported. However, messages that can be sent as signals on a band of energy can be far-reaching and flexible.

Light—The Fantastic Carrier

visible light waves

The role and importance of carriers can be illustrated by comparing **visible light waves** with radio waves. Both of these wave forms are on the energy spectrum. Visible light is further along the spectrum in the direction of speed. Light waves, therefore, have a higher frequency than radio waves.

wavelengths

Radio, television, microwaves, infrared and visible light are radiated waves on the energy spectrum. They all travel at the same speed of 186,000 miles per second. However, they differ in wavelength, the number of times they vibrate per second. Messages are carried on the wave length, from one vibration to another. The more vibrations per second, the more information that can be carried.

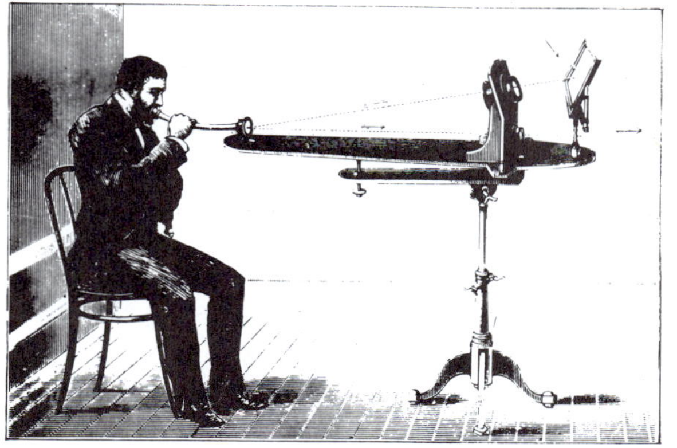

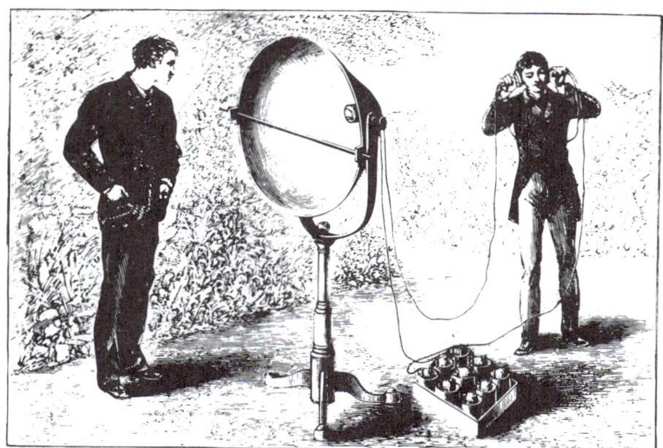

Until the invention and development of the laser, a system of light as a carrier of information was not feasible. The signals were easily lost because of dust, fog and other interference.

frequency

Visible light vibrates at frequencies up to a million times faster than AM radio waves. The radio part of the energy spectrum is crowded, because it is narrow and many people want to use it. In some frequency ranges, the stations may seem packed one on top of the other. The visible light spectrum has a million times more room than all the radio, TV, shortwave and microwave bands that we now use for communication. Light waves have great potential as carriers.

Losing Information—Entropy and Noise

Transmission of information by any media on any carrier is not 100 percent efficient. All of the signal does not get through. The transmission of information is improved by building in **redundancy.** This means sending more signals than are actually needed. Television stations send excess messages so that even sets with poor reception can receive a picture.

redundancy

Loss of information is usually a result of partial loss of a signal. Ink fades with time. A tape may lose some of its magnetism. The grooves on a record will become worn.

In early studies of information, research used a telegraph key. These early research efforts tried to transmit a signal that was clean and crisp. However, the signal always came out of the receiving set a little distorted. Something or some part of the signal was lost in the transmission.

signal entropy

This loss was called **signal entropy.** Signal entropy is the loss of usable information as it is changed from one form to

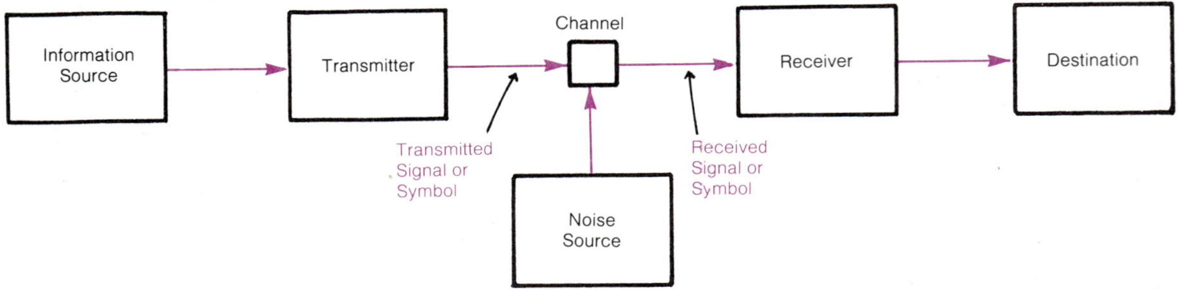

This general model of communication, developed by Shannon about 1950, and still in use, shows the parts of the communication system. Noise interferes with the signals and symbols as they are transmitted on the channel. Entropy takes place during each step from the information source to the destination.

another. The voice you hear in a telephone receiver is not as clear and distinct as the actual voice of the person. A flashing warning beacon at an airport appears less bright at a distance or when the weather is misty or foggy. These are examples of signal entropy.

Symbols may also lose some of their meaning in the process of translation. Signals and symbols, even those with minimum entropy, will mean little without a prearranged code. For example, if two people operating a telegraph are not using the same code, it will be impossible for them to understand each other.

We have only a few recorded messages made by ancient people. Many messages are in a code we do not entirely understand. We have trouble decoding the messages for three reasons: the materials may have worn away or decayed; we may only have part of the object on which the message was recorded; or we may be unable to break the code they used to understand the symbolic meaning of the words and phrases. Our experiences are not the same as the experiences of people many centuries ago. Some words may have very different meanings now.

noise

entropy

Signal loss is also caused by noise. **Noise** is a special term that indicates interference during transmission. Shannon's model shows the point in communications at which noise interferes. Information entropy and the loss of signal becomes more of a problem when interference or noise increases.

All communications systems have some noise that interferes with information transmission. Radios and telephones have static. A computer may have stray voltage. A photocopier may reproduce dirt on the lens. All of these interfere with transmission. If the static becomes too strong, the volt-

One-way flow of information is called open-loop communication. The heat currents rising from the fire cause the turbine fan in the chimney to turn. An energy transmission system causes the meat on the spit to turn. The hotter the fire, the faster the fan turns and the faster the meat turns. Since the fan does not control the fire, the system is an open loop.

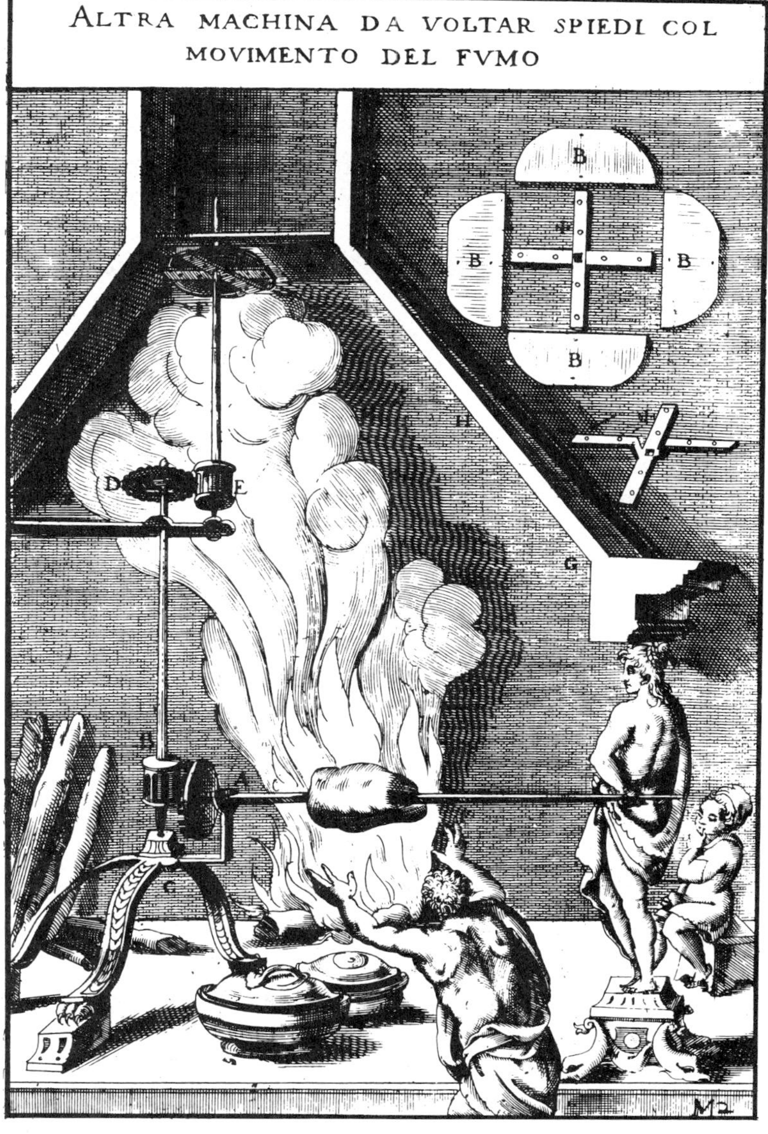

age too high or the copier too dirty, the transmission of the information becomes impossible. The signal is lost when the noise level becomes too high. The signal must remain higher than the noise level to be received. To maintain good transmission of information, a high signal-to-noise ratio must be maintained.

Talking Back to the Sender

Have you ever called someone on the telephone and found that you can barely hear them? This lack of contact with the "receiver" is a special situation in communicating. This situation is called an **open-loop process** of communication. The message sender gets no information back about the transmission. When you communicate with a machine, the same situation is called an **open-control loop.** Return information from the receiver to the sender is missing.

open-loop process

open-control loop

Most communications have some return contact between receiver and sender. This contact closes the communication circle. These situations are called **closed loop.** Return information from a receiver to a transmitter may be in many forms. A thermostat control on a furnace, the water level control on a toilet tank and the display screen on a computer terminal are all examples of closed-loop communication.

closed loop process

Feedback is the term used to describe the contact from receiver to sender. Feedback closes the loop of a communication or control process. It is simply information provided by the action of the receiver. This information indicates that the message was received. The sender can use the feedback to learn how to communicate better.

feedback

People's facial expressions and questions provide feedback on communications. We can then restate things or give examples to clarify our messages. This process is called feedback. Feedback happens only if some change, like the change in expression, takes place. Until then, we can only guess at the success of our explanation.

Feedback from and to machines is also based upon change. This change is actually a deviation from an existing situation to a new situation. Change or deviation from the set temperature in a kitchen range provides feedback to the heating system of the stove. A temperature drop in a room provides feedback to the thermostat as a change from the set temperature. The feedback causes a furnace to turn on.

Symbolic Communication

Printing, photography, lithography and electro-static reproduction are a few of the advances in communication. The general term **graphic communication** refers to the visual

graphic communication

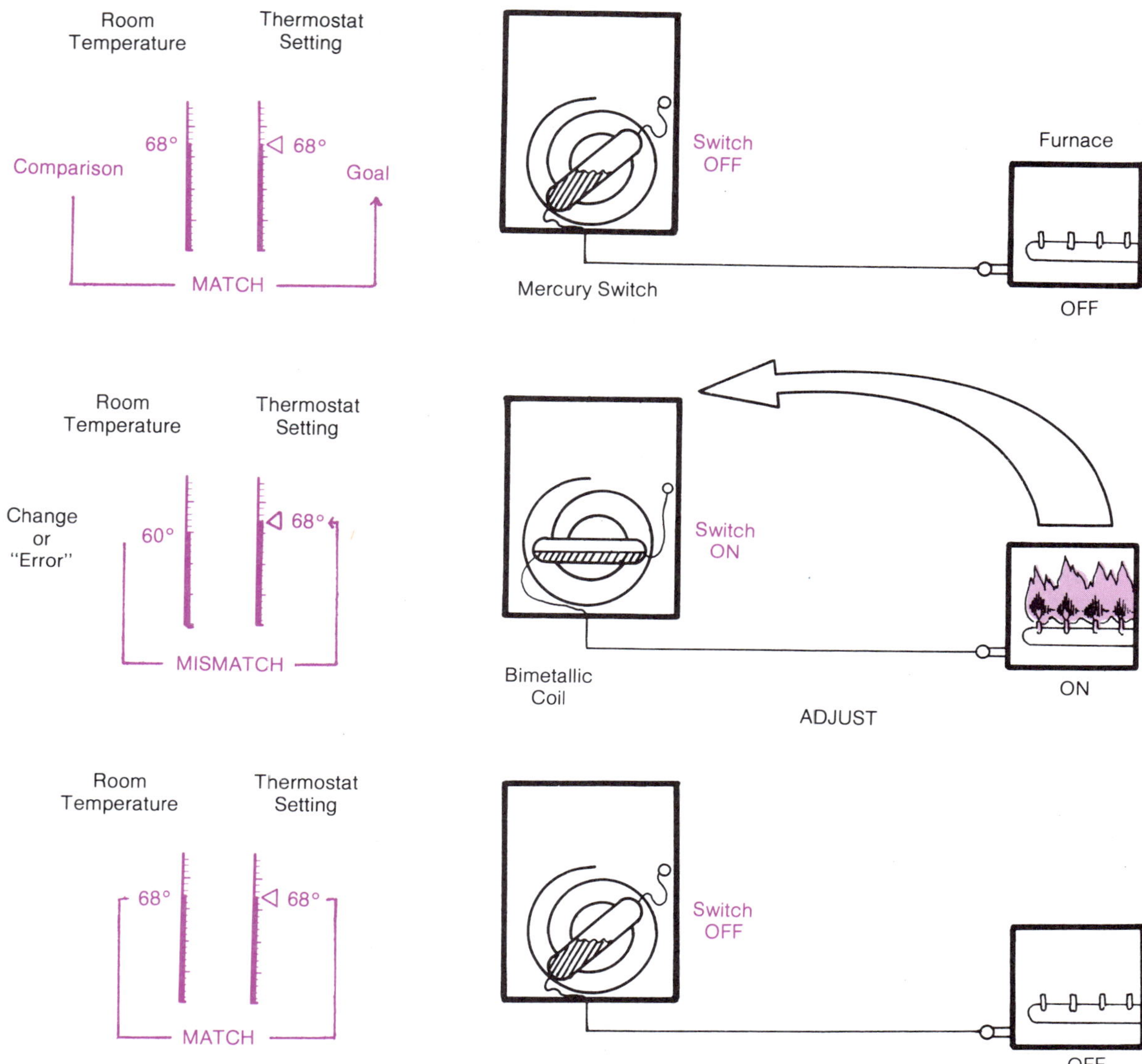

Temperature changes are received by the thermostat of an automatic furnace. The thermostat then sends a message (signal) to switch the furnace on or off as required. This sensing device provides "feedback" which makes the furnace a closed-loop communication system.

forms. Graphic communication includes the printing technologies. It also includes visual communication, such as sign language for the hearing impaired. Language forms developed from sign language can even be used for communication with selected animals.

Visual communication also includes developments in graphic design. Attempts to communicate with people in a variety of languages are often seen in road signs. A set of symbols common to most people is being developed. An example is the plaques placed aboard the Pioneer 10 and 11 spacecraft. Their visual message provides information about the place, time and origin of the makers of the craft. This example of visual communication was transported in tangible form. Other messages have been transmitted on energy waves. The transport of a videotape is a tangible form of visual communication. The broadcast of a television program is visual communication transmitted on energy waves.

videotape

broadcasting

Communication by Signal

telecommunication

A study of communication would be incomplete without some attention to what is called **telecommunication.** The prefix "tele," as used in telephone and television, means operating at a long distance. Telephone really translates to mean "long talk" and television means "long see" or "long

This drawing shows the communicator that E.T. used to send messages into space. The communication system, built from common parts and devices, actually operates in a manner similar to larger, more sophisticated systems that send messages into deep space.

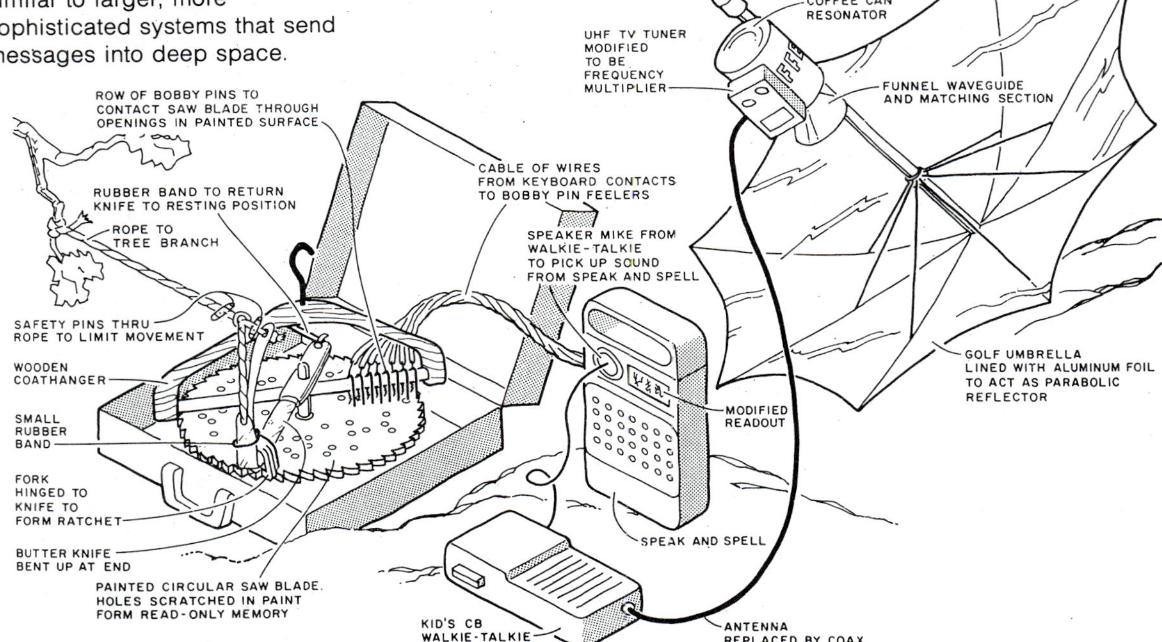

sight." "Radio" could have been called "teleaudio" or "long hear."

Listed here are some examples of telecommunication:

telegraph

- **telegraph**—an apparatus or system for transmitting signals in code by electric impulses sent by wire or radio.

telemetry

- **telemetry**—the process of measuring and recording television images and readings of instruments at long distances.

telephone

- **telephone**—a system or instrument used to convert speech to electrical impulses to be sent long distances by wire or other carrier.

telescope

- **telescope**—a device for viewing distant objects by means of lenses.

teletypewriter

- **teletypewriter**—a form of telegraph in which a typewriter sends a transmission that causes the corresponding keys of a receiving typewriter to operate.

television

- **television**—the process of transmitting visual images by electric impulses superimposed on radio waves to be projected on a luminescent screen (TV set).

Animals and humans can communicate. This chimp read the coded message on the board which says, "Sarah insert apple pail banana dish". The chimp then placed the apple in the pail and the banana in the dish.

In 1800, Alessandro Volta invented the electric battery. Volta's battery laid the groundwork for Samuel Morse's invention of the telegraph. The importance of Morse's work was not immediately apparent. This is often the case with new ideas. Other people use the breakthrough and find important applications that were not recognized by the inventor.

binary code

The obvious result of Morse's work was the telegraph itself, with its capability of sending coded messages over metal wire at the speed of light. But the telegraph's symbols also changed written letters of the alphabet to signal energy. These dots and dashes, along with the electrical pulses that caused them, were re-examined and used by numerous scientists 100 years later. This binary code was crucial for the development of communication and information systems. One such system was used in 1974 to send a radio space message from the observatory in Arecibo, Puerto Rico. The huge radio transmitter there stretches between mountains and sends the strongest signals leaving the Earth. Its message leaves our solar system less than six hours after transmission. The electronic blips, sent in blocks, can be translated into a visual checkerboard message.

This message about Earth and the people who live on it was encoded and transmitted into deep space from a huge radio transmitter in Puerto Rico. The transmission was an attempt to communicate with others outside our solar system.

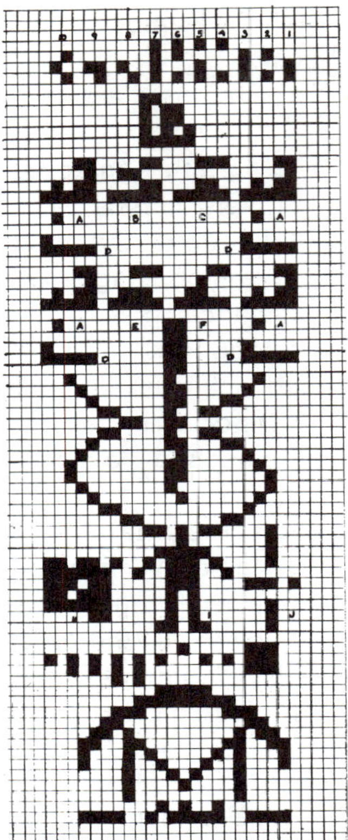

Communication Systems

Technological activities tend to evolve into systems and then into still larger systems. A **communication system** may be very simple and involve very few participants. An example is the buzzer or doorbell that many of us have in our homes. A visitor at the front door communicates to people in the house through a push button connected to a bell. A more sophisticated system is the telephone system that links a large portion of the people of the world through an intricate information network.

communication system

Communication systems may also be established between machines. A simple example is the system between the furnace and the control thermostat in a building. Automated production is a more sophisticated system.

synchronous satellite

There are also communication systems that use synchro-

Another message about Earth, its location and its people was placed on 6″ × 9″ gold-anodized aluminum plaques. This was attached to and traveled with Pioneer 10 and 11 in their journeys into space.

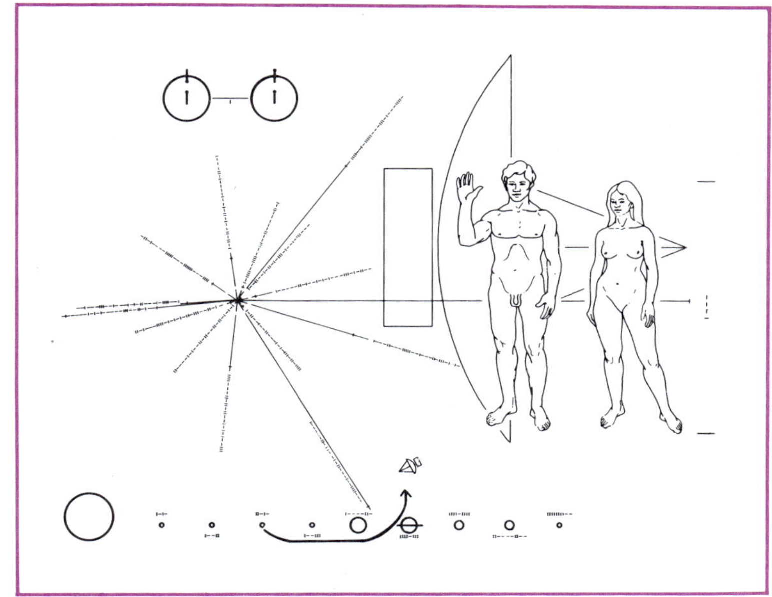

A communication satellite is being prepared for a launch into orbit around the Earth. This satellite was placed in a geosynchronous orbit so that it keeps pace with the rotation of the Earth, yet appears to hang motionless in the sky.

nous satellites. The satellites are placed in an orbit whose speed matches the speed of the turning earth. The satellites appear to hang motionless over fixed points on the earth. Three of these satellites can provide contact with nearly every point on the globe.

The above examples are only a few of the many systems in the communication arena. Information communicated by satellite may also be printed rapidly in thousands of newspapers. For this process, paper, ink and other materials must be produced and transported to the printing presses. A network for distributing the papers, handling money and even recycling the newspapers is needed. All aspects of the system are interdependent. Each element needs all the other elements to work. This interdependency may be evident only when the process stops. A strike affects the papercarrier, workers in paper mills, truck drivers and people who depend on the newspaper for information.

The exchange of money is another example of a communication system. The trade of surplus goods stimulated transportation systems. It also fostered changes in methods of recording the value of the goods traded. In earlier times, as

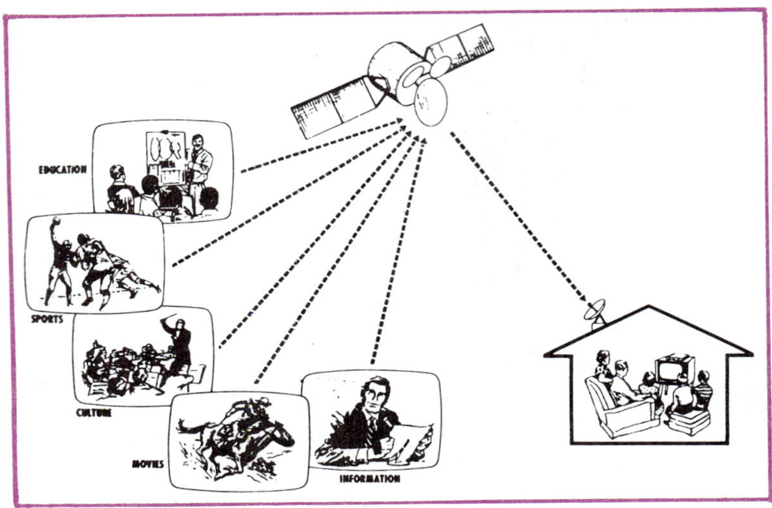

Satellites now provide direct-to-home transmission of television programs on separate channels. The signals are picked up by rooftop antennas.

medium of exchange

people traded farther away from home, they needed a common medium of exchange. Salt was an early form of money, as were shells and sand dollars. Coins were made to represent the value of the objects traded. Later, paper money was used to make the exchange of goods and services easier.

communication medium

The use of symbols rather than physical objects increased the possibility of complicated transactions. A system of credit and loans allowed people to buy goods and services easily, on the promise that they would pay later. Credit cards can now be used to obtain cash, as well as goods and services. Credit cards and cash are being replaced by electronic "blips." This electronic "money" allows machines to communicate with machines and make the money transfers. No physical information in the form of coins or paper will change hands.

Some people trust only gold or diamonds to hold true value. Money can be touched and counted. Money is put in a bank in trust that the money can be retrieved later. Credit cards are a means of making paper transactions. As the information found in money is changed to electronic messages, some people fear we may lose control of our money.

Before total electronic banking can happen, vast changes in the system are required. New machines are needed which will send and receive messages wherever transactions occur.

A U.S. Government satellite provides data to ships that allows them to sail north in the strong currents of the Gulf Stream and to avoid the currents when sailing south. The heat sensitive sensors on the satellite locate the warmest water in the Gulf Stream where the currents are strongest. They save considerable energy, time and money.

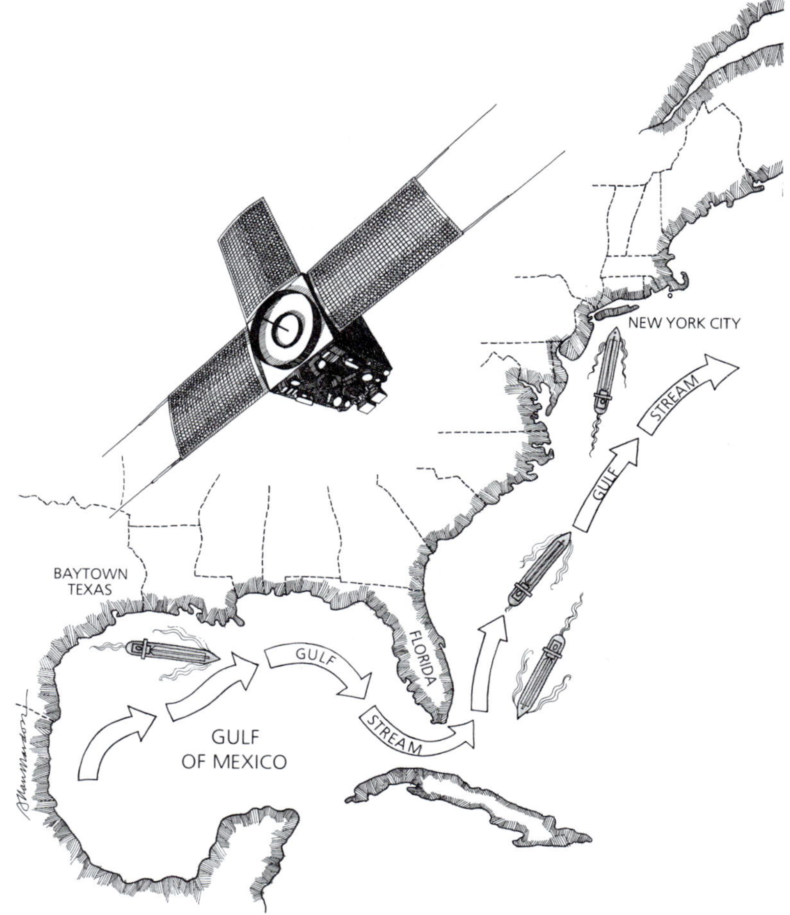

The marketplace and the banking system must change. Many people regard their money matters as highly personal and private. These attitudes will have to change or they will slow the development of the system.

Summary

The sending of symbols and signals occurs between two or more participants, machines, humans or both. Characteristics of the carrier, such as the frequency of an energy wave, determine how much information can be sent in a given length of time and the amount of redundancy needed. Transmitted information goes through a natural process of deterioration called entropy. The loss of symbols/signals due to en-

The expansion and improvement of communication systems promises new methods of information transmission. This Japanese prototype attempts to integrate information from newspapers and television for use in the home.

tropy affects the efficiency of the communication. The effect of signal loss is reduced by sending more than is needed or by choosing a more efficient carrier. Communication systems require feedback before communication ends.

Communication through the use of visual symbols follows the same rules and principles of signal transmissions. Symbols are very important as a connection between machines and humans. Signals are used in connections between machines and machines. Communication systems are an integral part of the other technological activities. Information must flow among the participants involved in all technological activities. Without a sophisticated communication system, many activities would come to a standstill.

All communication can be described through this general model. Information moves from a source to a destination through a transmitter and a receiver. The signals and symbols as they are carried on a channel will lose their fidelity because of some source(s) of noise. Feedback, as we will explore in more depth in the chapter on controlling may also be a part of communicating.

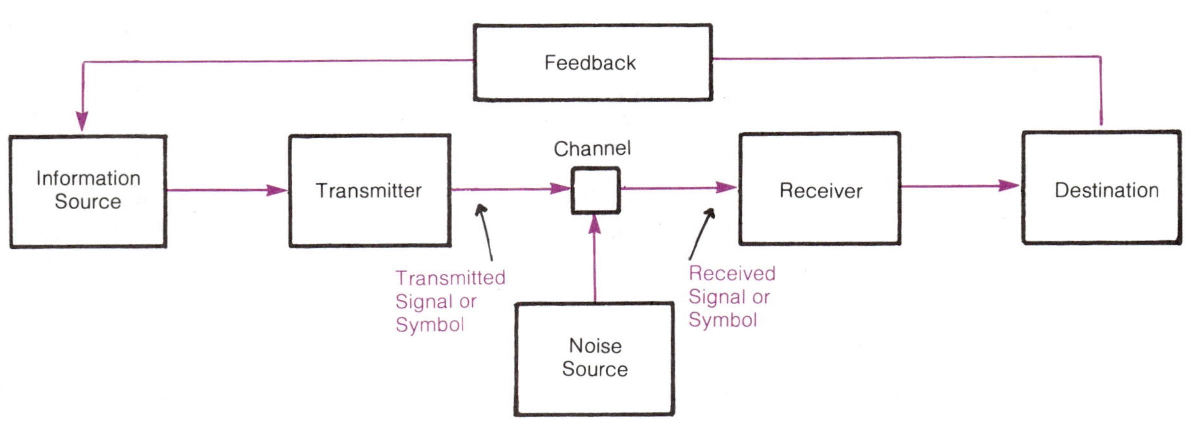

Key Concepts and Terms

binary code
broadcasting
carrier
closed loop-process
code
communication medium
communication system
communication vehicles
compatibility
dynamic media
entropy
feedback
fiber optics
frequency
graphic communication
information
language systems
laser
media
medium of exchange
multiple messages
noise
open-control loop
open-loop process
receiver
redundancy
replication
sender
signal entropy
signals
symbols
synchronous satellite
synthesizers
telecommunication
telegraph
telemetry
telephone
telescope
teletypewriter
television
transmitter
videotape
visible light waves
wavelengths

One of the very first production assembly lines was, in fact, a disassembly line. This packing and canning company used many modern innovations. Conveyors similar to these were used by Henry Ford in his mass production of the Model T.

Chapter 11 Producing

production

Production includes any activity that yields or creates a product. Have you ever wondered where your blue jeans came from? All over the world, people wear the denim pants that originated in the United States as work clothes. Several different technological activities are involved in producing blue jeans.

The jeans you may be wearing started in a cotton field. This field may have been in the warm regions of the United States. Or perhaps it was in another cotton-producing country. The cotton may have been planted, hoed and harvested completely by people-power. However, in the United States, several machines were probably used. The cotton was cleaned, ginned, dyed and otherwise prepared for use. The fasteners were made from metals or plastic. All of the required materials were transported to one place by the most economical means. This may have been by ship, train or plane. Weaving machines made the cloth. The fabric was cut according to patterns that have been developed over many years. These patterns are based on the body measurements of many people. These measurements were averaged and patterns made in several different sizes and shapes.

The cut pieces of cloth were machine-stitched, probably on an assembly line. Some people put in zippers while others put on pockets. The process continued with each machine or person following a standard part of the process. The sewn garments were then riveted at stress points to make them wear longer. The garment was then pressed.

Permanent press jeans were treated with chemicals designed to resist wrinkling. The jeans were heated to a given shape, treated with chemicals and then cooled so that the shape was "permanent."

Energy was required at all stages in the harvesting, refining, extracting, transporting and assembling of the jeans materials. Information was needed in operating the machines, choosing the sizes and shapes and printing the labels. The labels indicate size and care of the fabric, as well as who made and inspected the garment. The jeans were then packaged and transported to the stores. The cloth remnants were recycled.

Although the production of jeans is only one example, notice how the materials were produced. The metals were extracted from the ground, the cotton was harvested, the chemicals were refined and the machines and people were transported to one location for manufacturing. Orders, purchases, machine operation, developed patterns, labels and cost accounting were needed.

Most of the products you use are produced by a complex set of activities. Sometimes a product is made by only one person who produces all the materials needed. Often, the production process is highly specialized. The person who plants the cotton may never see what happens to it after it is harvested. Those who cut the materials may never meet the person who designed the patterns, and so on.

process

Extracting Minerals

Our planet contains the materials and energy needed to produce the products we use. The process of taking those natural materials from the earth, air and sea is called **extracting.** Extracting includes mining materials such as coal, salt and diamonds. It also includes drilling and pumping for materials such as gas, oil, water and sulphur.

extracting

mining

The mining of raw materials such as flint, salt and copper dates back many centuries. Before the 16th century, the major source of fuel was wood. In England, however, wood became scarce because it was used faster than it could be replenished. Before wood became scarce, mining was considered an affront to nature. When the need for an alternative

Mining in the middle ages was primitive by current standards. This ingenious lift could be raised or lowered depending on which lever was pulled. Pulling the left lever fed water into the front half of the wheel, lowering the lift. Pulling the right lever raised the lift.

A—Reservoir. B—Race. C, D—Levers. E, F—Troughs under the water gates. G, H—Double rows of buckets. I—Axle. K—Larger drum. L—Drawing-chain. M—Bag. N—Hanging cage. O—Man who directs the machine. P, Q—Men emptying bags.

fuel source became strong enough, mining coal became an acceptable method of extracting fuel.

Early mines were relatively small, shallow and dangerous. Mines were made deeper after new ways of building support structures were developed. But deep mines had problems with ventilation, water and transportation. The problem of removing water, getting sufficient air to the workers and of lifting the materials to the surface of the ground required the development of additional new machines and structures.

Modern mines may reach several thousand feet into the earth. The deepest mines are the gold mines in South Africa that go down more than 12,000 feet. This is deeper than humans have ever been before.

longwall mining

Another method of mining is **long wall mining.** In this way, coal can be removed mechanically in large cuts, allowing the earth to settle and fill the open space. There is no need to build piers and supports in the mined sections, so mining costs are reduced.

strip mining

Open pit or **strip mining** is used to obtain gravel, coal, copper, diamonds and other materials. A great deal of topsoil is moved and huge ugly holes are often made. New laws in some states require that mining companies restore the land before they move on to another site. Trees and grass must be replanted and the land restored for other use. Open pit and long wall mining require less complicated support structures.

A modern deep mine requires a large system of support services to keep the mine operating. This illustration shows the elaborate transportation system for moving mined materials.

Deep-pit, open mines are practical where there is only a small amount of rock and soil covering the ore or coal. This mine is in the Ruhr Valley, West Germany.

Air shafts and protection from cave-ins are not needed as in deep mining.

Drilling and Pumping

Far beneath the surface of the earth are deposits of the hydrocarbon remains of plants and animals. These are in the form of **oil.** Over millions of years, the remains of animals and plants piled up on prehistoric forest floors. Eventually, these remains were covered by soil during the later stages of earth's development. Pressure, heat and time formed oil from these organic materials. Drillers try to reach the oil and pump it to the surface so that it can be refined for use. This is often a difficult effort. It requires costly structures and high-level technology.

oil

Just as the scarcity of wood in 16th century England stimulated mining activities, the scarcity of oil has caused new developments in drilling and pumping. New methods have been developed to get the maximum amount of oil possible out of wells that were previously abandoned.

Two general approaches are most used. One approach is to pump heated liquids or solvents into the wells to liquify oils that are too thick to pump. The second method is to pump water into the well to float the oil within reach of the pump intakes.

The idea of pumping liquids into a well to help get the oil is not a new idea. In fact, it was first used more than eighty years ago by a man named Herman Frasch. Frasch wanted to mine sulfur but could not reach it because of sand and water. He

Early Chinese used well drilling techniques as shown here. Similar techniques were used for many centuries throughout the world.

used hot water to melt the sulfur. Once melted, the sulfur was easy to pump from the well. Today, wells of this type are the primary source of sulfur. Other materials extracted through drilling and pumping include water, natural gas and helium.

Oil is often extracted and stored long before it is refined for use. Large storage tanks have been made by forming cavities in large salt domes. Water is pumped into the salt dome to dissolve the salt and make the cavity. The salt water is then

Conventional pumping techniques can remove only about one quarter of the available oil from a well. But if water is pushed under pressure through the rock, oil can be forced out of pores and cracks. This helps to double the amount of oil that can be recovered.

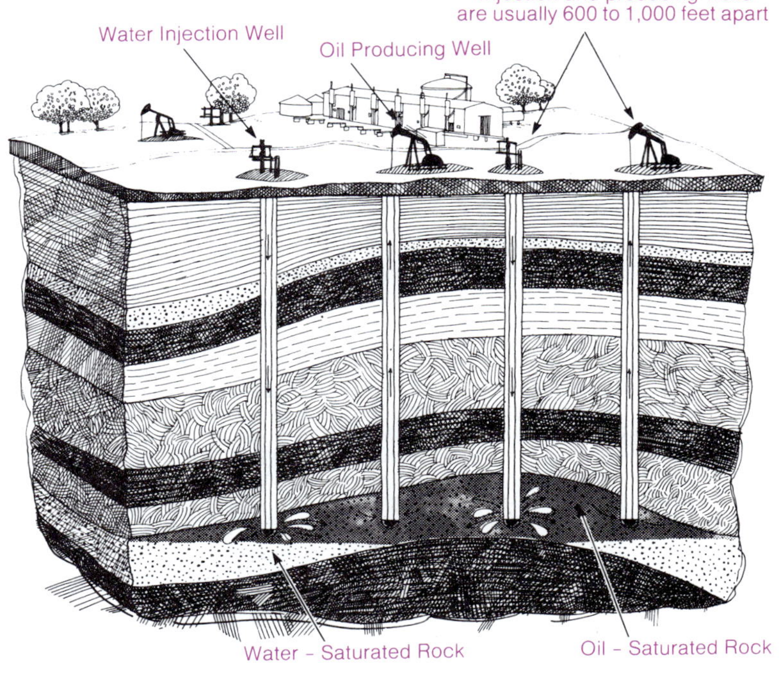

forced out to allow room for the crude oil. This storage technique is used to insure that an adequate stockpile of oil is available during periods of shortage.

Harvesting Food

harvesting

Harvesting satisfies many of our needs for food, shelter, and clothing. If you have ever picked fruit from a tree, you were engaged in a harvesting activity. Most fruit orchards, even the largest ones, still require hand picking. Tomorrow's fruit trees, however, may be developed, grown and trimmed so that the fruit can be picked by a machine. This machine could also be equipped with pruning and spraying attachments.

ecology

mechanization

Mechanical harvesting could increase the productivity of the orchards and increase the yield of the harvest. But, it could also create problems that include ecological impacts, increased energy requirements, a need for larger orchards and fewer employees. These problems are already evident in many farms that use mechanical harvesting techniques.

New developments in apple trees and picking machines could increase future harvests and make the gathering of the fruit more efficient.

Each crop and animal on a farm can be considered a small production system. The plants and animals convert materials into new, more usable and valuable forms. A cow eats grass and grain and converts it into milk. A hen converts feed into eggs. A field of corn converts the nutrients in the soil, water and sunlight into usable grain. Cattle, sheep, hogs, chickens and goats take in plants, grains and other feed and convert them into increased body weight. This increase adds to their value as a usable product.

forage crops

Forage crops are plants that are raised as food for animals. These plants convert nutrients to increase the size of the plants themselves. Most forage crops are members of two important plant families, the **grasses** and the **legumes.** Important members of the grass family are rice, wheat, sugarcane, corn, millet, sorghum, oats, barley and rye. Peas, soybeans, peanuts, chickpeas and a variety of beans are members of the legume family.

grasses
legumes

The grain combine was named because it combined into one operation five jobs that were once done by hand. The machine (1) cuts the stalks replacing tools like the sickle, (2) feeds the stalks through the machine, eliminating the need for carrying and hauling, (3) threshes the grain from its hull, replacing the flail, (4) separates, eliminating the hand discarding of the stalks, and (5) cleans, replacing the process of winnowing.

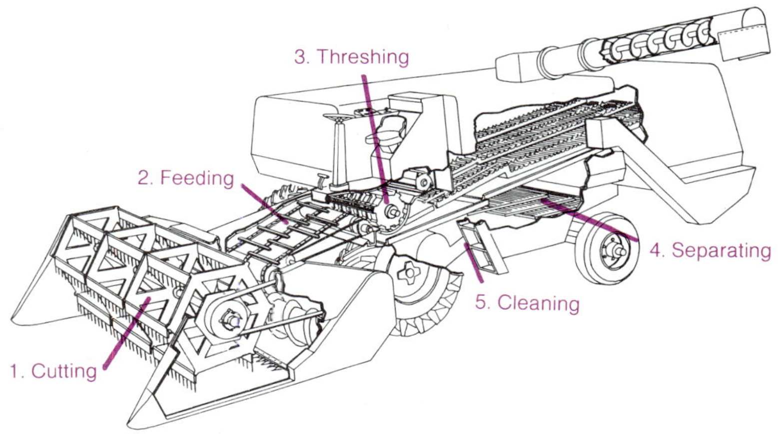

Extensive systems of harvesting technology have been developed to plant, cultivate, harvest and use forage and grain crops. One example is the self-propelled harvesting machine or **combine.** This machine combines the reaper, a machine that cuts grain, and the thresher, a machine that separates the grain from the rest of the plant. The combine separates the grain from the straw without damaging the grain. This bulky machine must move over relatively soft soil. It must operate efficiently while bouncing over uneven surfaces.

combine machine

It is difficult to machine harvest tomatoes and other fragile vegetables and fruits. The problems have led to attempts to change the properties of the fruits and vegetables themselves. For example, scientists are trying to develop square tomatoes with strong skins that can withstand rough handling by machines.

Farming machines are usually specialized equipment that can do only one task. Consequently, much money must be invested in a number of machines. As the number of such specialized machines increases the cost of producing foods gets higher. Thus, farming has moved from small operations to large operations that require a lot of money and equipment. This change from hand labor to expensive mechanization is evident when early farm machines are compared to modern farm equipment.

agribusiness

This horse-drawn walking plow dates back to the 1880's. It was the major machine that opened up the heavy sod of the Midwest and plain states.

Modern farming methods require huge, costly equipment that consume large amounts of energy.

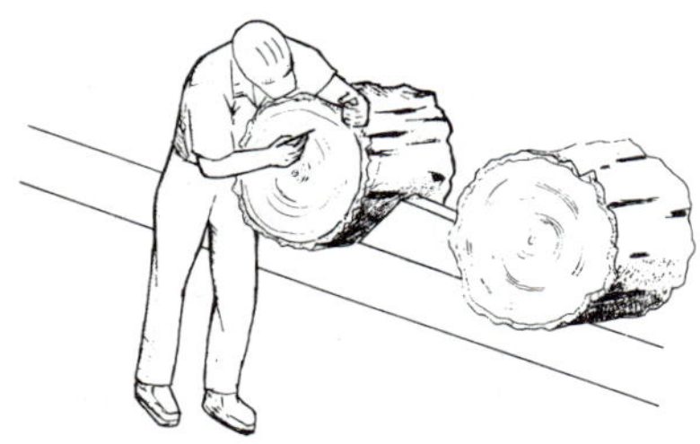

Both of these trees are about 30″ in diameter. The tree at the left is about 450 years old, the one at the right only 60. Experiments to produce fast-growing, giant trees are a major research activity of the forest industry.

Modern farming also requires a large amount of agricultural chemicals for fertilizing the crops and for controlling weeds and insects. These chemicals also increase the costs of production.

tree farming

Large scale tree harvesting is relatively new. Trees have been cut and used for many years but farming techniques of planting and tending have only recently been used. One company has just completed a twenty-year effort to develop "superseedlings" of softwoods such as Douglas fir. About one dozen of the finest trees were selected from nearly 100,000 acres of forest land. A few ounces of seeds from these trees were planted in a greenhouse and carefully nurtured. Each year the best seedlings were saved. Finally, after many years of work, enough new seedlings were ready. The company used these superseedlings in areas that were being harvested for lumber. Superseedling trees can be grown seven times faster than naturally grown trees.

fish farming

Harvesting is not limited to the farming of land crops. **Fish farming** is also a harvesting activity. Artificial structures are placed underwater to provide surfaces on which algae can grow. The algae, in turn, attract certain fish populations that can be harvested in highly efficient ways.

floating fish factories

Sophisticated harvesting techniques, especially those used by floating fish factories, threaten certain species of fish. **Floating fish factories** are ships designed to collect the fish, process and preserve them in one continuous operation.

Plants, such as kelp, are also harvested from the sea. Shells are gathered to be used for buttons. Oysters are "seeded" by adding sand. Then, the cultured pearls are collected and made into jewelry.

Many natural materials must be processed before they can be used for human needs and wants. That process is called **refining.** Foodstuffs, fuels and many other materials can be refined to make them more useful.

refining

Refining Metals

manufacturing

reduction

Most metals that are used in manufacturing and construction are refined. Except for a few select metals, refining is essentially the process of driving out excess oxygen. This kind of refining is called **reduction.** The metal is reduced to a more pure form.

All metal ores that contain oxygen must go through some form of reduction process. Some reduction processes are quite simple, such as reducing iron ore to iron and tin oxide to tin. Other metals such as uranium and cobalt are more complex. All of the processes, however, need large quantities of heat to complete the materials conversion process from the ore to the metal.

Iron is one of the most useful and important metals. It is the basic ingredient of steel. Without steel, industrialized society could not operate.

Many metalic ores are metals that have oxidized (taken on oxygen). The metal is extracted from the ore by a process of reduction. The ores are heated by a carbon based fuel. The oxygen combines with the carbon from the fuel to form carbon monoxide or carbon dioxide, leaving the molten metal behind.

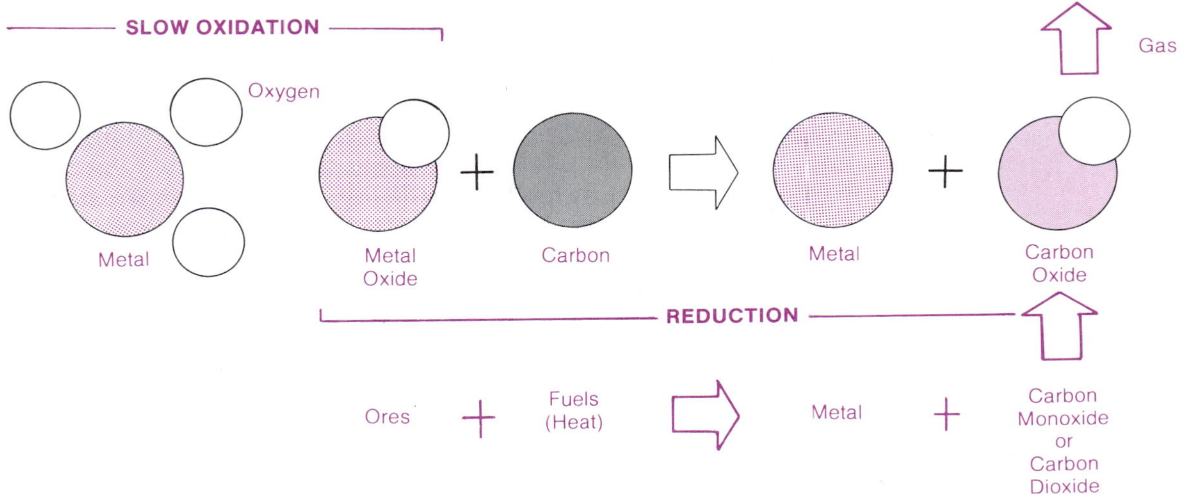

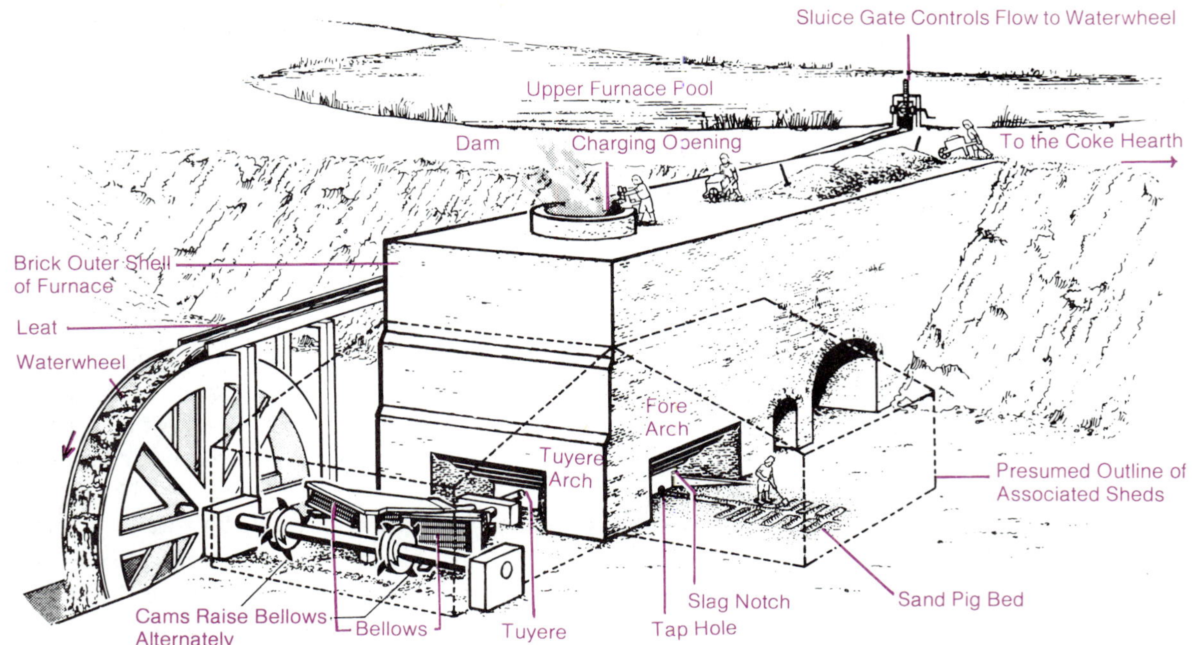

This historic blast furnace operated in England in what is viewed as the beginnings of the Industrial Revolution. Modern blast furnaces operate in much the same manner but with far less human labor.

blast furnace

A **blast furnace** is used to drive off oxygen from iron ore. This converts the ore to iron. Such a furnace can be as much as 100 feet high. To melt the iron in the furnace, temperatures have to be in excess of 1537°C. Once a blast furnace has been started, it must be operated continuously. The materials are fed in at the top and the molten iron is drawn off at the bottom. If the furnace is shut down, time is needed to heat everything back up to the proper temperature.

Metals are not the only materials that are refined for later use. Ceramic materials, such as Portland cement and limestone are also reduced or refined. Limestone is an important ingredient for concrete. To make Portland cement, clay and limestone are mixed together and heated in a kiln. These materials are then ground to a fine powder and bagged for future use.

Refining aluminum requires a different process. Bauxite, the ore of aluminum, shows little effect from direct application of heat. The ore melts but the aluminum does not separate. After years of research, a man named Charles Hall found a practical way to obtain aluminum from bauxite. He passed electricity through the melted ore. The melted aluminum settled to the bottom of the furnace where it could be drawn off. Before Hall's process was used, aluminum cost more than $500 per pound. Today, aluminum costs about one dollar a pound.

crude oils
hydrocarbons
paraffins
naphthenes
aromatics
fractional distillation

Refining Fuels

The refining of oil produces a range of usable fuels, as well as materials for manufacturing. The **crude oils** consist primarily of hydrocarbons. The three main groups of **hydrocarbons** are the **paraffins,** the **naphthenes** and the **aromatics.** Oil refining requires two types of processes. **Fractional distillation** is the physical separation of the oil into several useful forms. The second type of process includes cracking, reforming and polymerization.

The refining of crude oil is a boiling process in a tall tower. Each of the many products boil-off at different temperatures. Gasoline is one of the first to boil off, while asphalt is left behind as a residue. Each product is considered to be a fraction of the crude oil. The process is referred to as fractionating or breaking the oil into fractions.

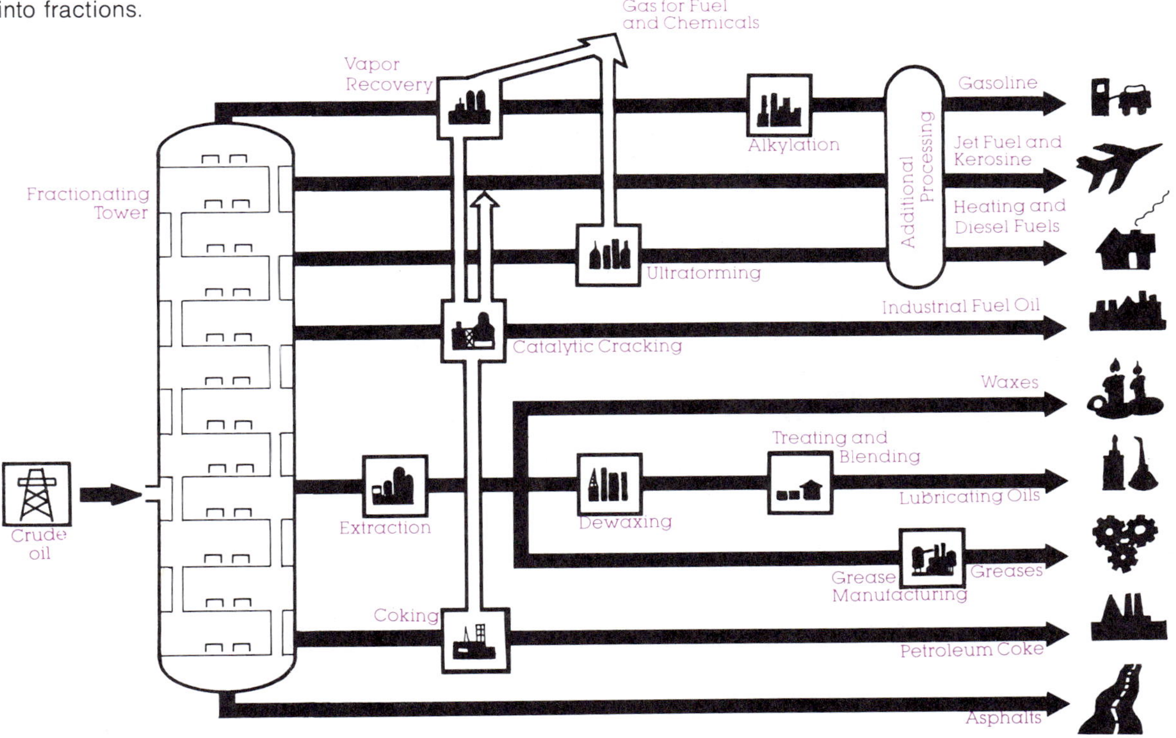

distillation In fractional distillation, the crude oil is heated to boil off the many liquids. The tall columns at an oil refining plant are sophisticated distilling vats in which the oil is heated. The lighter parts of the oil, called **fractions,** boil off first. These vapors rise in the column and are drained off into holding tanks. Liquids that boil off next are caught at a different height in the column. The oil is usually sent through several columns. Sometimes it is recycled more than once before the distillation process is complete. This separation is a lot like skimming the fat from a pot of chicken soup. The fat is lighter and rises to the top.

Refining processes also change low octane to higher octane fuel. This process is called **reforming.** The **polymerization process** joins molecules to form more complex molecules. Many plastics are formed through polymerization processes. Additives may be used to increase the stability or other properties of the material.

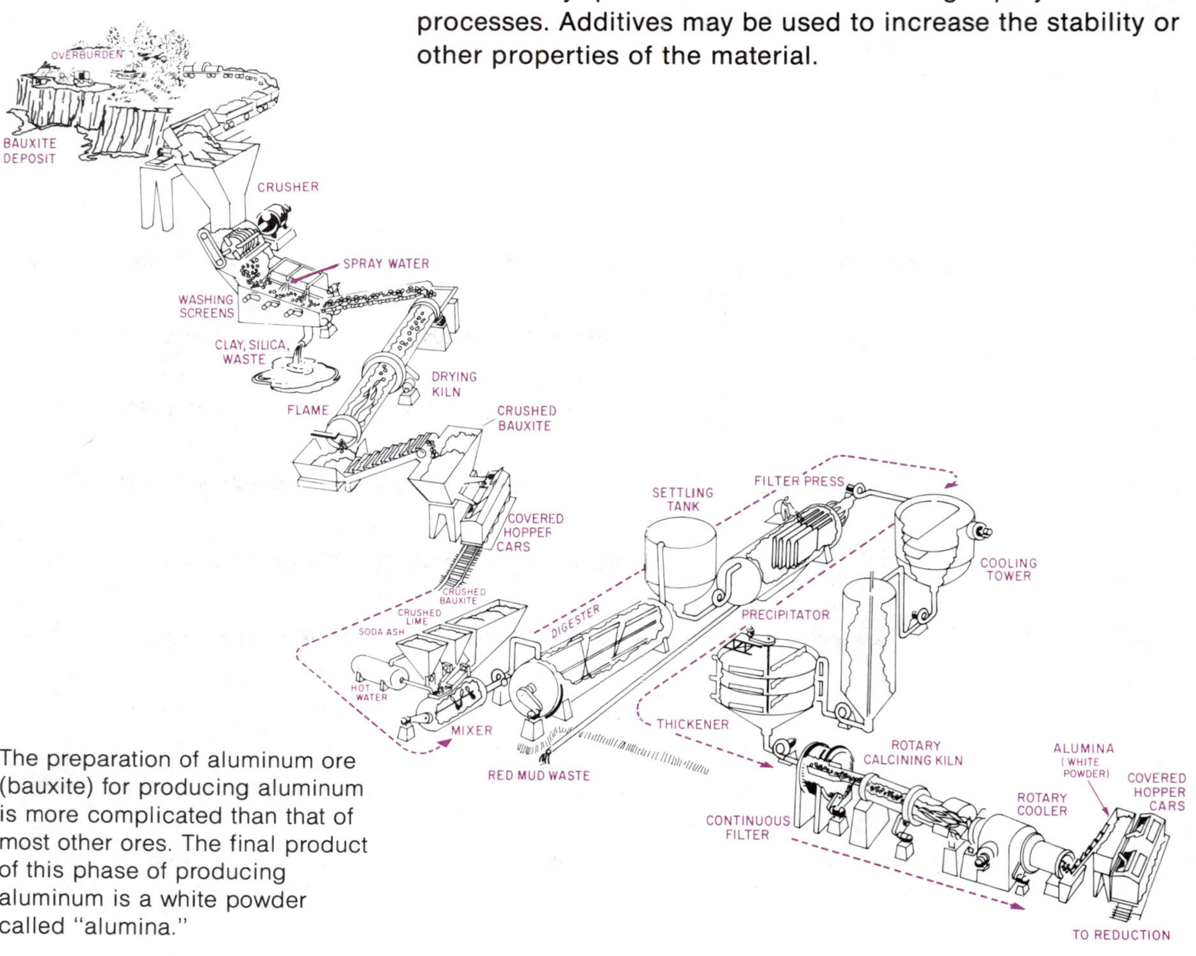

The preparation of aluminum ore (bauxite) for producing aluminum is more complicated than that of most other ores. The final product of this phase of producing aluminum is a white powder called "alumina."

Most of the coal that is mined today is used in its natural state. Coal can also be refined to produce **synthetic natural gas.** This synthetic gas has many of the same properties as natural gas and can be sent through the same pipelines. Through a **coal gasification process,** coal and hydrogen react together to produce methane.

synthetic natural gas

coal gasification process

Refining Foods

Foods are refined to remove impurities or change the food to be more desirable. Sometimes, substances are added to improve the shelf life, the nutritional value and the appearance of the food. The milling of grain into flour is an example of food refining. In many parts of the world, this grain is wheat, but flour may also be made from rice, rye or barley.

Different flours may be blended to provide the desired characteristics. Flours for bread, pizza and cake are all different. In addition to blending, flours may be enriched by adding

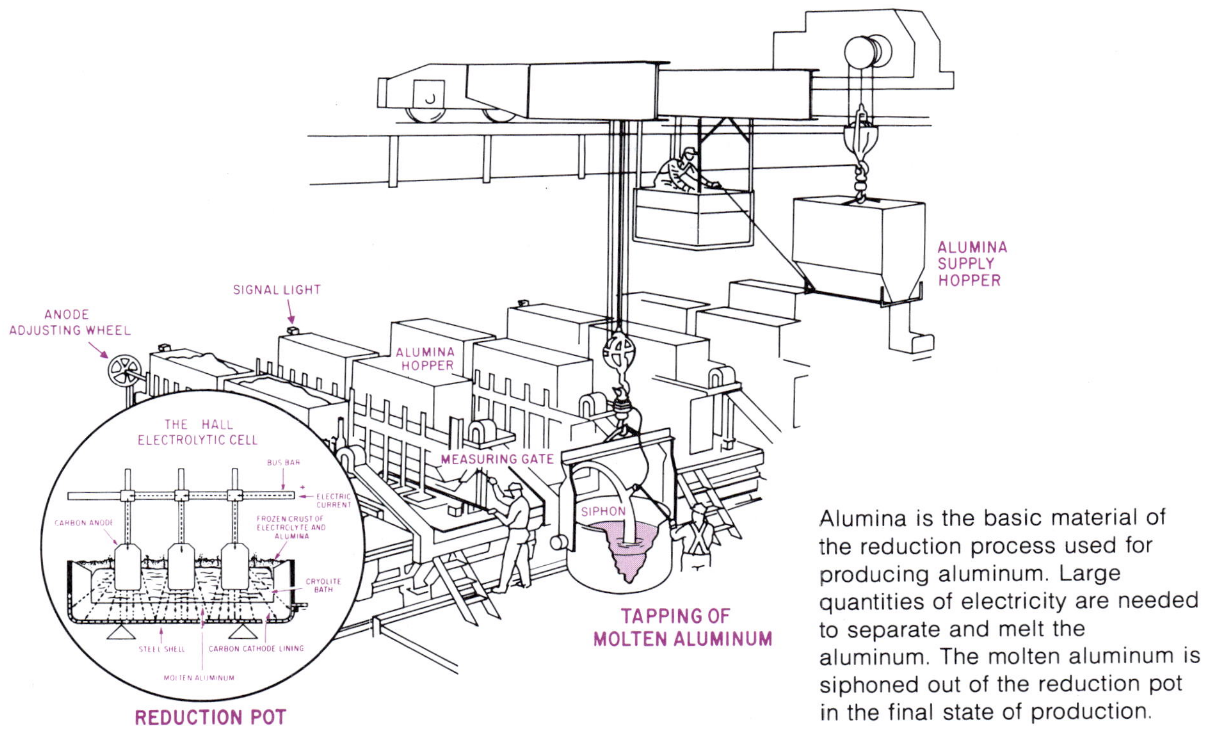

Alumina is the basic material of the reduction process used for producing aluminum. Large quantities of electricity are needed to separate and melt the aluminum. The molten aluminum is siphoned out of the reduction pot in the final state of production.

The parabolic solar collector is used to obtain high temperatures. Some systems based on the same principle use many reflectors and can reach temperatures in excess of 5000 degrees.

vitamins that may have been lost during the refining process. Rolled oats, cornmeal and even sugar are refined in essentially the same way.

Producing Energy

In the earlier chapter on Energy, special attention was given to the "prime sources" of energy. We mentioned only a few of

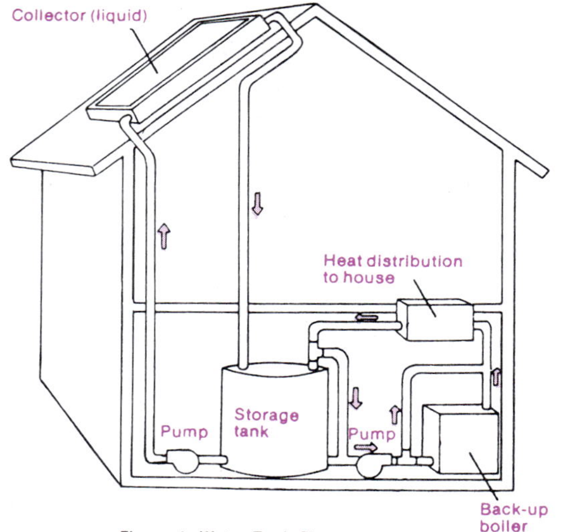

Figure 1. Water Tank Storage

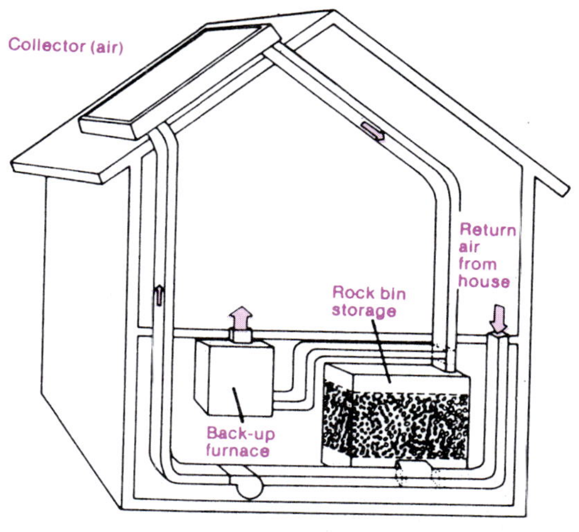

Figure 2. Rock Bin Storage

The temperature of sunlight is capable of heating water for home use. Solar-heated water systems are popular in climates with numerous "sun" days (southern climates); but can also be effective in climates where "sun" days are limited.

Because the Earth's crust is so thick, only limited use of geothermal energy has been made for producing electricity. If new techniques can be developed to reach and tap the thermal energy of the molten rock far beneath the crust, the problems of energy shortage would be solved.

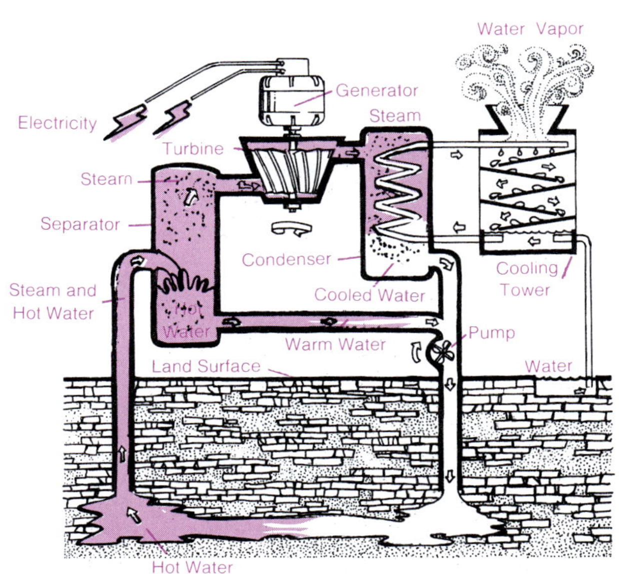

generating

the processes and devices that can be used to generate useful forms of energy. In this section we will expand on these earlier concepts. Our focus here will be on ways in which energy from these sources can be changed to a more usable or desirable form.

hydroelectric power plant

turbine generator

Selected energy devices are used to convert prime energy into secondary sources that are readily usable. These energy conversion devices include solar energy units, geothermal devices, windmills, tidal powered devices, hydroelectric power plants, nuclear reactors, fossil-fuel power plants, steam turbine generators, magnetohydrodynamic generators and fuel cells.

Humans, other animals and plants are not included. These are biological sources of energy. New processes are being developed to use these energy sources. These processes are still in the experimental stage.

solar energy

solar cells

Solar energy is used to generate heat or electricity. **Solar cells** are devices that convert the light of the sun directly into electrical energy. During the past few years, the number and kinds of solar collectors have increased significantly.

Because there is a considerable amount of heat in sunshine, it is fairly efficient to use that energy directly for heating water or warming air in a building. The system shown here heats water that is then fed into the existing hot water system for a home.

Another prime source of energy with considerable potential for direct conversion is the heat of the earth. Geothermal energy can be compared to the fantastic heat of volcanic lava. Geothermal techniques harness this heat for practical purposes. Approaches that capture the heat that escapes through the surface of the earth are used. Sophisticated approaches drill deep into the earth for heat.

geothermal energy

Geothermal energy is generally thought to be nearly unlimited for all practical purposes. The problem is how to extract the energy from deep within the earth. Natural hot springs can be used to generate energy. The conversion of the heat for generating other energy forms is accomplished three ways. **(1)** We can change the water to steam. This is done by allowing the pressure to drop and using the resulting steam to drive a turbine generator. **(2)** We can maintain the pressure on the water so it will not become steam. Then we use a heat exchanger to extract part of the energy from the hot liquids. **(3)** We can use both the steam and the water to drive a hybrid turbine generator. These have two parts, one that can be driven by steam and one that can be driven by water. If such approaches are successful, a reservoir of energy could come within our reach that would dwarf all the present fossil fuel deposits combined.

wind power

The **wind** also has great potential as an energy source. Prior to the development of the steam engine, windmills provided more energy for the factories than other sources. Windmills must have some way of orienting themselves into the wind to maximize their power. They must have some way to regulate how much of the wind they attempt to harness. Without a braking device, even the strongest machine would beat itself to pieces in a powerful windstorm.

Energy from water power is available in two major forms, mechanical and thermal. Mechanical energy exists in the movement of the waves, tides, and currents of the oceans. Thermal energy exists as the difference between the warm temperature of the surface water and the cold temperatures of water deep in the oceans.

Modern windmills like this Darrieus rotor operate on the same principles as older windmills.

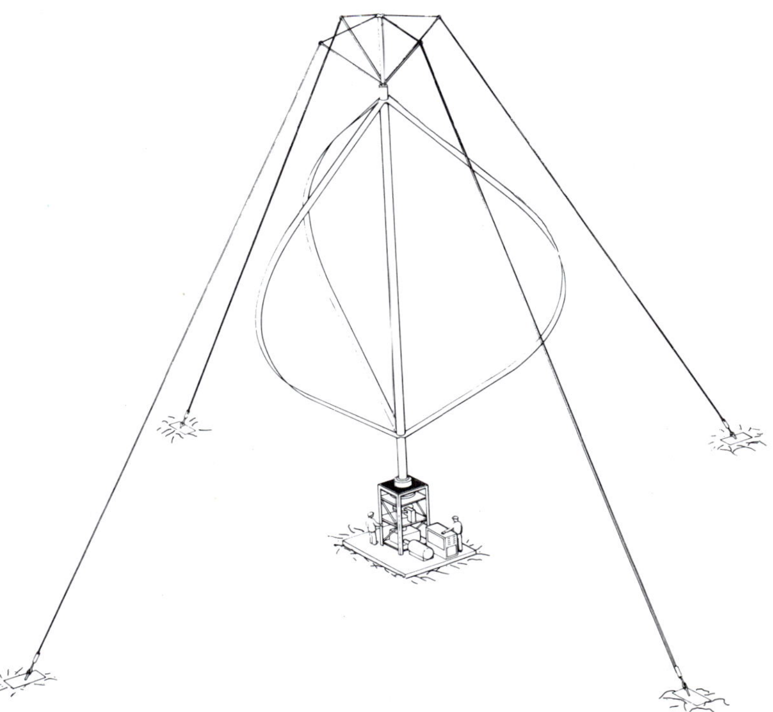

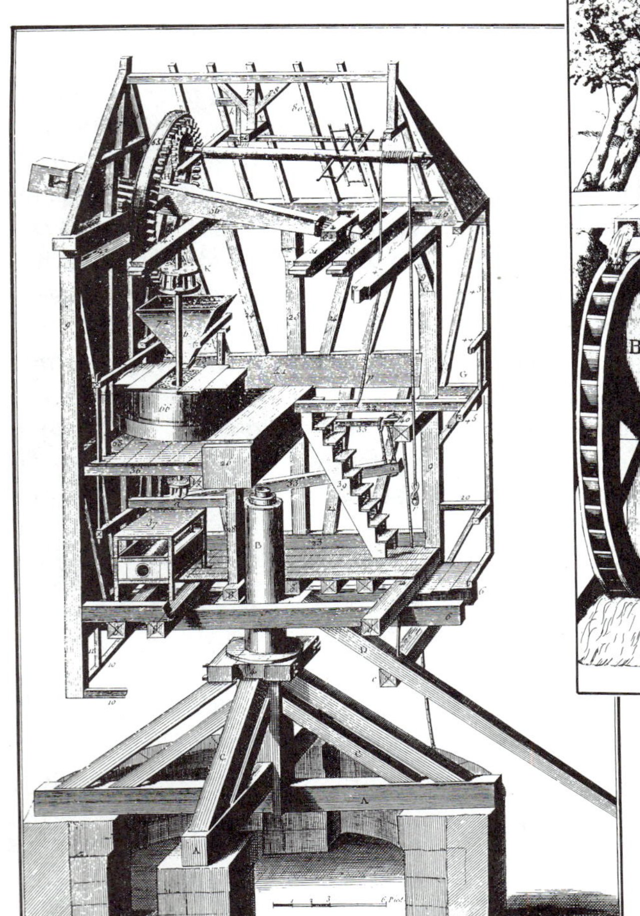

The use of waterwheels and windmills to provide energy producing products, such as flour, became popular in the middle ages. The waterwheel and the windmill also provided power during the early phases of the Industrial Revolution.

water power

Water power was an early prime source of energy. Waterwheels were often used to convert water power to do work. Recently, inventors have begun to develop devices to capture energy from the ocean tides and waves. One approach to capturing and converting mechanical energy from the waves is through the use of large flexible rafts that move like a hinge. As the raft flexes, pistons operate to pump water through turbines. This, in turn, generates electrical power. Another approach is through the use of wheels and similar devices. One such device looks like a waterwheel that has

lifting foil

been extended like a long vertical conveyor. It is called a lifting foil. The foil looks much like a continuous venetian blind. It is turned by the movement of the tide as it moves toward the land and again as the tide moves out again. A similar version of the lifting foils might be used some day to harness the moving water in rivers and ocean currents.

Hydroelectric power plants are the most direct means for using water to produce energy. The generation of energy is similar to the method used by steam-powered turbines. In hydroelectricity, the driving force is the moving water. The water turbine is a logical development from early water-raising machines. The waterwheel takes advantage of the weight of the water, and increases efficiency with a long drop.

nuclear fission
reactors

The two major ways to produce nuclear energy are fission and fusion. **Nuclear fission** splits unstable atoms. The nuclear generating plants now in use are called **reactors.** They use fissionable materials, such as uranium, that can be split to generate huge amounts of heat. This heat is tapped and converted into steam to drive conventional electrical generators.

Waterpower became very important with the development of high speed turbines that could capture the energy of water falling from great heights. The turbines turned large generators that produced electricity.

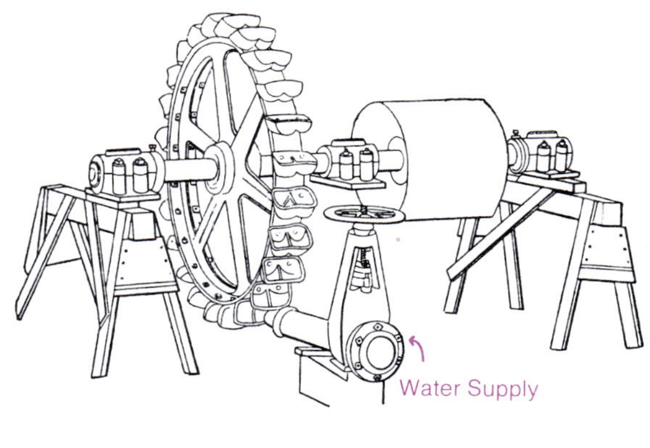

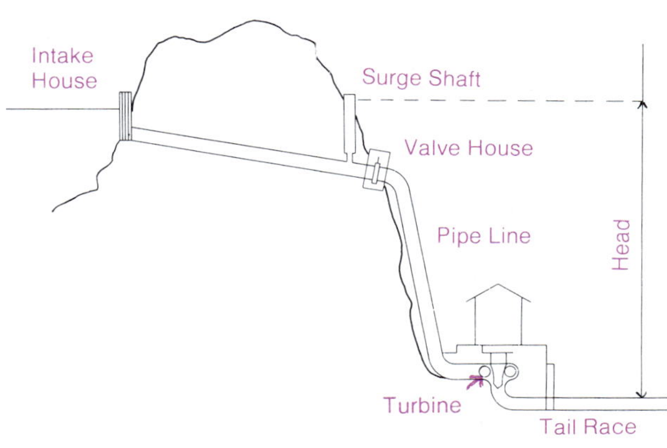

This is a model of a proposed "Tomac" nuclear fusion reactor. If and when such a reactor is successful, a miniature version of the sun will be held in a magnetic bottle. The heat generated by this device would produce steam to drive electric generators.

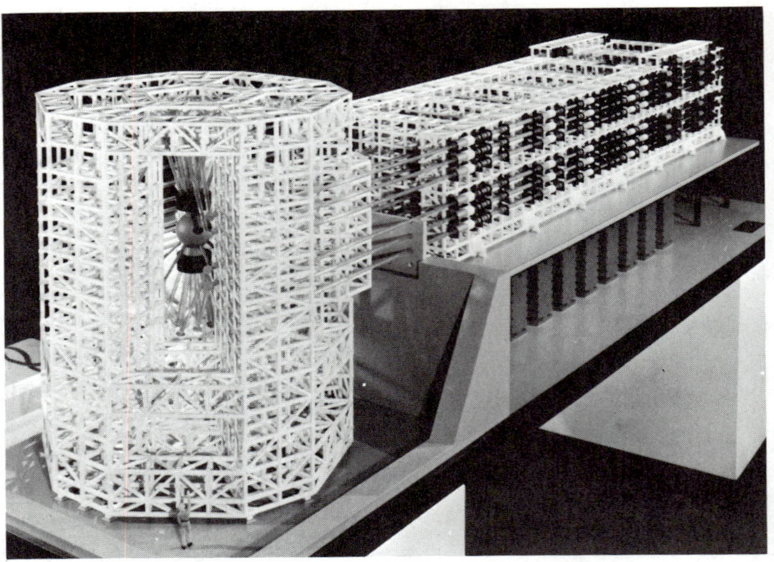

nuclear fusion

Nuclear fusion is in its earliest stages of development. It is generally accepted that fusion reactors could produce vast quantities of energy. However, many problems and much work lie ahead. The major problem is in initiating and controlling the 100 million degree temperatures created by the fusion process.

fossil-fueled power plants

Fossil-fueled power plants are primarily large furnaces that are used to convert coal, oil or natural gas into heat. The heat is used to convert water to steam. The steam is used to drive a steam turbine. The steam turbine is connected to an electrical generator. Fossil fuels supply approximately 75% of the energy used to produce electricity.

The oceans hold tremendous amounts of thermal energy. One approach to converting this energy is to use stationary and floating structures. These are designed to take advantage of the difference in temperature of the surface and deep waters. A liquid of low-boiling point, such as Freon or propane, is cycled through a machine. This machine works like a refrigerator in reserve. The liquid is first cooled by the deep waters and loses energy by giving off heat. The liquid is pumped to the surface where it takes on large amounts of energy and is vaporized under pressure by the warm water. The high-pressure vapor then operates much like steam, to drive an electrical generator.

Floating structures hold considerable potential as floating factories. A ship of this type could use the electrical energy generated to produce ammonia. Ammonia is an important ingredient for fertilizer for food crops. Currently, much of our ammonia comes from natural gas a fossil fuel. The floating ammonia factory would capture energy from a new, replaceable source and conserve our scarce, non-renewable fossil fuels.

ammonia

A new energy producing device is in the experimentation stage. It is called a **magnetohydrodynamic (MHD) generator.** The MHD generator takes advantage of the electrical properties of a super-heated gas called a **plasma.** Air is heated to almost 5000° F through several steps and then passed as a plasma through the magnetic field of a supermagnet. Free ions are captured from the plasma by the magnetic field. The ions cause electrons to flow, producing electrical energy.

magnetohydrodynamic (MHD) generator

plasma

Fuel cells are also being developed as an alternative to fossil fuels. The fuel cell operates on the same principle as a

fuel cells

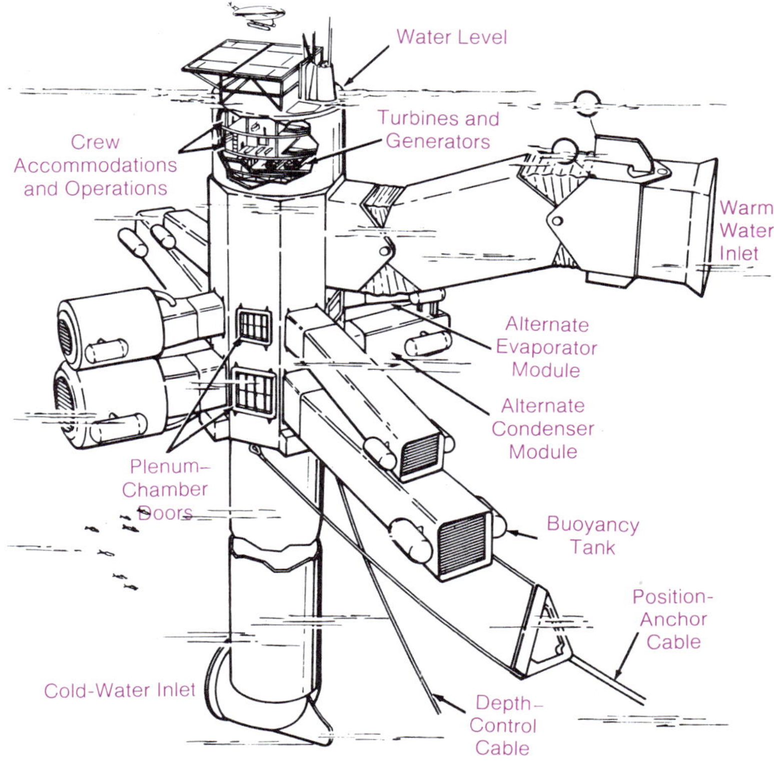

This proposed Ocean Thermoelectric Powerplant would use the energy difference between the deep cold water and the surface warm water in the ocean to produce electricity. The system would operate like a huge air conditioner that runs backwards.

This ocean going factory would use the heat from the sun to convert nitrogen from the air and produce ammonia for fertilizers. Twenty such ships would produce enough ammonia to save 475 billion cubic feet of natural gas a year.

wet cell battery. In this case, however, natural gas can be converted directly into electrical energy. The hydrogen content of the natural gas is the energy source for the fuel cell. At the present time, there are limited applications for fuel cells as energy converters. Although the efficiencies of some fuel cells approach 50%, their manufacture and operation is still too expensive for general use. Costs may go down as people learn more about fuel cells.

Manufacturing

Manufacturing can be thought of as the activities through which useful and saleable products are made. The end result of manufacturing is a product that is ready to be used. Obviously, some manufacturing processes produce only certain materials and parts of a final product. For example, automobile tires are manufactured to be used as parts of a completed automobile. The preparation of the materials for the

This device is a magnetohydrodynamic (better known as an MHD) generator. Hot gases become ionized and conduct electricity when they are forced at high speeds through a strong magnetic field. The ions cause electrons to flow, providing electrical energy.

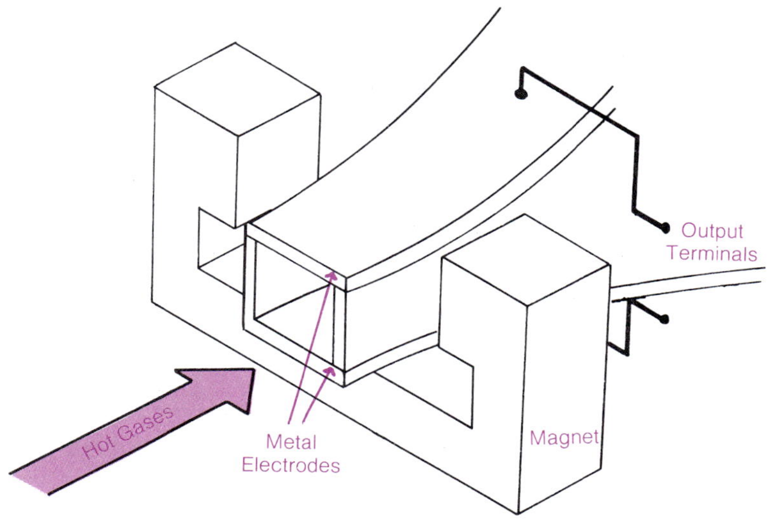

Fuel cells, originally designed nearly a hundred years ago, now take many different forms. All fuel cells capture energy in the form of oxygen ions released in the oxidation of conventional fuels. Fuel cells provide a direct conversion of energy and avoid the wasteful detour of first generating heat to be converted into useful energy. Such direct energy conversion reduces the loss of usable energy, or entropy.

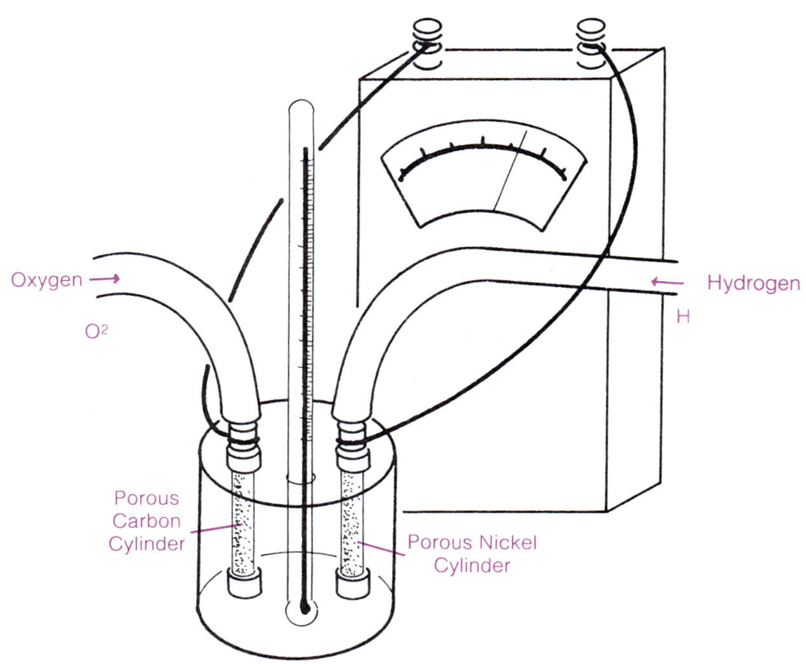

tires may include extraction and refining as well as manufacturing.

Manufacturing falls into three types: (1) the custom producing of items, (2) the intermittent producing of "batches" of products, or (3) the continuous production of products in mass. Custom manufacturing can operate on a small, medium or large scale. Intermittent manufacturing usually operates on a medium or large scale. Continuous or mass production can operate only on a large scale.

custom manufacturing

Custom manufacturing may range from small activities to very large ones. An individual making a one-of-a-kind leather purse is an example of a small custom activity. The production of a large airplane or ocean liner is custom manufacturing on a large scale. In the custom manufacture of individual items, there may be concern to make no two items exactly alike. Often, the uniqueness of an item is valued.

intermittent manufacturing

Intermittent manufacturing is used when there are too many products to be made by the custom method. The prod-

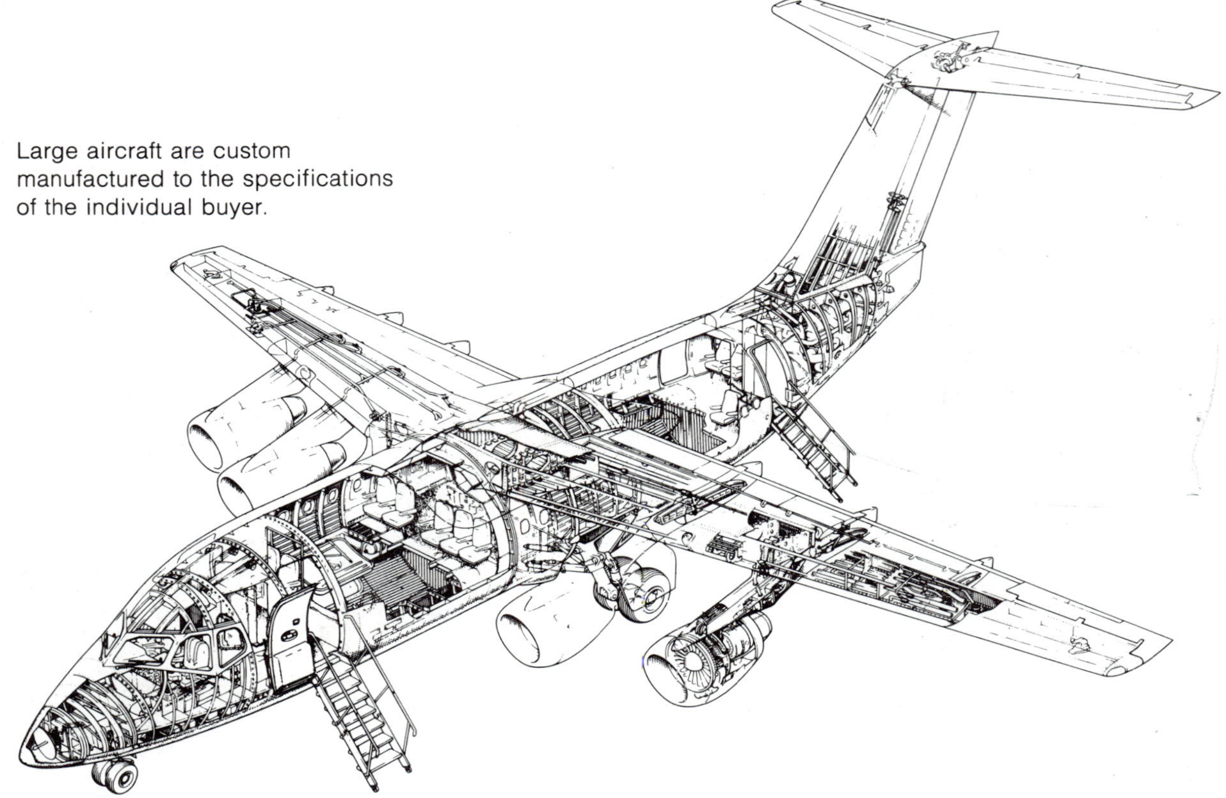

Large aircraft are custom manufactured to the specifications of the individual buyer.

ucts are processed in quantities called "lots" or "batches."

lots
batches

Lots refer to a group of individual items that are made one at a time. **Batches** are quantities of materials or products that are made all at once and then separated into individual products. For example, paint is mixed by the batch and then poured into appropriate sized containers for marketing. The label of a can of paint will often have the identification number of the batch from which the paint was taken. Bakery cookies and bread are usually mixed by the batch and then portioned out for baking as individual items.

mass production

Continuous or **mass production** is used when the products or parts are to be made identical and a very large number is needed. Mass production by continuous means is possible only if there is a mass market to buy the product. Products such as automobiles, refrigerators, radios, toasters and telephones are mass produced on large assembly lines.

feedback
control
automation

Continuous manufacturing is also used to process oil, chemicals, foods and metals. Continuous manufacturing is possible only if a fairly sophisticated level of feedback and control has been achieved. A production line must be a working system, not just a large number of unrelated operations. For example, it is possible to mass produce a million loaves of bread with a thousand bakers working at a thousand ovens. But, the process is not continuous. It could not be automated into a self-monitoring and correcting system.

Recycling

recycling

If we continue to use Earth's materials for production and construction, we will eventually exhaust our supplies. Buckminster Fuller, however, saw the earth as a spaceship. All the materials that were here when the earth was brand new are still here, but in different forms. Fuller says the problem is not in running out of something, but in learning how to reclaim, recycle and reuse the resources that are available.

The problem of recycling may not be as direct and simple as Fuller describes. Even after products wear out, many of them can be used again. Glass bottles, for example, can be reclaimed for their glass. The glass can be used to make new bottles or mixed with road paving materials.

Reclaiming materials such as glass, aluminum, iron and paper can save money. A large part of these savings can be

A continuous manufacturing process can be used to produce an endless steel slab from molten metal fed into a ladle and oscillating mold. This approach gains more acceptance and use as smaller steel mills become more common.

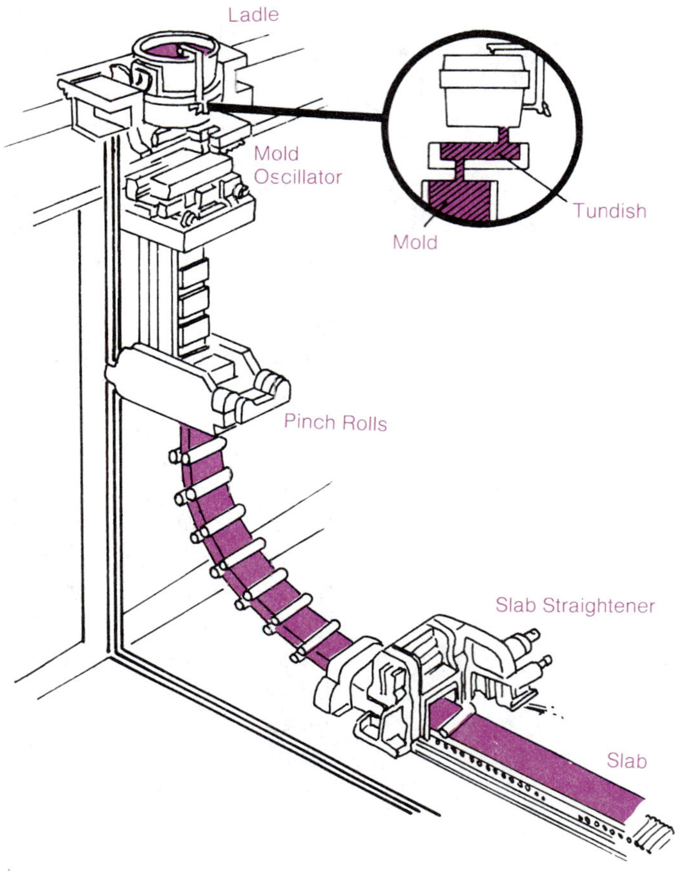

used to make a new product. For example, it takes much less energy to make new cans out of melted ones than it does to make more aluminum from aluminum ores. The Earth's energy problems will make reclaiming of materials more and more important in the future.

Everyone complains about taking out the garbage. Garbage is something we would like to do without. It takes money to get rid of it, and it smells. But, what if we consider garbage to be a resource? We know, for example, that garbage can be burned. In well-designed incinerators, the garbage that is thrown away each year could generate heat that is equivalent to burning 400 million barrels of oil. One out of every ten barrels used in the United States could be replaced with "garbage oil."

The concept behind recycling is to "close" the system. The

scrap, waste or refuse from the system can be reintroduced and used again. Glass bottles are recycled regularly to reclaim their materials for use in making other bottles and glass products. Worn-out automobile and truck tires ground up into pieces have been used with asphalt to build roadways and sidewalks. The rubber has a nice bounce when you walk on it and rain drains right through, leaving the walkway free of puddles. As we become more creative in reusing the scrap, waste and refuse that come from our industries, farms and homes, we will be able to reduce the amount of materials and energy needed to make some products.

Research has shown that the weeds, food and garbage thrown away by most families and the human waste that is regularly flushed down the toilet can be converted into methane gas. The gas can be produced in sufficient quantities to reduce each family's energy bill. Your home could be designed to capture your family's waste products so that they could be used in a methane converter. This converter would produce gas to supplement the fuel you burn for heat. Such an approach would be "closing" the system of your home waste environment so that the refuse could be used as a raw material. The idea is not new. Farmers have used animal wastes to fertilize crops for years. Animal "chips" are a major source of fuel in many developing countries.

Wood is a major fuel in many areas of the world and attention is required to find more efficient ways of using it. In these countries, large amounts of time are spent gathering and transporting wood. This time will be even longer in areas where the demand becomes greater than the supply. If wood continues as their cooking fuel, these people could benefit by more efficient stoves.

In all of the previous examples of recycling, two central ideas occurred again and again. Waste and refuse can be considered as a potential resource. The system of operation, whether large or small, can be closed by feeding back into the system any part of the garbage, scrap, waste or refuse that can be put to productive use. Recycling, therefore, is another practical case of establishing a self-sustaining technological system.

Summary

Production is a system that takes materials and energy from the earth, sea and sun and converts them into useful prod-

The model of production is a more complicated version of the black box model used to describe machines and processes. Goals determine the inputs of materials, energy and information. These are changed by machines, processes and humans into output of new products of materials, energy and information. Similarly, production systems generate waste, scrap and refuse. Some of this will be usable for recycling and some will cause problems of pollution.

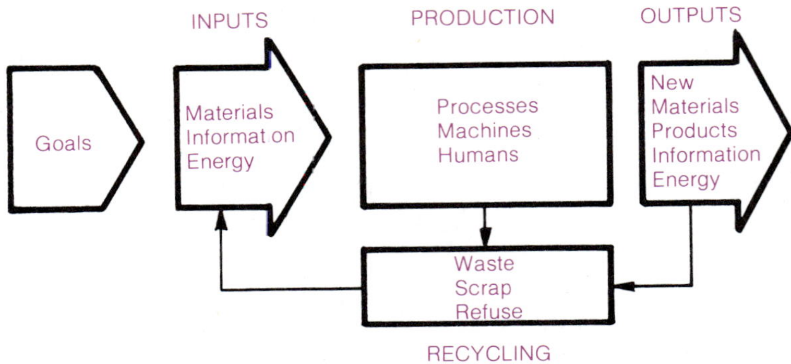

ucts. The system can be closed by recycling wastes to be converted into new forms of usable materials and energy.

Production activities include extracting materials and fuels, harvesting food and fuels, refining materials, foods and fuels, generating energy and manufacturing products. Producing allows many of the other technological activities to operate by making available the tools, devices and machines that are used in constructing, communicating and transporting.

Key Concepts and Terms

agribusiness
ammonia
aromatics
automation
batches
blast furnace
coal gasification process
combine machine
control
crude oils
custom manufacturing
distillation
ecology
extracting
feedback
fish farming
floating fish factories
forage crops
fossil-fueled power plants
fractional distillation
fractions
fuel cells
generating
geothermal energy
grasses
harvesting
hydrocarbons
hydroelectric power plant
intermittent manufacturing
legumes
lifting foil
longwall mining
lots
magnetohydrodynamic (MHD) generator
manufacturing

mass production
mechanization
mining
naphthenes
nuclear fission
nuclear fusion
oil
paraffins
plasma
polymerization process
process
production
reactors
recycling
reduction
refining
reforming
solar cells
solar energy
strip mining
synthetic natural gas
tree farming
turbine generator
water power
wind power

Summary—Unit Two

In this unit we have shown how the model of the six elements of technology have grown and fit within the four systems of technology—constructing, transporting, communicating and producing. The four systems depend upon each other and do not work as isolated activities.

The major activities of technology are constructing, transporting, communicating and producing. Structures are important to the activity of constructing. The five types of structures are for support, shelter, containers, direction and transportation. Constructing and structures provide environments for shelter, movement and work.

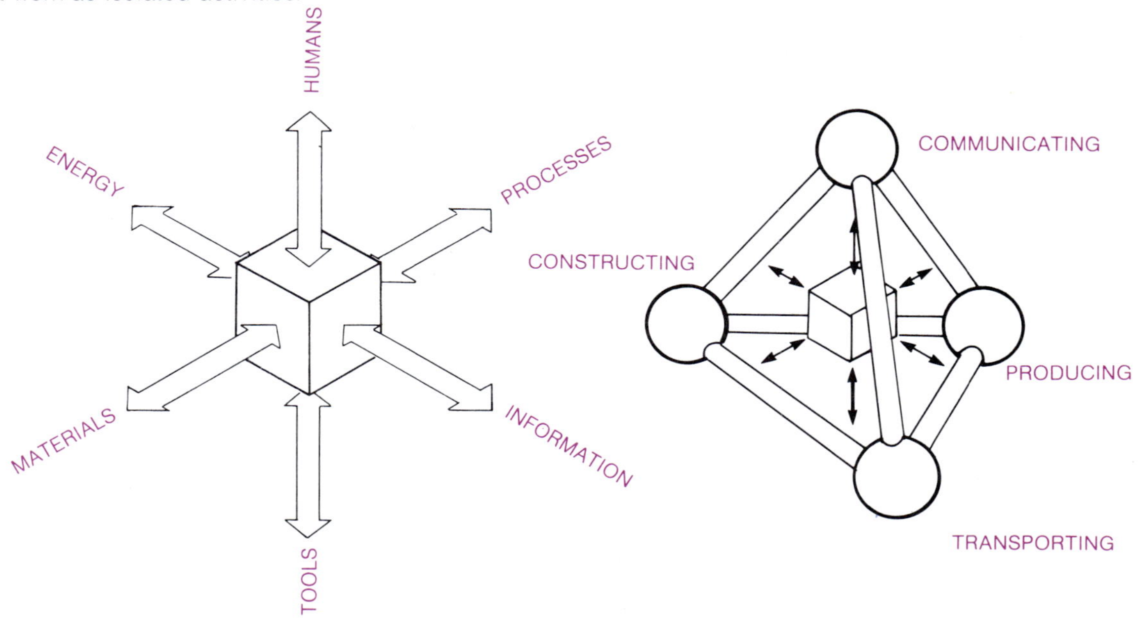

The important elements of transporting are vehicles, ways, loads and purposes. When the method of transportation allows movement in several directions, it is difficult to maintain control over vehicles. Ways are paths for transporting that include one directional movement to six directional movement.

Symbols, signals, codes, messages, media and participants are major elements of communicating. Participants may be either humans or machines. Understandable messages require prearranged codes with symbols that humans can understand and signals that machines can understand. The concepts of entropy, redundancy and feedback are also important in communication.

Production is a system that converts materials, energy and information into useful items. The activities related to producing include extracting, harvesting, refining, generating, manufacturing and recycling. Through producing, we convert resources into the six elements of technology and the four activities of technology. (The activities are constructing, transporting, communicating and producing.) All of these activities can support changes and improvements in technology.

UNIT III The Changes of Technology

Introduction

The focus of this section is on the manner and processes by which technology grows and develops. People's desire to know more and improve products and processes has expanded our horizons. Products have been improved through development and design. Development has extended the elements and activities of technology so that planned change can occur.

Changes in manufacturing and production have led to automation and cybernetics. Large scale technology systems have been made self-directing and self-regulating in order to meet the demands of mass production. The control and development of technological activities have evolved with the development of the computer, miniaturization and increased capability of computers.

Exploring, a technology supported activity, has taken on new dimensions, new vistas and new meaning during the twentieth century.

Chapter 12 Exploring

exploration
direction
development

The desire to know is a major characteristic of the human species. Early humans explored their surroundings on foot. Later, they developed wheeled vehicles and boats, which increased their range of travel. As journeys lengthened, it became very important to maintain a sense of direction. New tools and instruments had to be developed.

Which Way Are We Going?

Early in human history, the sun and stars were known to move in predictable ways. The sun always came up and set in the same general pattern day after day and year after year. When the seasonal variations were taken into account, the sun became a very predictable and dependable tool.

sundial

Many of the early direction-finding devices used the sun. A **sundial** uses the shadow cast by the sun shining on a vertical pointer to indicate the time of day. Although they were not as accurate as modern day digital clocks, sundials served the purpose. The sundial could also tell early explorers which direction they were heading. The Vikings took advantage of this knowledge and developed a device called a **ship's bearing dial.** The device provided orientation to the north. It had a moveable pointer that enabled sailors to know in what direction the ship was sailing. The pointer could also provide a bearing or direction that the ship should maintain.

ship's bearing dial

Except for some accomplishments of the Phoenicians, the

The Vikings used a simple but reliable navigational device. The ship's "bearing dial" allowed long ocean voyages like those from Norway to Greenland. The shadow of the sun at noon pointed to the north. Other directions could then be determined by rotating the pointer around the 32 point scale. The ship could be steered by using a preset bearing or direction. This method, as well as more modern ones, is known as "dead reckoning."

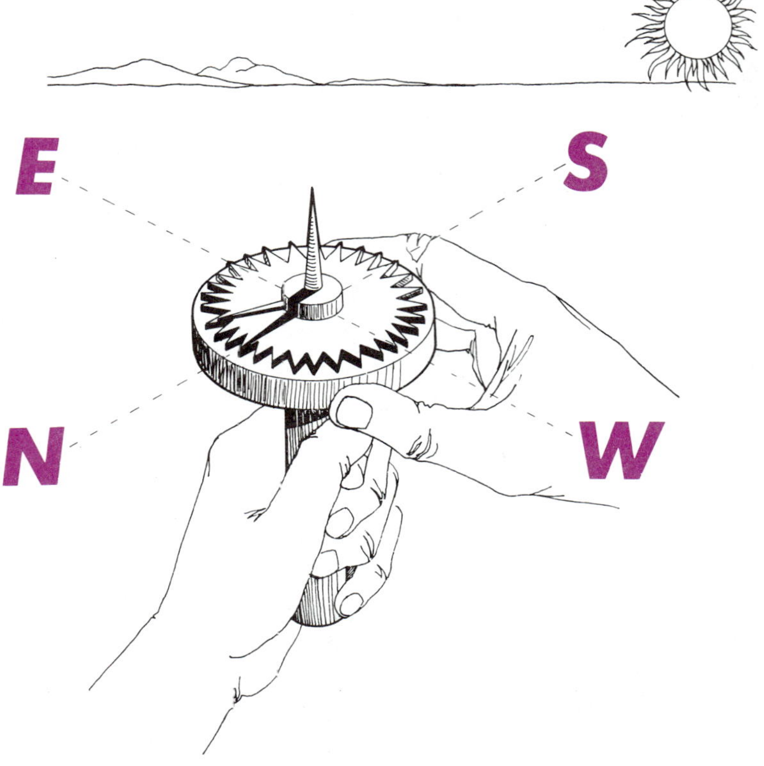

Polynesians and the Vikings, exploration of the oceans continued without major developments for two thousand years. The invention of the **hinged rudder,** between 1000 and 1200 A.D., opened the way for safe ocean travel for all people. Prior to that time, all ships were steered with one or two oars. This arrangement made it very difficult to maintain direction on the open seas.

hinged rudder

The hinged rudder was attached to the boat below the surface of the water. The control of even large ships was now within the power of one person, the pilot. The ship could also perform the difficult maneuvers needed to "tack" into the wind. Its direction could be better controlled in the changing currents of the ocean. Eventually, bigger and stronger ships were built with better masts, sails and rigging.

The magnetic compass, which may have been developed by either the Arabs or the Chinese, provided the much needed direction indicator for sailing on the open seas. Before the compass, sailors had to depend upon sightings on

Early ships were piloted by "steering oars" similar to the ones used on this Viking craft. The hinged rudder was introduced into western Europe from China, where it had been invented centuries earlier. The rudder replaced the steering oars because of the mechanical advantage and control it provided. The advantages of the rudder made it possible to build larger ocean-worthy ships for world-wide navigation and exploration.

the shore or the sun and stars to keep track of the ship's location and direction.

The compass works because the earth acts as a large magnet. The compass has a magnetic pointer that is suspended so that it can turn freely. Unless it is interfered with by large metal objects or other sources of magnetism, the compass will align itself with the earth's magnetic field.

magnetic north
true north

The compass needle will always point north and south. But, the north on a compass is different from the north on a ship's bearing dial. The compass points to **magnetic north** and the ship's bearing dial points to **"true" north.** There is a difference of about 1000 miles between the two "norths." This problem plagued early explorers.

Where Are We?

bearing

cross staff
astrolabe
quadrant
sextant

Devices that determine direction are very important tools of exploration and discovery. But, as early explorers sailed the oceans, they found that merely knowing their bearing or direction was not enough. In time, a series of devices was developed that helped to determine location. These included the **cross staff,** the **astrolabe,** the **quadrant** and the **sextant.** All four devices work on the same principle. They measure the angle or height of the sun above the horizon.

All four of these devices measure the angle of the sun or a star and the horizon. Sightings taken of the sun at midday made it easy to determine a ship's position north or south of the equator. Fixing of position in this manner was helped by the development of data charts that allowed sightings at other times of the day or night. Electronic equipment is now used based on similar principles. Sightings on navigational satellites can determine, within yards, the actual location of a ship at sea.

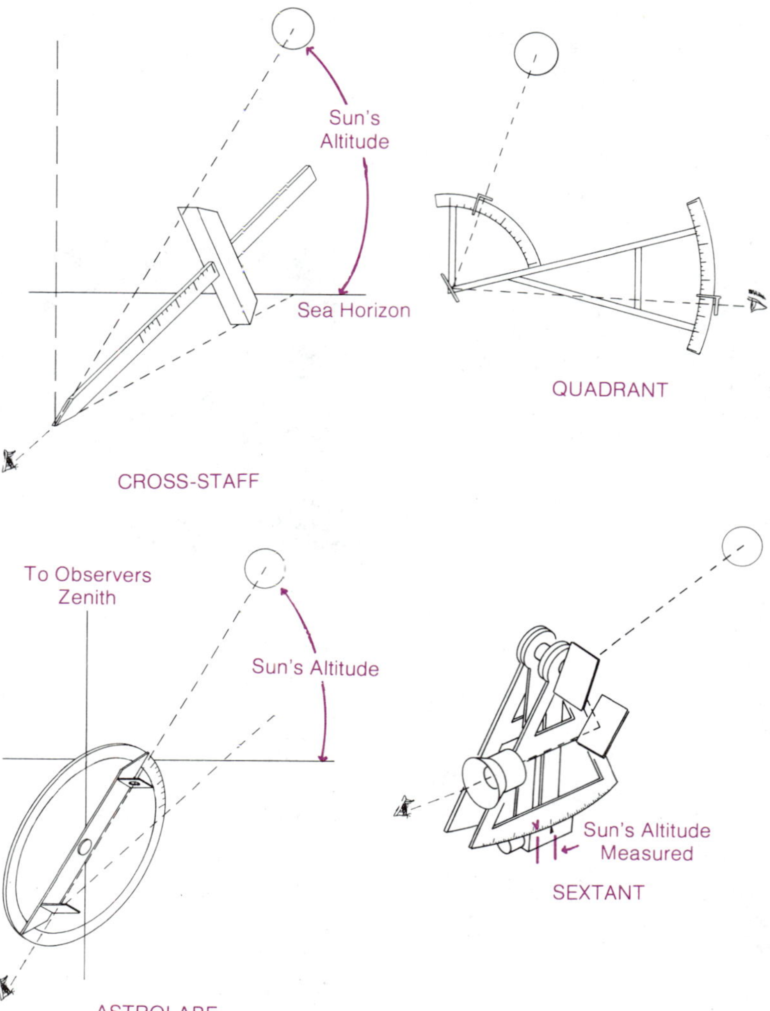

The compass was far more useful when teamed with an astrolabe or sextant. The compass indicated the ship's direction and the astrolabe found its position. But, the relative position of the sun at given times changed as ships proceeded east or west. Because the sundial was not an accurate timekeeper, another device was needed.

The development of the clock provided the necessary information. The clock allowed explorers to measure how far they were from their starting point. Before leaving their home port, sailors set a clock at 12:00 when the midday sun was directly overhead. A sighting was taken the following day

The first compass, called a "south-govenor", was developed by the Chinese over 2000 years ago. A ladle or spoon was carved from a piece of lodestone or magnetite and placed on a bronze plate. The round part of the plate represented the heavens. The square part with the directional markings represented the Earth. When the ladle was spun it would always come to rest pointing to the south.

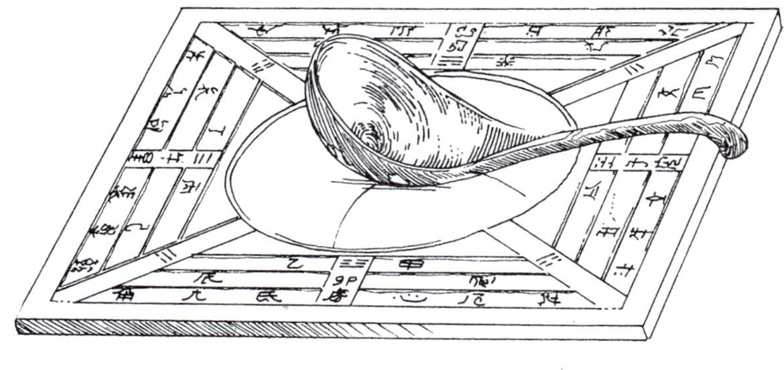

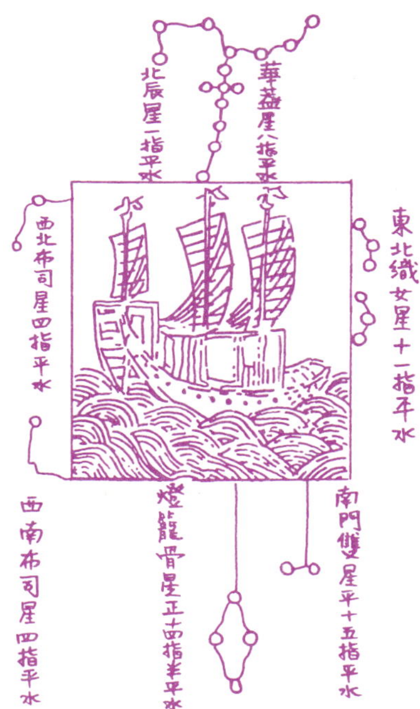

The compass made it possible to sail the oceans and navigate over long distances away from the sight of land. This star chart was developed by Zheng He, a Chinese navigator, between 1405 and 1433, during his voyages to countries as far away as India, Arabia and Africa. Can you find the Big Dipper, the North Star and the four stars that make up the Southern Cross?

when the sun was directly overhead. If the clock said 1:00, it meant that the ship's relative time was one hour different from the time at home base. Because the sun takes 24 hours to go around the earth and the ship sailed westward, the sailors could determine that they had gone 1/24 of the way around the earth. By combining this information with a reading of an astrolabe or sextant, the ship's position could be accurately located on a map.

The three related ideas of **direction, orientation** and **time** serve as the basis for navigation, even today. Passenger planes, oil tankers, spacecraft and satellites all use these three kinds of information. But, modern, more accurate measuring instruments have been developed.

Where Have We Been?

A logical outgrowth of early navigation was the improvement of simple hand-drawn maps into accurate maps and charts. These early maps and charts, which took several forms, were the first **documents** of space. One of the first was the **stick chart,** used in the South Pacific to fix the position and distance between the islands. Some of the sticks also interpreted the wave pattern and currents of a given area of the ocean. The Portuguese developed another type of chart called **rotarios.** The rotarios listed the directions, distances and hazards of a particular course. They were much like the flight maps used by present day pilots. Although these maps were not accurate by today's standards, they were helpful to the early explorers.

As navigation instruments improved, more accurate and

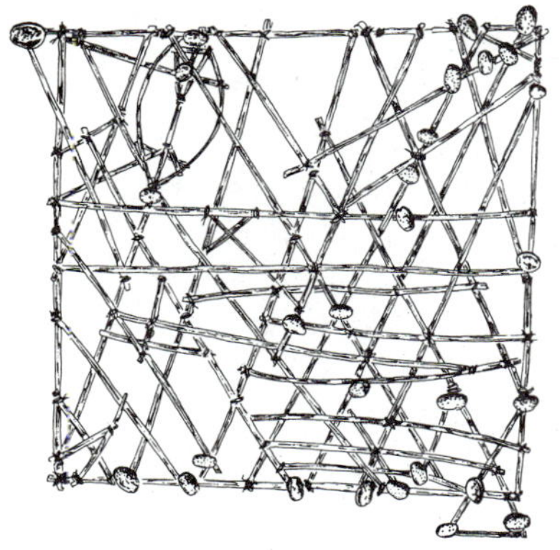

This is not an abstract sculpture. It is a map used by early Polynesians to sail among the islands of the South Pacific.

useful navigation charts and maps were needed. The first maps were drawn on flat paper. It was necessary to develop ways by which the features of a round world could be illustrated on a flat surface. The method most commonly used today is a **projection.** The technique was developed by Gerard Mercator in 1569.

What Is Out There?

Many scientific and technological achievements culminated in landing the Viking crafts on Mars in 1976. The rockets, the guidance system and the computers were used primarily to transport other discovery devices to the surface of Mars. The cameras, the sampling devices and the sensors became the long distance replacements for the humans back on Earth. The spacecraft provided the opportunity to transport a complicated system of exploration tools into space. Until the development of the rocket, exploration of space was limited to observation and analysis through land-based telescopes and radio equipment.

receiver

refractor telescope

reflector telescope

The exploration of space started long before the invention of the rocket. Early observers of the heavens were limited to the use of the unaided eye as a **receiver** until the invention of the telescope. The earliest telescopes, such as the kind that Galileo used in 1609, were of the **refractor** type. Approximately fifty years later, Sir Isaac Newton invented the **reflector** telescope. It was a great improvement. Most astronomers

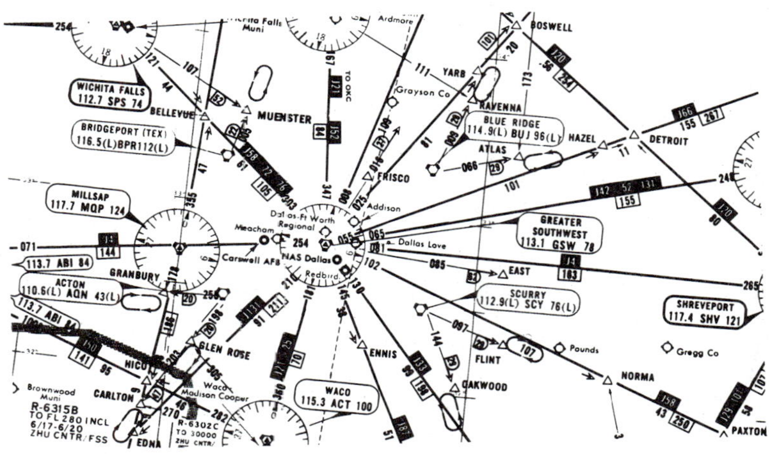

This modern aviation map marks important places and the airways between them to help pilots navigate from one point to another. Early maps called "Rotarios" used the same approach for ocean navigation.

The telescope has been the primary exploring instrument for astronomers since the days of Galileo. Galileo did not invent the telescope but he used its capabilities to help change the way people looked at the universe.

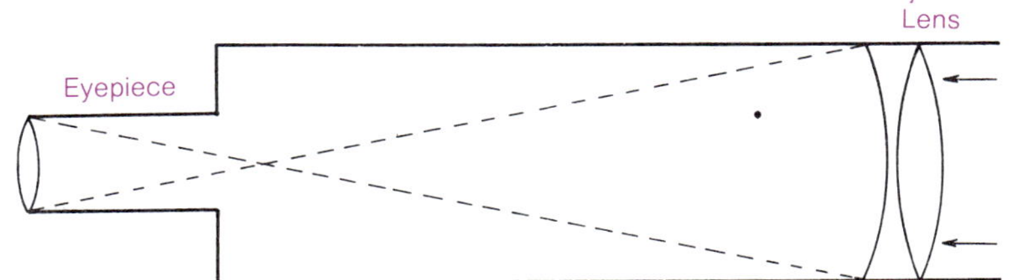

The refractor: This simplest of telescopes employs the same principle as binoculars and opera glasses. The objective lens forms a small image that can be observed through the magnifying eyepiece.

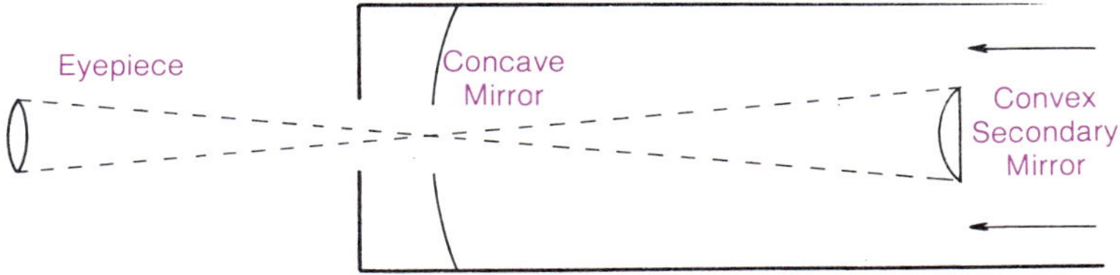

The reflector: This telescope, used most often by serious astronomers, works on reflected light. This illustration is of the Cassegrain-type telescope. It reflects light from the concave mirror to form a small image on the convex mirror. The image is viewed through a hole in the concave mirror.

still use the reflector telescope which, for nearly three hundred years, has been the major device used to explore outer space.

radio telescope

In 1931, the first **radio telescope** was built by Karl G. Jansky while doing research on the causes of static on early radio sets. Since that first accidental discovery, radio tele-

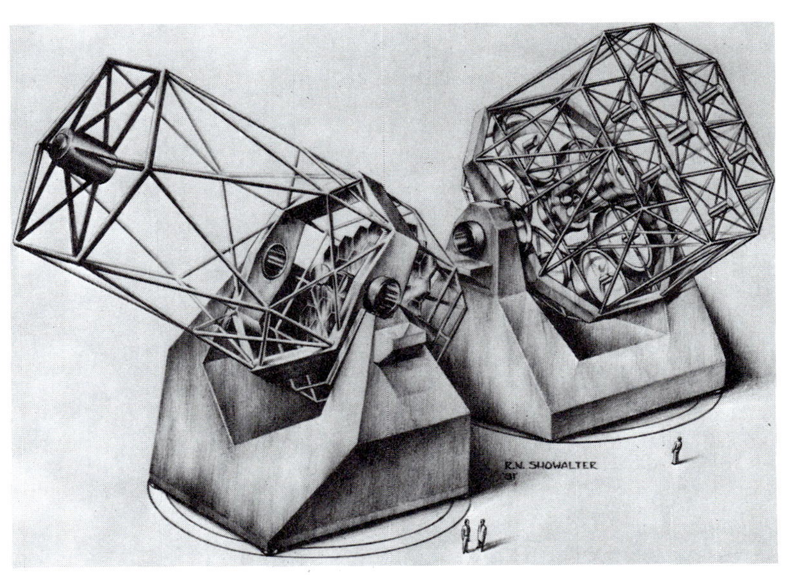

These modern telescopes use multiple mirrors to work as one giant mirror. They explore much deeper in space than ever before possible. This approach provides a telescope far larger than could be manufactured with a single piece mirror.

Jansky's radio telescope, the first ever to be built, was of modest size but tremendously important for the changes it introduced into astronomy.

scopes have been developed that receive and collect energy from pulsars, quasars, radio galaxy and hydrogen clouds. These devices are now able to pick up energy that may have been generated in the creation of the universe billions of years ago.

Radio telescopes such as these in Socorro, New Mexico have reached enormous proportions and can be linked together in a VLA (Very Large Array) as though they were one device. The map illustrates the size of the VLA if it had been built at different places around the country.

What Is Down There?

Some of the earliest exploring activities required travel on the oceans. Eventually, people became curious about the seas themselves and what might lie below their surface. Undersea

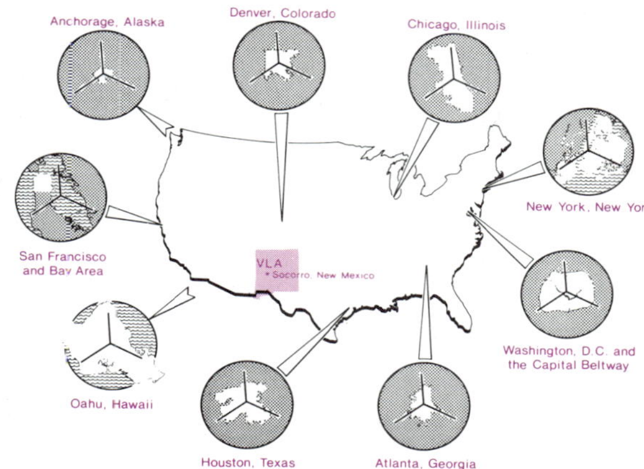

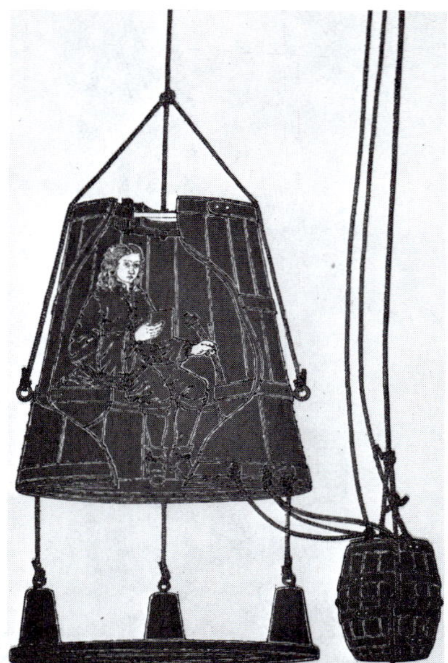

The first diving bell was a large bottomless barrel in which a person could descend to modest depths. The barrel at the right provided an anchor for the air supply lines.

exploration was held back for years, due largely to the lack of suitable technology to support human life in that environment.

Only in the past few decades has exploration of the seas begun to blossom. Because stronger metals have been developed, it is now possible to construct deep sea exploration vehicles that can withstand the extreme pressure of the ocean depths.

Maps of the ocean floor and currents have recently been developed. The hills and valleys of the earth's crust are fairly well known. However, exploring under the earth's crust still poses many problems. Seismographs have been used to chart the earth's structure. Attempts have been made to chart movement of the earth's layers and to predict earthquakes. However, we probably have a more accurate understanding of the make-up of the solar system than we do of the earth itself.

What Is In There?

microscope

The **microscope** is the counterpart of the telescope. Telescopes are used to explore outer space. Microscopes are used to investigate micro(small) space. Early microscopes were optical in nature. But in recent years, new electron-type microscopes have been developed. Explorers of microspace now have powerful new tools of exploration.

electron microscopes

Electron and x-ray microscopes enable us to explore the structure of molecules themselves. Recent developments in the use of lasers indicate that it may soon be possible to see even more clearly into the microspace of our world.

research

Some of the most interesting microspace exploration is currently being done by medical researchers. New developments in electronic instruments have made it possible to learn much more about how the body functions. Ultrasonic beams, X-ray scanners and infrared photography can now be used instead of surgery to determine the causes of health problems.

How Old Is It?

The exploration of time has two dimensions, the past and the future. The exploration of the past is a search for evidence.

Microscopes play important roles in the exploration of objects usually not seen with the human eye.

The optical microscope was used for more than three centuries. It magnifies images using a beam of light reflected off a mirror through the specimen to be studied. This device can magnify an object up to 1000 times its normal size.

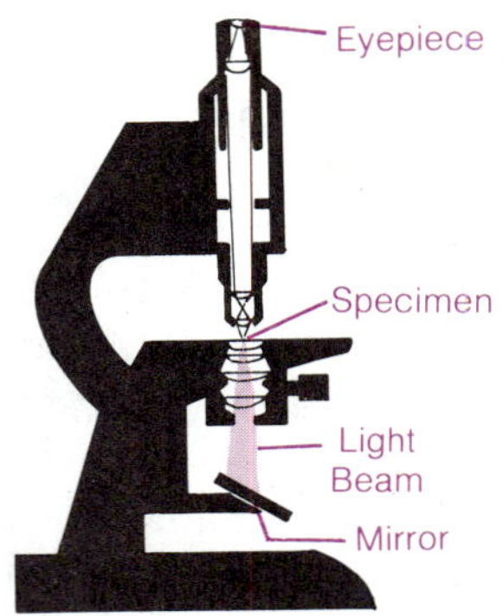

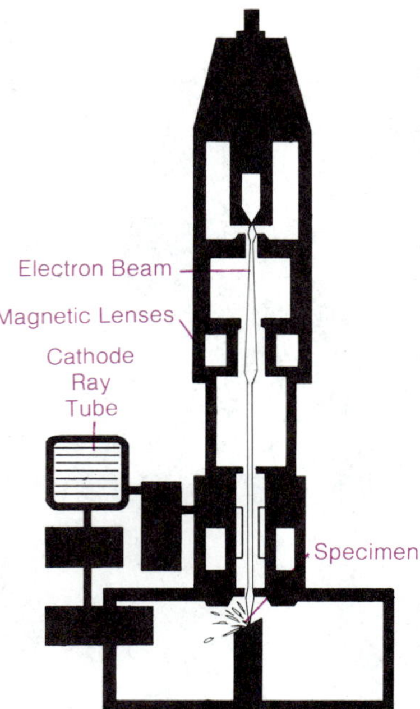

The transmission electron microscope uses electron beams to replace light rays. Objects can be magnified 100,000 times normal size.

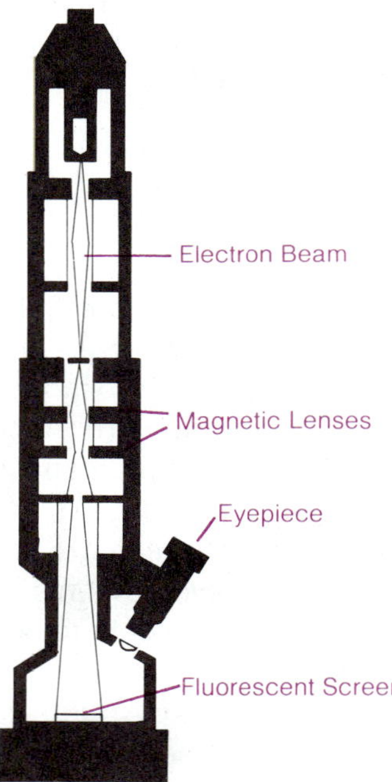

The scanning electron microscope (SEM) provides a different technique for generating images. Three-dimensional images are possible. The microscope uses a scanning beam of electrons to trigger the release of the specimen's own electrons. These electrons can be captured photographically.

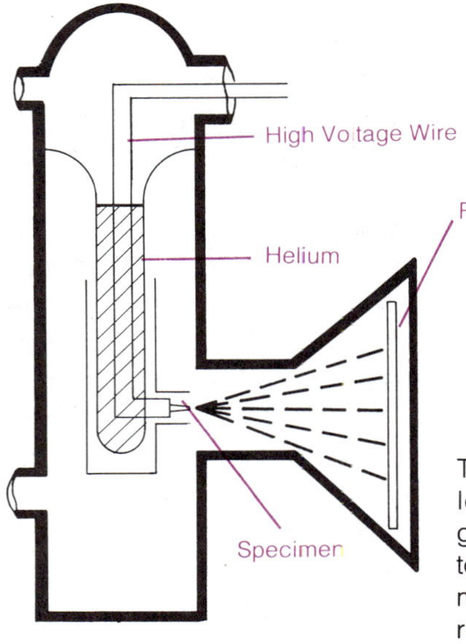

The field ion microscope has no lens but uses a cloud of helium gas to capture images. It is used to study the atomic structure of metals. A large electric current is run through metal specimen 1000 times smaller than a human hair. The electrical energy in the specimen causes helium atoms to shoot out from the specimen. When capture on film, the atoms are magnified a million times or more.

Historians search for evidence in the technological remains of past activities. These artifacts are products that give us clues about how humans lived centuries ago.

Historians and archeologists often use special instruments to help determine the age of artifacts. One such instrument measures the amount of radioactive carbon that remains in

Scanning devices are designed to explore inside the human body for diseases. These devices can also be used to conduct research. The photographic records of the specimen below can help us understand what really happens during the burning of coal.

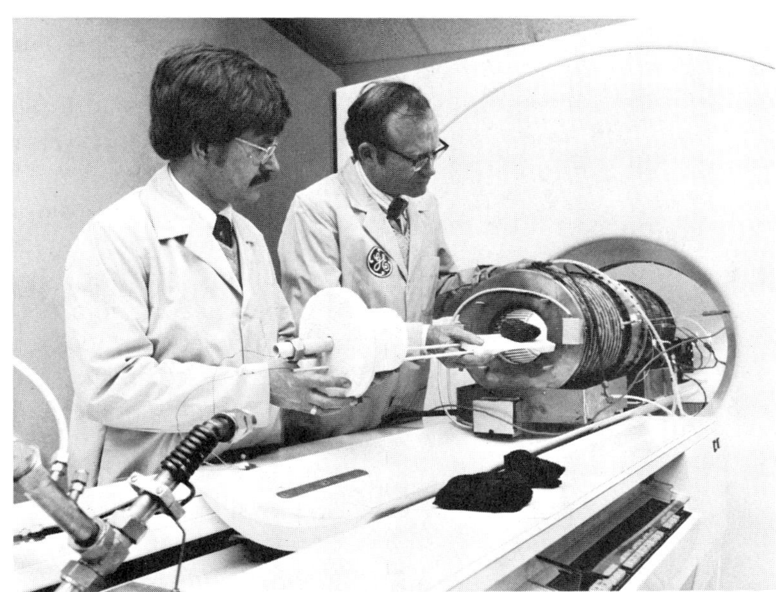

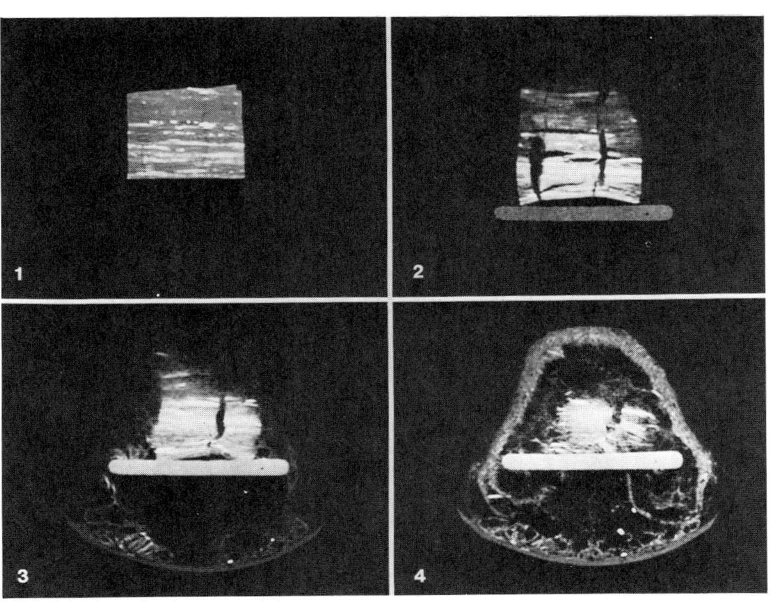

the object. It is known that radioactive carbon atoms decay at a predictable rate. Archeologists determine where the radioactive carbon atoms are in their decay sequence. This information provides a fairly accurate index of the length of time that the material has been decaying. For example, it is relatively easy to determine the age of a prehistoric campsite by sampling the remains of the charred wood. The wood would have been dead at the time of the fire or slightly before then. Thus, we can determine the date of that fire by finding the age of the wood.

The archeologist is a central figure in historical research and exploration. But archeology has only evolved into a scientific profession in the last 100 years. Before then, technology had not yet developed enough to provide the important tools needed in this field.

The archeologist now has many technological tools. The airplane is used to find the best possible site for a dig. The camera is used to document the "dig." The electronic mine detector is used to find metal objects. The X-ray spectrum analyzer can determine the chemical composition of an object without harming it. The radioactive sampler can determine the age of objects through radioactive carbon dating.

forecasting

Exploration of the future is often called **forecasting.** Forecasting also depends on technological tools, but differs somewhat from other exploring activities. The weather forecaster, for example, begins each day by gathering information about local conditions. Barometric pressure, wind direction, humidity and temperature are recorded. Satellite photos are examined for evidence of upper wind patterns and major storm areas. Data from upwind stations indicates what kind of weather may be moving into the area. Based on all the available information, the forecaster makes a prediction of what will probably happen.

The key processes of this type of forecasting include (1) information collection, (2) analysis of data, (3) identification of probabilities and (4) forecasting the possibilities. Most future-oriented exploration follows a similar sequence of action.

future-studies

The weather forecast uses mostly current information. If the forecaster needed to predict the weather months in advance, then records from past weather events would be used. Weather seems to occur in cycles. Long-range weather forecasters try to determine these cycles or trends, to help trend analysis

them in their predictions. This method is called **trend analysis.**

analysis

The computer is an important tool in trend analysis. The computer makes it possible to analyze satellite, seismographic and production data to predict likely areas for drilling new oil wells. The computer also can be used to analyze data from accidents and breakdowns of vehicles, structures or other products. This data can identify patterns and help predict potential trouble spots. These patterns or trends are important as feedback for making improvements and reducing the likelihood of future accidents.

delphi method

historical analogy

Forecasting can also be done using other approaches. The **delphi method** is an approach that draws information from a group of experts. These experts indicate what they expect to occur. A set of procedures is followed to come up with a kind of consensus. Another method, the **historical analogy,** makes comparisons of past events and sequences. This approach forecasts similar trends and patterns. The major source of data is what has happened in the past. Both approaches use a limited amount of high quality data.

causal model

A different type of forecasting uses models. A **model** is a more sophisticated kind of forecasting tool to include more powerful relationships. Causal models are not easy to develop. They attempt to describe all that is known about an event, usually in the form of mathematical formulas. Models

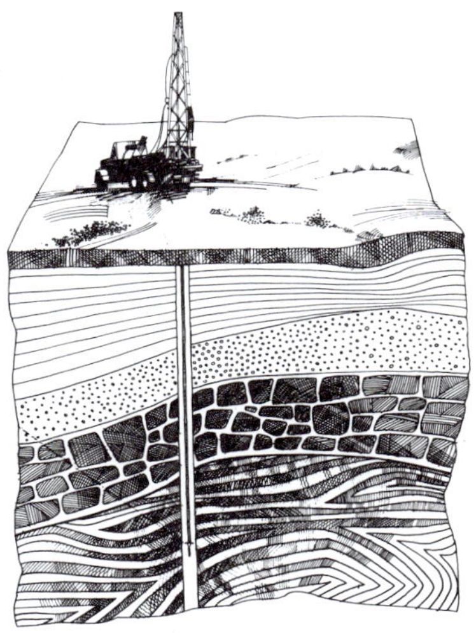

Exploring also includes the search for scarce materials, such as uranium. This metal is used for research and generating nuclear energy. After a geologist discovers a promising site, a drilling rig is set up. This rig takes samples from rock layers several hundred feet beneath the surface. After the drilling is complete, instruments, including a geiger counter, is lowered into the hole to check for the natural radiation of uranium. Almost all the usable uranium in the U.S. is located in the West.

are revised over and over again as more is learned about the system. As more and more information is incorporated, the accuracy increases. A well developed model may make it possible to identify future possibilities. The model, therefore, can be very helpful in the exploration and even discovery of an event, before it happens. Even well developed models, however, cannot predict with certainty. As the model is revised, the probability of being right increases.

probabilities

tectonic plates

The surface of the earth moves in segments called **tectonic plates,** that slide past each other. Scientists have developed a model of this movement from the understanding of what has happened in the past. The model is helpful in searching for minerals, such as uranium, and for fossil fuels, such as oil.

At some points of contact, the movement of the plates is very easy. At others, the movement is very difficult and great tension develops. These tensions are the source of earthquakes. Understanding these tensions and learning to monitor them was an important step. As actual earthquakes are monitored, the occurrence of the next possible earthquake in that locale is forecast.

Forecasting the future is important for planning our present actions. If we think the climate may become very cold, we might store reserve food supplies and fuel.

Why Explore?

People explore for many different reasons. Some explore simply because they want to know something new. Others explore for a more practical purpose. Some may explore to find materials or knowledge that are worth something economically.

A major product of exploring is new knowledge. The procedures of exploring can be controlled in specific ways. If the exploring follows these procedures, it is called **research.** Research can be very theoretical or very practical.

basic research

pure research

applied research

Basic or **pure research** is research in which the primary aim of the investigator is more complete knowledge or understanding of the subject under study. No concern is given to how the knowledge will be used. **Applied research** is directed toward using knowledge. Products which have specific commercial objectives may be the intended outcome.

Research is "exploring that attempts to produce new knowledge, of both a pure and a practical nature." In its early stages, the laser had no practical applications. The laser was described as ". . . a solution in search of a question."

Research to find new products or better processes is included here.

Research procedures must be replicable. The person must record exactly what is done under what conditions. These procedures must be precisely described so that another person can follow the same procedures and get the same results.

Summary

Exploring appears to be a basic activity for humans. As they explore people have a need to move about. This in turn requires technological devices for navigation. The devices help people to determine their direction of travel, to calculate their exact position to document where they have been.

Space exploration was initially limited to using receiving devices that collected energy that reached the Earth's surface. The rocket eventually allowed humans to transport themselves as well as exploration devices into space. Other types of space exploration include efforts to study and chart the ocean floors, the interior of the Earth and the microspace of molecular and atomic matter.

New knowledge and developments often emerge from exploring efforts that started in very different directions. The discovery of gunpowder resulted from experiments in alchemy during the search for the formula of immortality. The Chinese alchemists used sulphur and saltpeter as ingredients in their experiments. These substances, when combined with charcoal, resulted in an explosive mixture—gunpowder or "huoyao" (fire) medicine) as the Chinese still call it. The Chinese put gunpowder to work in many ways to include its use in rockets and fireballs as early as the year 1000 A.D. Historical records show that in 1221 A.D. 17,000 fire arrows were used by Song dynasty generals in a single day during a twenty-five day defense against invaders from the north.

Exploration adds to our knowledge. Part of this knowledge can be applied and put to use in developmental efforts. We shall see in the following chapter how experience and development can also add to our fund of knowledge.

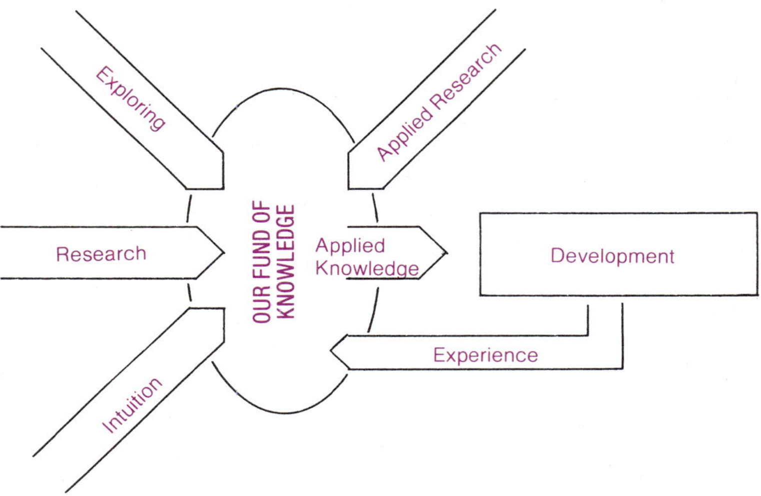

Humans also explore within the dimension of time. Specific technological devices help us explore the history of events while other devices and procedures help us explore possible future events.

Exploring is important to technology because it produces new knowledge through research. New practical knowledge comes from applied research. Applied research is intended to solve specific problems. Pure research is conducted to produce more knowledge without concern for the immediate practical use of that knowledge.

The next chapter considers how the knowledge generated through applied and pure research is used to support the work of people involved in development.

Key Concepts and Terms

analysis
applied research
astrolabe
basic research
bearing
causal model
cross staff
delphi method
development
direction
electron microscopes
exploration
forecasting
future-studies
hinged rudder
historical analogy
magnetic north
microscope
navigational chart
orientation
probabilities
projection
pure research
quadrant
radio telescope
reflector telescope
refractor telescope
relative time
research
rotarios
seismograph
sextant
ship's bearing dial
stick chart
sundial
tectonic plates
time
trend analysis
true north

The industrialization of space is expected to be one of the major developmental efforts of the coming century.

Chapter 13 Developing

The development of new technologies has a tremendous impact on people's lives. Many things that we take for granted today were merely dreamed of a generation or two ago.

The Wright Brothers' test flights began in the early 1900s. Since that time, flight has developed to speeds faster than sound. Aircraft and spacecraft can now travel thousands of miles in a day. Developments in technology continue to evolve at a faster and faster rate. Amazing things have already happened during your parents' lifetime. But even those will be overshadowed by many new technological developments during your lifetime.

In the world of medicine, some injured or diseased body parts can be replaced by ones manufactured from synthetic materials. As new medical techniques and devices are developed, the age of the "bionic human" may become a reality. Thousands of people with medical problems are now being helped by technological devices. It is likely that millions more will benefit in the coming years. It is quite possible that you will live to be more than 100 years old. What will you be doing on your 116th birthday? Do you think it is possible that your children will live to be 150 years old?

Discovering, Inventing, Innovating

Often in our history, people learned how to solve a problem before it was known why the solution worked. Many techno-

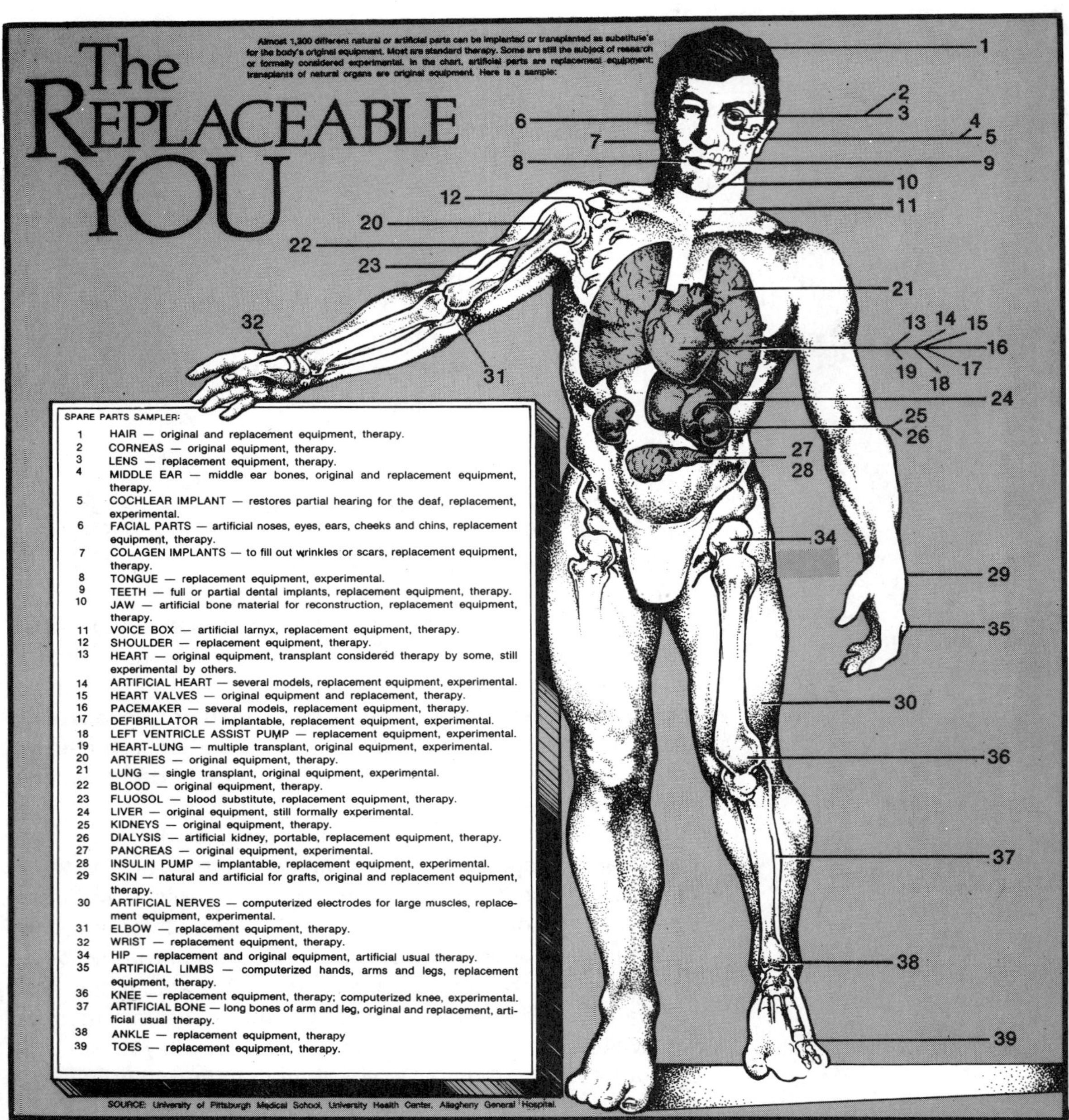

Almost 1,300 different natural or artificial parts can be implanted or transplanted as substitutes for the body's original equipment. Some are still experimental but most are currently available.

logical advances have come about by accident, or as side effects of working on unrelated problems. Most advances, however, have occurred as a result of systematic efforts.

Development is the planned use of knowledge from research and experience, for the purpose of achieving a desired goal. The three parts of development are **discovery, invention** and **innovation.**

discovery
invention
innovation

A *discovery* is new insight into the uses of something that already exists. An *invention* is something that a person makes that did not exist before. An *innovation* occurs when a person takes several things that already exist and brings them together to produce a new device or process.

lodestone

magnet

If you found a piece of interesting rock, it would not be considered a technological discovery. But, it would be a discovery if you found that small pieces of other materials were attracted to the rock and you used the rock to collect those pieces. You could call the rock a **lodestone** and the material "iron." It is an additional discovery if someone rubs a sliver of iron on the lodestone and the lodestone picks up other iron slivers. You could call the lodestone a **magnet.**

An iron magnet floating on a small piece of paper or wood in a bowl of water will always point the same way. The person who noticed this invented the basic compass. Someone later may have decided to shape the iron into a pointer so that it could turn on a pivot point rather than use a messy bowl of water. That person also decided to put a ring around the outside of the needle with marks for North, East, South and West. These additional steps were innovations or improvements to the original invention.

In Search of Answers

research and development

Technological development has a place and function within a larger system of **research and development.** The research and development system uses existing knowledge to solve practical problems. This knowledge, in turn, can generate new knowledge about how practical problems might be solved more easily. It is a circular, closed-loop process. Development is a system of using and generating knowledge that recycles almost continuously in search of better ways of operation.

Suppose your mother makes the best pie crust around.

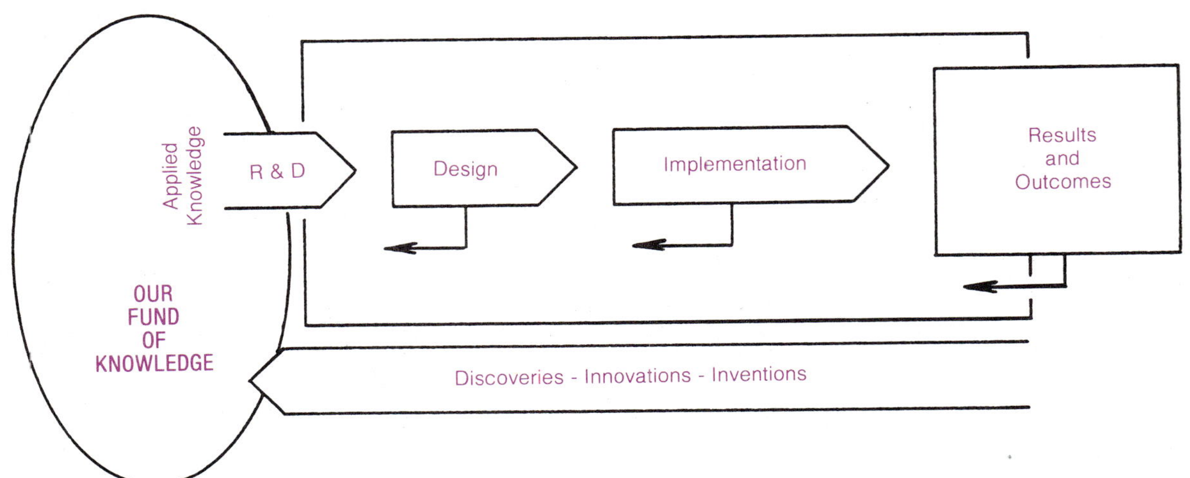

Developing is a process that uses experience as a type of feedback. Closed loop systems continually improve ways of solving problems.

One day, a friend takes one of her pies home to his mother. She is vice president in charge of product development for a national food industry. She thinks the pie is the best she has ever tasted. She wants to develop a new line of these pies.

quality control

The quality of the pies must be retained, even though they are mass produced at the rate of 10,000 per month! A set of quality control procedures must be developed. A crew of food technologists watches your mother's work. They analyze every material and process she uses. This crew gathers an accurate measure of her knowledge of pie-baking. Observations of her behavior and statements that reflect her experience and intuition are included, along with measures of ingredients. The number of strokes she uses in mixing and the temperature of the ingredients are all recorded. They take the knowledge back to the factory laboratories. They can soon duplicate her pie in the lab.

supply and demand

The next step is planning. What processes are best done by machines and which ones by people? What machines are needed? Where can large quantities of the ingredients be obtained? What is the effect of substituting one ingredient for another? What problems in production can the team anticipate? What is the best size and shape for the pie? What kinds of pie are already on the market? What kinds will sell best? The list will be long. This planning procedure gathers together the best knowledge available before the pies can go into production.

Several alternative procedures might look good. Tests will be conducted for each procedure, and results will be compared. Hours and hours of carefully controlled and recorded work will go into making pies in the laboratory. The pies will be carefully evaluated. Some may taste great, but fall apart easily. Others may stand up to being transported and sitting on a shelf, but taste like cardboard. Some taste great and can be delivered in good condition for sale. But, a single pie will cost ten dollars!

developing
applied research

Another example will show the developing process in action. The first step of development is **applied research.** This is a special kind of controlled problem solving. When people, as well as instruments, were the cargo of a space ship, a reentry heat shield for the manned space capsules was needed. From instrument flights, scientists and engineers learned that a capsule entered the earth's atmosphere at an orbital velocity of about 17,000 miles per hour. Because of the high speed,

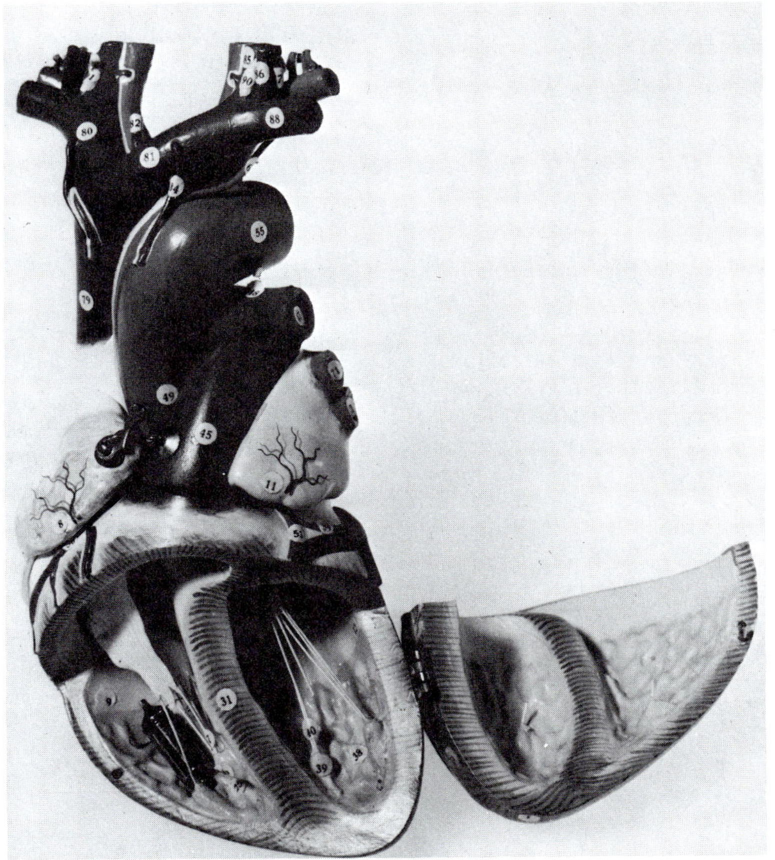

The invention of the pump provided a model to understand the heart. An artificial heart has been developed and is being used. The artificial heart is actually a very dependable special purpose pump.

Prototypes of inventions or improvements on existing inventions are often crude looking and do not look to be important. This rough working model was James Watt's improvement of the steam engine. By adding a separate cylinder for condensing the steam, he was able to significantly increase the power output of existing steam engines. For this development, he was mistakenly given credit for inventing the steam engine.

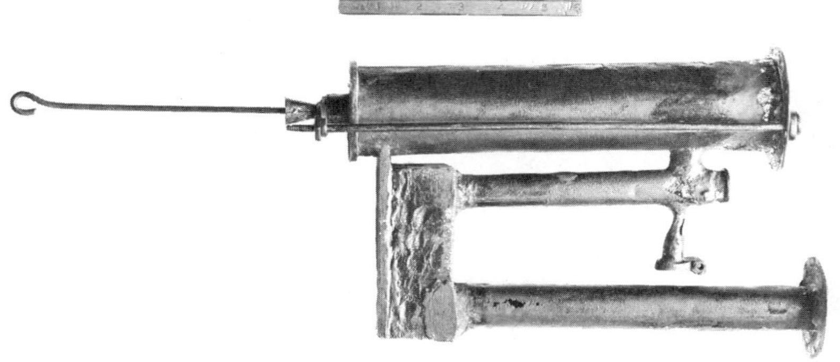

the capsule would be subjected to tremendous air friction. This friction could generate enough heat to melt the aluminum alloys and other structural metals then in use. Special alloys were needed to withstand the high heat and strain. The heat generated by the air friction would be severe enough to raise the cabin temperature to dangerous levels. Elaborate methods of insulating the cabin or carrying off the heat with cooling devices would make the craft too heavy.

The space scientists and engineers faced a new research and development problem. They had to find a method of protecting the astronauts and capsule from heat during reentry. This method should not create excessive weight and should fit in with the structural requirements of the capsule.

Early manned spacecraft used a heat shield that burned away during re-entry to get rid of excess heat. The space shuttle required materials that would last through the stress and wear of several re-entries.

Insulating tiles developed for the space shuttle require minute inspection to insure that the heat will not reach the body of the shuttle. The material is so efficient in getting rid of heat that you can hold a very hot piece without being burned.

The solution to the problem was the development of a new material. It was lightweight and strong. When subjected to severe heat, the new material slowly burned away. As it disintegrated, it carried dangerous heat away from the spacecraft. This type of material is **dynamic** because it changes its characteristics during use.

dynamic

Although the first phase of development was complete, more problems remained. A method was needed to produce the new material at a reasonable cost. Techniques were developed, however, and the material was successfully used on dozens of manned space missions.

Later, the space shuttle was developed. It would be reusable. The same spacecraft would fly several times. The earlier heat shield material could be used only once, so an entirely new material was needed. Again, a solution was found. Engineers found that silicon tiles would do the job. They would be glued to the spacecraft. A new glue was required. Do you see how one technological development often leads to another?

Why Develop?

Development has a wide range of results and outcomes. This activity is a part of almost every technological system. The efficiency

drive to do things more efficiently and at less cost is at the heart of technological change. The goals of development include:

- Improvement of existing products
- Development of new products
- Improvement of existing processes
- Development of new, more efficient processes
- Solution of current and predicted problems in production, communications, transportation and construction
- Reduction of costs (there are often significant human and environmental "costs" that cannot always be measured in dollars)
- Development of new uses of existing materials, products, devices, or machines
- Adaptation of products to new markets
- Improvement of customer and public relations
- Exploration of diversification or expansion opportunities
- Assurance of adequate materials supply
- Assistance to management planning
- Abatement of dangers to people and property and the creation of commercial products from wastes
- Assistance in standardization
- Development of prestige and acceptance for the company and its products (through advertising).

Any given project may achieve more than one of the above goals. The desired outcomes or results usually take one of three general forms. The results are new or improved **products, processes** and/or **systems.**

Product Development

products

Development may have several specific outcomes. The three general categories are new or revised products, processes and systems. The most evident and often the most spectacular results of development are the end products. These may include new machines, tools and toys, a more comfortable fabric, more efficient buildings, bigger spacecraft, smaller artificial hearts or more powerful computers.

New products, such as organic computer chips, artificial butter and non-caloric sugar, may take years of research and development.

There are three types of products that can be developed:

consumables

- **Consumables**—these products are made to be used up. Examples are new food products, new fuels and new materials.

tools and devices

- **Tools and Devices**—these products are improvements over earlier tools and devices or are new designs to serve new needs.

structures

- **Structures**—these products include the buildings in which we live, play, work and travel.

Consumables

The most important consumable is food. What we eat is largely dependent on the technology available. Early and important food breakthroughs were originally technological improvements—better ways to grow, dry, preserve and store food. Most foods we eat are a product of nature. However, they require intricate technological support systems.

Some foods are synthetic in form. Margarine is the result of long efforts to duplicate the taste and consistency of butter. Artificial orange juice, sweeteners, vitamins, dried soups, instant coffee and baby formula are only a few of the many foods that are products of developmental work.

Another type of important consumable product is fuel,

limited fuels
renewable fuels

such as coal, oil, gas, thermal energy, solar energy, atomic energy and others. There are two very different categories of fuels: **limited** and **renewable.** The "energy crisis" is, in a sense, a shortage of limited fuels.

Obviously, renewable fuels hold considerable promise. At the present time, however, we do not know how to use the safer, renewable forms to meet the world's enormous energy demands. Developmental efforts are aimed at generating

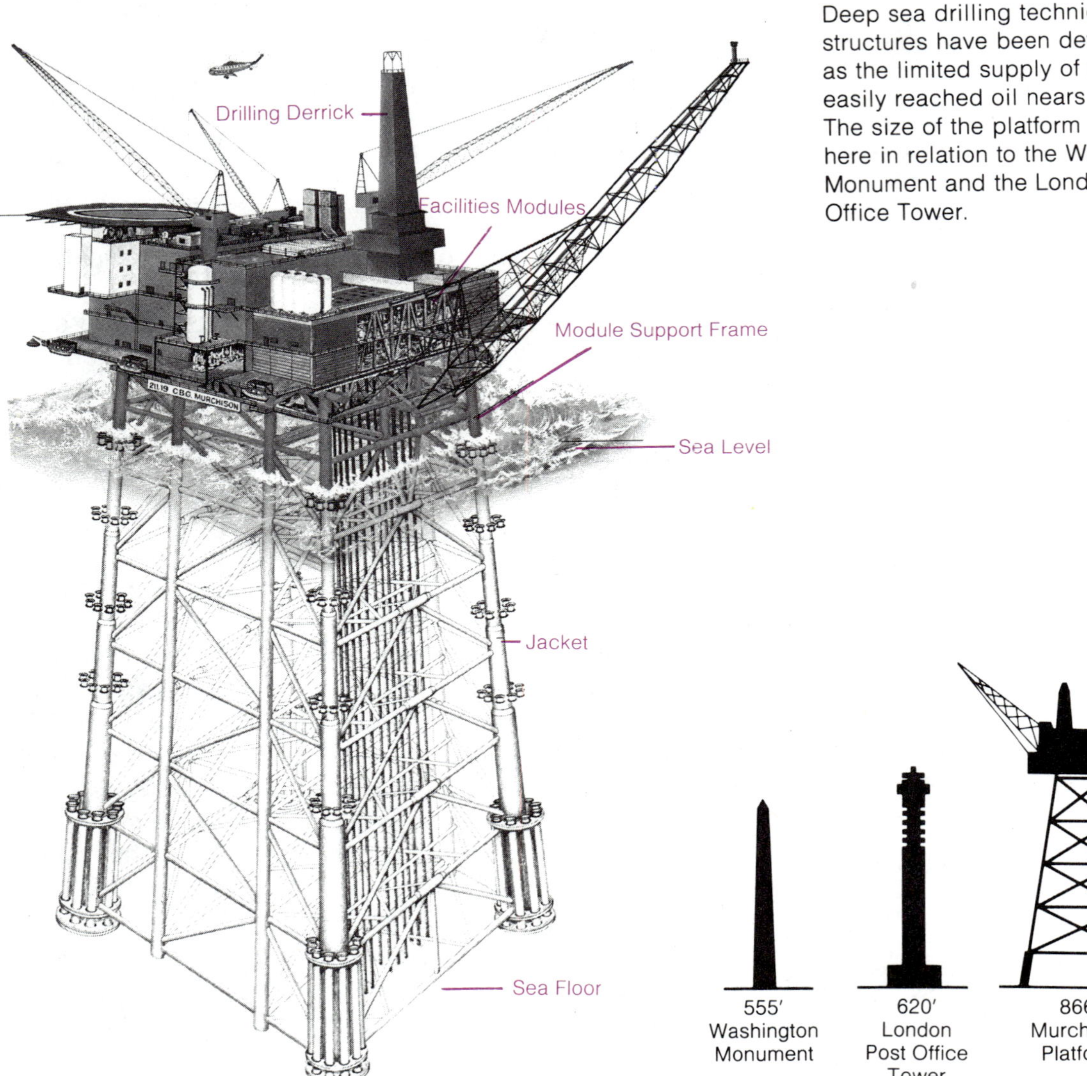

Deep sea drilling techniques and structures have been developed as the limited supply of more easily reached oil nears depletion. The size of the platform is shown here in relation to the Washington Monument and the London Post Office Tower.

more fuels. Other efforts are aimed at developing more efficient use of fuel. But, what happens to the ecological balance if we continue to alter the form of so much energy? The issues are much more complex than just questioning whether we will have enough of a supply.

A third type of consumable product is the materials we use to manufacture other materials and products. This category includes timber, cotton, wool and minerals. Oil, gas and coal are also included. They are considered as materials for the manufacture of plastics, fertilizer, dyes and other products.

Consumable materials also fall into the categories of limited and replaceable. Minerals such as copper, tin and aluminum exist on earth in fixed amounts. Timber and cotton are examples of materials that are replaceable.

This distinction between limited and replaceable materials becomes very important as development efforts are planned for the future. The question is not only when supplies of limited materials will run out, but how they should be used.

The industrial robot was developed from knowledge accumulated from over a century or more. The high investment required for the development of and the cost to produce these robots makes them rather expensive.

Should we make and operate automobiles or should we make fertilizers to increase the food supply? Developmental efforts are needed to make materials that can be recycled or quickly recovered as basic materials. Research and development has also been concerned with recovering limited materials for reuse.

renewable materials

Information about the consequences of our continued demand for products can be gathered by research. The questions about what products are most important and what consequences are acceptable are human issues. People will have to decide—hopefully by peaceful means.

consequences

Tools and Devices

The second major category includes devices, tools and machines we use to produce products, devices and other tools and machines. These tools are especially important because they allow us to continue to produce the things we need and want. Without tools, we would have a hard time replacing the materials we use. We could not replace the television sets or tools that become broken. Without tools and machines, we would be limited to the food gathering methods of prehistoric people. Almost all of our time would be spent just surviving. We would be without many of the devices, tools and machines that make our work lighter and our lives more enjoyable.

It is important to look at issues such as supply of materials, the priorities as to how materials should be used and the effects of continued or increased use. Should developmental activities focus on tools and machines that use limited energy forms or materials? What will happen if everyone in the world gets two cars and a kitchen full of electric gadgets?

effectiveness

Structures

The third major category is the structures we use in work, play and travel. Buildings in which people live represent most of the structures in the world. Office buildings, factories, schools and hospitals are a few of the structures in which we work. Stadiums, coliseums, arenas and gymnasiums are a few of the structures we use for play. Bridges, tunnels, stations, terminals and roads are structures for travel. Each of these structures has evolved as new approaches and materials were developed.

Sheltering structures are even now in short supply in many

areas of the world. As the population grows, these demands become even greater. Developmental efforts to provide low cost housing for many people are certainly important. Shelters that are aesthetically appealing and express individual preferences are a luxury enjoyed by only a few people.

cost

In general, most developmental efforts have been directed toward designing and producing housing that uses the materials of the area, is low in cost and can be maintained with a minimum of effort. The energy issue has stimulated development of new products such as efficient insulation and structures for deep sea drilling.

Development of structures for shelter, support, containing and transporting has expanded as people look for improved approaches and explore new environments.

Process Development

processes

Processes are also improved through developmental efforts. Innovations are made in the (1) converting processes, and the (2) constructing, transporting, communicating and producing processes. Process research and development is similar to product research in that it takes place at two basic levels: **(a)** improving existing processes and **(b)** developing new processes.

Often, process research and development follows closely on the heels of product research. A new product may be better, but it must also be produced more easily or in larger quantities. There are many incentives for improving a process. The prime incentive is that of reducing production costs. This may help keep a product competitive, return a better profit or lower the cost to the consumer. Improved product quality and reliability is also achieved by improving the process.

For example, it is not enough to produce cheap potato chips, if ten percent of them are burned in the process. Customers remember the few burned potato chips. They will probably buy another brand, even at a slightly higher cost. The company with bad chips will need to experiment with different ways to select the potatoes, different handling techniques, different oils for frying, and so on. Efforts will continue until a process is found that consistently works.

Continuous and Automated Processes

Improvement of the production processes is often the key to better products and higher profits. Before automation was introduced, most products in the industrial nations were made by the **continuous production process.** In the continuous process, people, tools and materials were arranged so that everything came together at the right time and place. The products were cranked out one after another. The process is similar to the way fast-food restaurants make hamburgers.

continuous production process

automation

Automation was developed as a more refined version of the continuous production process. Machines are substituted for human workers. Products can be made faster, cheaper and often with more accuracy.

net profit

Automation has several important consequences. Originally, when more human labor was used, the need for new products was determined first. Then, a method of production was developed and the items were produced until they no longer sold. Today, the cost of automated equipment, such as computer controlled robots, is so great that millions of products must be made for a net profit. The producer must be reasonably sure that the new product will sell in a big way. Producers now advertise to create a market.

Systems Development

The third category of development is the systems of interrelated products we develop. For example, our automobiles, roads, bridges and fuel require a system of products and services we use almost daily. The telephone system is made up of many interrelated parts that help us send and receive information. Our television system is becoming similar to the telephone system, as two-way communications capabilities are developed. These systems of technology have far-reaching consequences.

Suppose food technologists develop an automated system to produce pies in quantity. The cost of production is low enough to allow selling the pies at a good price. A package is designed and tested. Perhaps they test-market one package with your picture on it and one with a picture of your grandmother. Maybe they try different colors. An advertising campaign is developed. A marketing system, a transportation system and a feedback system are also developed to test how people like the pie.

People must communicate with suppliers who have the ingredients. Dealers will be located to sell the pies. People who sell and service the production machines are involved. Accounting systems for the exchange of money are needed. The production system must be coordinated with the communication systems of telephoning, mailing and exchanging money. Transportation systems are involved in getting material together for production and moving goods to the marketplace. Construction systems are needed for building shelters, transporters and the like. All these systems must be monitored. Attempts are made to improve quality and efficiency all along the process. Developmental activities do this monitoring, testing and improving of systems all down the line.

Ways of Thinking and Developing

For a few highly creative geniuses, the processes of discovery, invention and innovation seem to come easily. Sometimes, the discovery or idea about an invention is made quickly. Often, the discovery is followed by the slow developmental work required to show others the idea. For most people, however, breakthroughs come only after a lot of hard work and frustration. The person with a unique idea may also have trouble finding any support for developing the idea. People may reject the idea simply because it is new.

Over a long period of time, a set of procedures has been developed that is helpful for those who do developmental work. A number of procedures can be used in designing, implementing and improving ideas and devices. These procedures can be considered as a few specific examples of tools to use in development.

designing

Designing is an activity that uses aesthetic, technical and scientific principles. The intent is to describe a device, a process or a system. The description is done in sufficient detail so that the proposed product can be made. The description is also complete enough to allow the designer to patent the idea. A **patent** is a legal statement identifying the ownership of the idea or device. The product may first be made as a model. This is especially true if construction of the real thing is expensive.

patent

This scale model of the space shuttle was put through extensive wind tunnel tests. Thousands of trials were made to determine how the shuttle would separate from the rocket. Real-life testing would have been impossible.

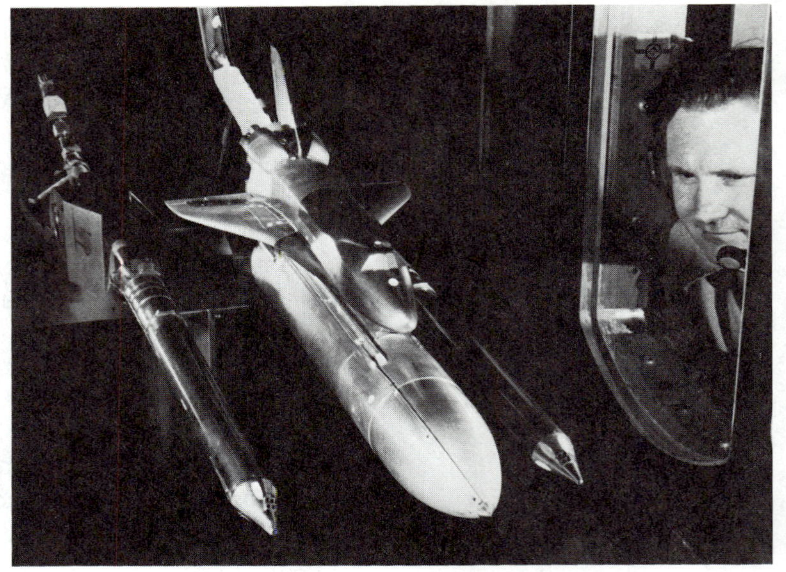

Producing on a Small Scale

Often, a new design or method is tried out on a smaller scale. This cuts down the cost of producing the design for testing. Alternative solutions can all be tested and compared without using as many materials and resources. Sometimes, the operation can be studied from models.

The use of physical models helps humans to decide what action to pursue. This press, when completed, will be as tall as a three-story building. The people in the picture would be able to stand upright in the claw of the arm used to feed materials into the press.

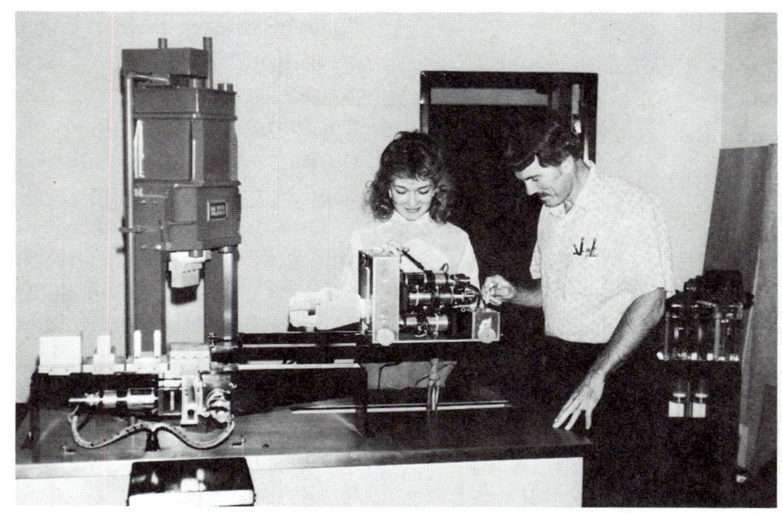

A small model such as this coffee dryer from Colombia, South America, can be used in development. The operation of the dryer, the procedures, and maintenance and repair routine can be tested. Potential problems can be located.

Models of the space shuttle were essential for testing how it would function. Real life testing in space was impossible until the actual launches and flights took place.

model

Physical models help us make calculated guesses about how and whether something will work. A physical model of a new airplane allows us to estimate more accurately how well the real plane will fly.

A symbolic model that uses mathematics helps us predict how the product may perform when it is completed. Calculations can show how many vehicles can move across a planned bridge before the bridge is built or how an automobile will react during a crash. Obviously, the use of a model in designing and in the early stages of implementing can be of considerable help. This may result in changing the design of

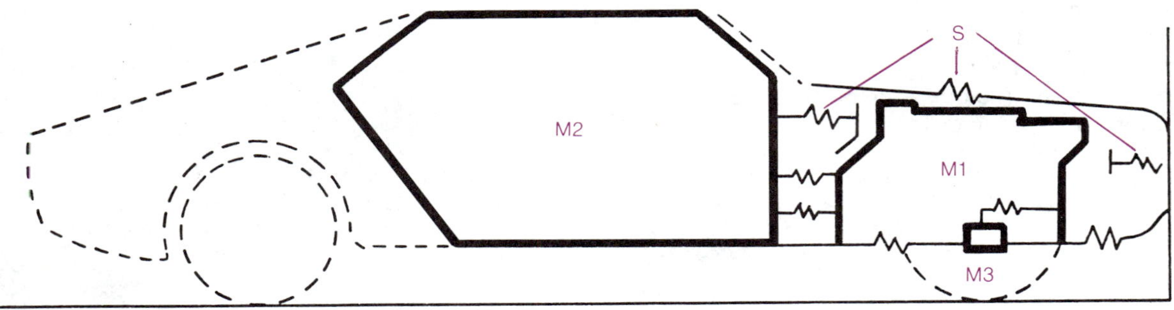

A mathematical model of an automobile allows for computer testing of safety ideas and body changes without the expense and delay of building and crash-testing physical models or prototypes. The parts of the car important in the crash are M1: the engine, M2: the body and M3: the front end. Each of these masses must be treated separately. The amount of crushing that key parts of the car can withstand are represented as springs and are shown as jagged lines (S). The size of the springs can be easily included in the mathematical model.

prototype

new airplanes, bridges and cars before they are built. The resultant savings in money, time and lives can be great.

Prototypes are a natural outgrowth of models. In some cases, the words are used interchangeably. A full size model of a device or machine is often called a **prototype.** These full size replicas may or may not actually work. For example, a large aircraft may be built to actual size, but it will not fly. Such a prototype is made to show what the finished machine will look like. It can be important in the final decision of whether funds for production will be made available.

Other prototypes can be described as working models. Such prototypes actually perform the way a finished product is expected to perform. Before a new television set is produced, a prototype is built to allow extensive testing. Obviously, the manufacturer would want to know if the new design works before investing the money required for automated mass production.

Different production processes require different prototypes. The working prototype of an aircraft may be the first vehicle in the actual production process. An airplane is often custom manufactured. Each aircraft is made individually with time given for extensive testing of each product. Televisions are most often mass produced by intermittent or continuous methods. A fully functioning prototype is required to determine how to set up the manufacturing lines. Usually only a few televisions are thoroughly tested after manufacture, however.

Prototypes are built to incorporate the best ideas found in the first stage of the modeling process. The work of the Wright Brothers is a very early example of prototyping a product after extensive modeling for scientific and technological

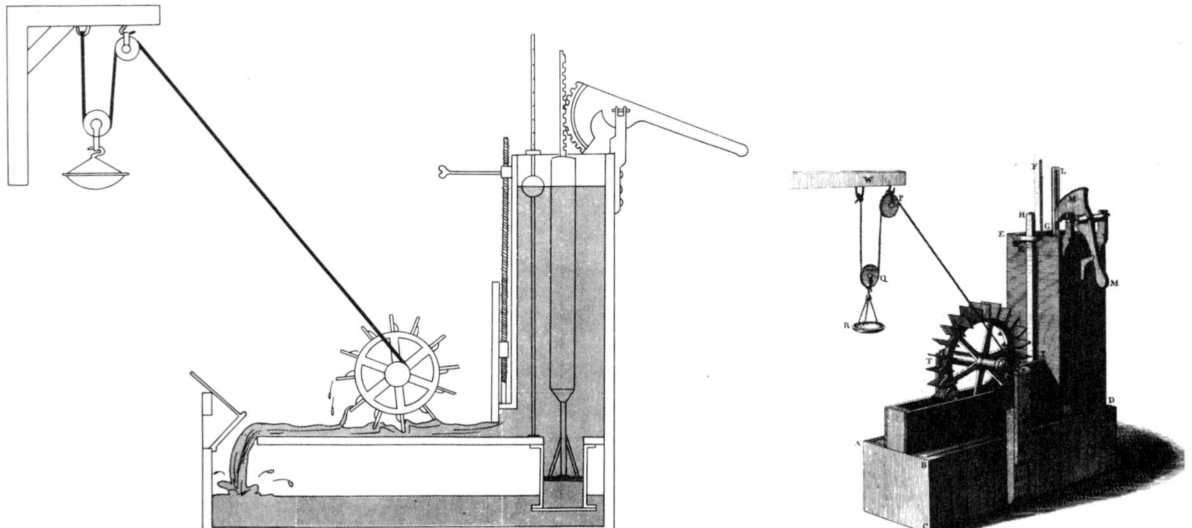

One of the earliest models used in developmental effort was built by Robert Smeaton to determine the efficiency of waterwheels and the best shape of the paddles.

purposes. They worked for years to build a flying machine. They found that developing an efficient wing shape was very important in making the airplane fly. To find that best shape, they built a wind tunnel. They made small versions of many different wing shapes. These models were placed in the wind tunnel and the lift upward, caused by the air flowing over the wing, was measured.

The Wright Brothers could not be satisfied with only the results drawn from models in a wind tunnel. They proceeded to construct a full size prototype of the aircraft they intended to fly. These prototypes were flown much like giant kites to determine if the full size versions of their ideas could actually

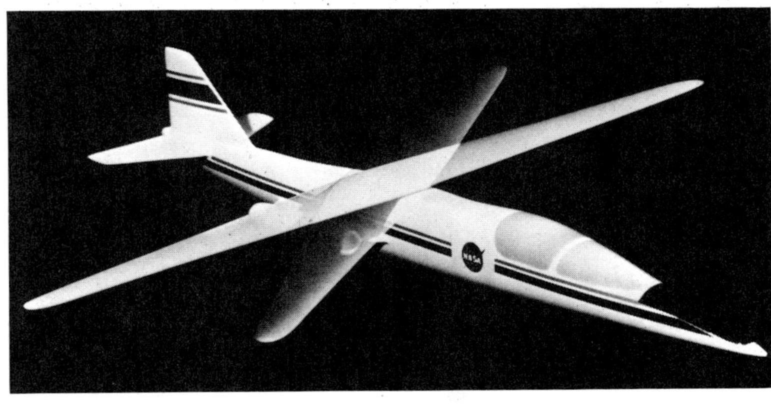

NASA has proposed this new airplane. It uses a long wing for maximum lift at low speeds. The wing would pivot to reduce drag during high speeds. Such a design concept requires extensive development and testing.

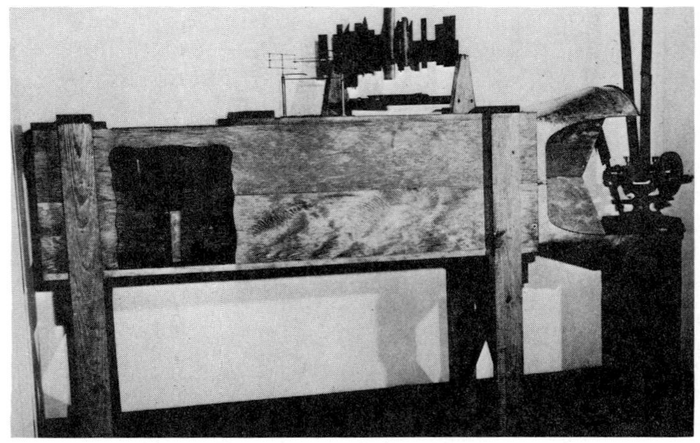

The experimental work of the Wright brothers was one of the first methodical research and development (R&D) efforts. It generated new knowledge that was needed to build a successful flying machine. The Wrights went through a long process that included (a) testing of miniature wing models while riding a bike and comparing the efficiency of those shapes to an established standard, (b) building the first wind tunnel so that the testing of wing shapes could be more accurately controlled, (c) flying larger scale models of their airplane designs as though they were kites, (d) flying their improved designs as gliders, and (e) after many changes and several years of recycling these steps, the launching of the first successful airplane at Kittyhawk, North Carolina in 1903.

fly. Only after extensive work of this type did they trust the plane enough to actually ride in it. Even then, they tested non-powered aircraft before adding the engine. Carefully, the development of the airplane evolved. Their first successful flight was the culmination of years of developmental efforts. They also drew upon research and development of several other people before them.

Optimization: Finding a Better Way

A developer is seldom lucky enough to find the best answer to a proposed problem and then use the solution without changes. Ordinarily, new procedures must be developed continually to improve the old ones. One improving activity is optimization.

optimizing

Optimization is the process of seeking the most favorable condition or solution to a problem, or a goal. Each time you tune your radio to a station, you are involved in the process of optimization. You move the dial back and forth until you get the best sound available. In this case, the goal or criterion is high quality of sound. The dial setting is the variable you

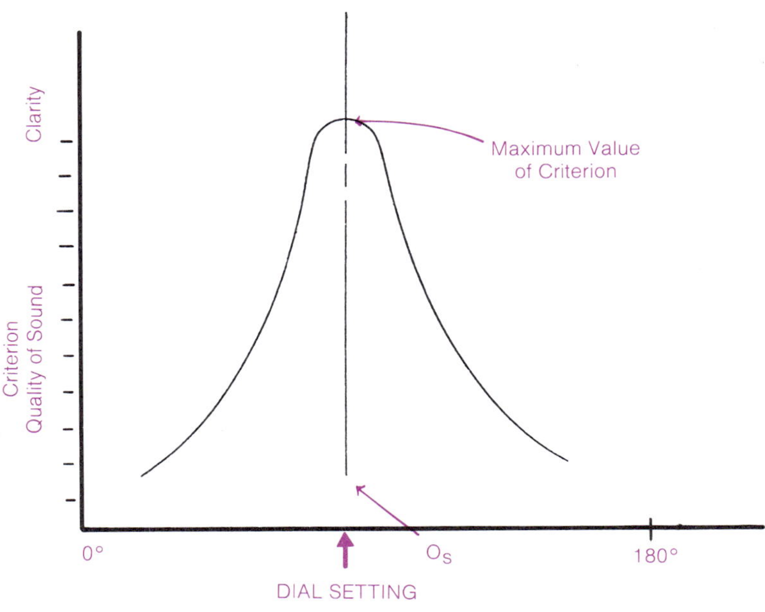

Tuning your radio is an example of optimization. You search for the optimum setting (Os) that will give you the clearest sound.

maximizing the criterion

criteria

manipulate to get the best sound possible. This process is called **maximizing the criterion.** The clarity of sound changes as the dial is changed. The optimum dial setting for the maximum sound is in a very narrow range.

Most problems of optimization are not this simple, however. Usually, there are two or more conflicting criteria. A television set provides an example. The sound and picture of your set, although controlled by the same dial, do not always match.

The best setting for the picture and the best setting for the sound may occur at different places on the dial. You must choose a setting that is a compromise between the two competing criteria. The setting you choose will depend on what is important to you about the program you are watching. If you are watching a musical show, you might place the dial closer to the optimum sound setting. If you are watching a documentary film on art, you may choose a setting for optimum picture clarity.

Decision making within technology is often similar to optimization. Many times there are competing goals and criteria. If you decide to build a clubhouse, you might consider such criteria as size, appearance, strength, cost, safety and durability. For any solution, these criteria will differ. A clubhouse that looks nice may not be secure and private. The same house may be acceptable in size but unacceptable in cost or

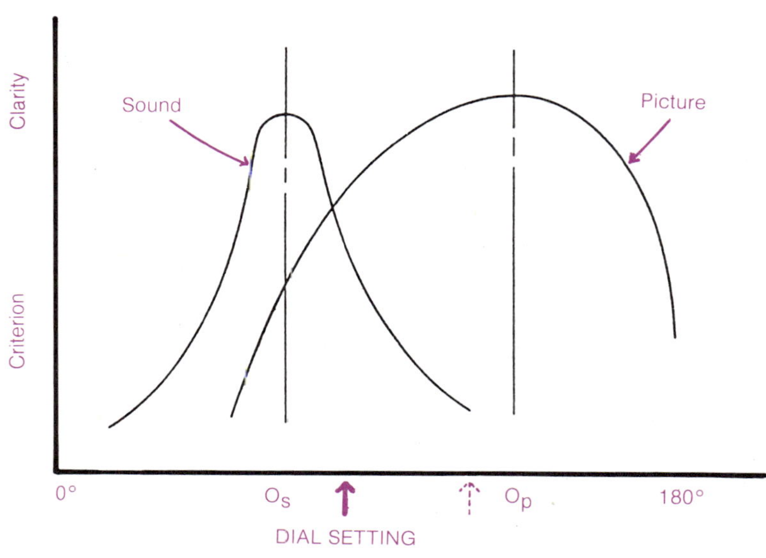

The dial setting that gives the best signal for sound (Os) differs from the optimum setting for the picture (Op). The operator must choose whether sound or picture (criterion) is most important.

safety. You are faced, therefore, with searching for the best balance or an optimum compromise. The necessary give-and-take required to reach a compromise is called the **trade-off process.**

trade-off process

As you consider the trade-offs, you must make a decision. You can decide what is a good design and what is not, based on personal preferences. You could use an authority's established criteria for what makes a good design. There are different opinions as to what these criteria should be.

Criteria For Design of a Product

- Low Cost
- Simplicity
- Reliability
- Usefulness
- Ruggedness
- Functionality
- Attractiveness

Optimization procedures improve the design of products and systems. Realistically, a product design must be economical. A product designer has described it this way:

> An economical design is an optimized design—getting the most for the least—it is a design without excess—a design with every part exactly sized and shaped for its intended purpose. No material is used that isn't needed. The material is placed and shaped to function at the maximum allowable stress level throughout. Weight, number of parts, number of assemblies, amount of machinery, complexity of shape are all minimized. The least expensive (but most appropriate) materials are selected and fabricated in the least expensive manner with the least amount of manual labor. In general, economy in design is the development of a product to do only and exactly what it is designed to do, and no more, with the least expenditure of materials, resources, production effort, and labor. . . . (Harrisberger, 1966).

Hexagon shaped washers? This approach to making washers promises to be a more economical method. It wastes less materials and under some conditions, this washer would appear to hold better than standard round ones.

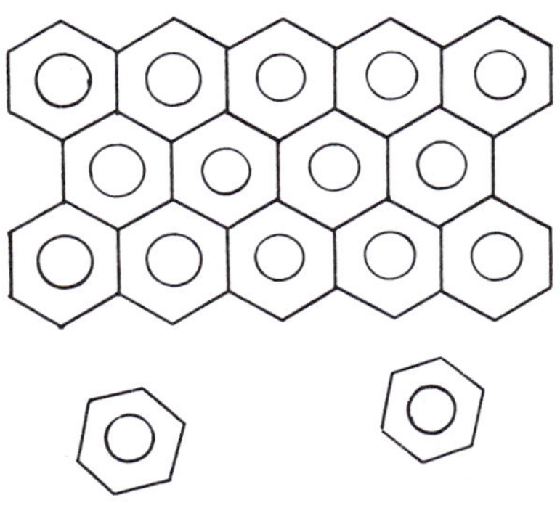

Efforts to optimize a product do not need to be sophisticated. Some very successful efforts have been very simple in nature. The standard round washer was redesigned to minimize the scrap that is made during manufacture. The result is a six-sided washer that is cheaper to make. Sometimes, a hexagon washer holds better than a round one. The design is simple, but resulted from hard work and much experimentation.

Recycling or Iterating

iterating

Iterating is the process of developing and improving an idea, process, product or system by sending it through a development stage more than once. The procedure is similar to combining recycling and feedback. Iterating is used in computers. Loops placed in the programs cause the computer to recycle through selected operations.

An idea is iterated by recycling it back through one or more of the steps of development. An idea may be returned "back to the drawing board" and moved again through the designing, implementing and improving phases. Or, an idea may be recycled through the optimization step only.

Iterating has sometimes been incorrectly labeled a "redesign." Iterating is actually a part of the normal design, development, and problem-solving cycle that attempts to find the best possible answer to a problem. Recycling an idea for its

Another clever but simple development combines standard devices with a special component to make a new tool. This device easily drills holes in things too large or awkward for conventional hand-held drilling.

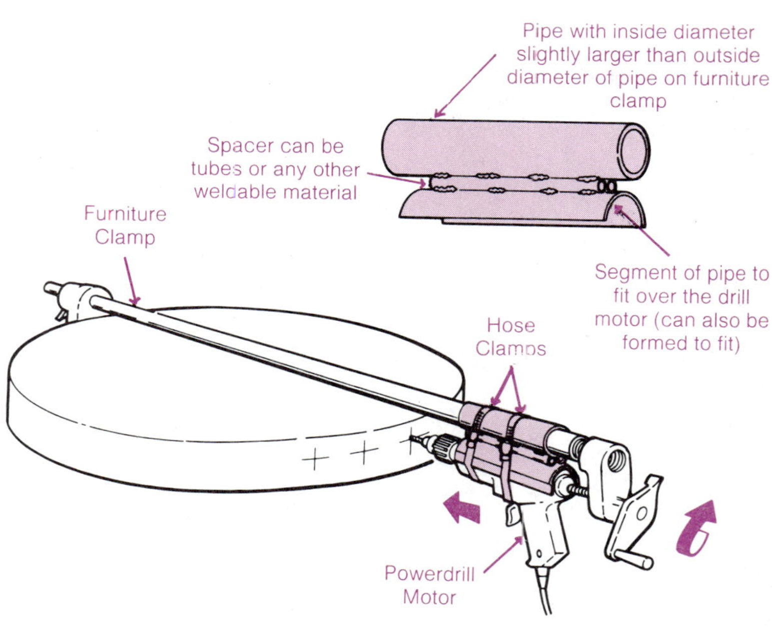

One goal of technology is to do more, with less. All homes have roofs and many of them use shingles. An interesting idea for development could be a shingle that also functions as a solar collector.

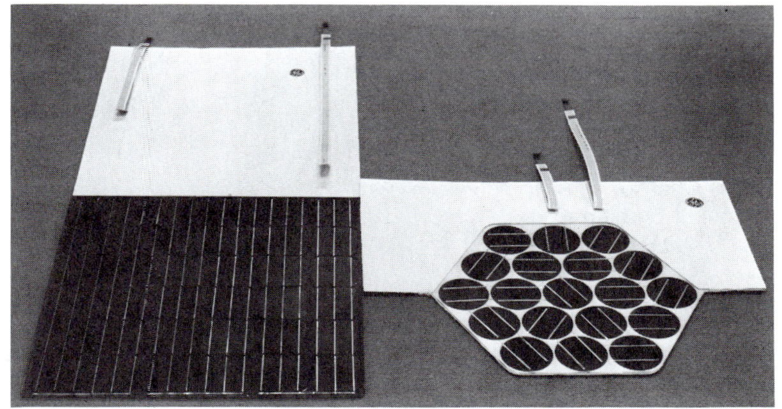

improvement through iteration techniques is part of the on-going design, problem-solving process.

Valuable feedback can be acquired through the process of iterating. The failure of a product to perform in an idealized way provides valuable information about changes that could be made in the process and/or the product. The difference between the best possible operating product and the product as it now exists can be viewed as a mismatch between the desired goal and the system output. Automated or cybernetic machines correct their operation based on the degree of error that is sensed and sent back into the control system. The designer uses the mismatch or error of the ideal and the real product as information that can be sent back into the design process. This information is required for the improvement of the product. Eventually, the development team agrees that the error between the real and the ideal product is as small as possible. The iterating process is then stopped and the product is ready to be used, constructed or mass produced.

Summary

Development is the planned use of knowledge from research and experience for the purpose of achieving desired goals. The goals of developing are varied. The general intent is to design new or to improve existing products, processes or systems. Products include consumable products, tools, devices and structures. Development focuses on new and improved processes for conversion of materials, energy and information. Other developments improve processes used in

The activity of development expands our understanding of technology. Technology not only includes the application of our knowledge, tools and skills but also the knowledge of how to apply what we know to solve practical problems.

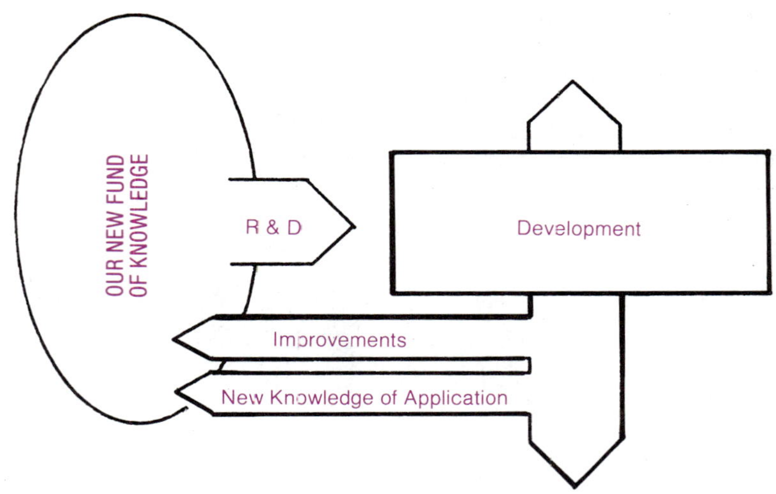

constructing, transporting, communicating and producing. Large-scale efforts integrate the products and processes that have been developed to improve systems. The technological systems include constructing, transporting, communicating and production as well as exploring, developing and controlling.

People engage in a range of developing activities such as discovering, inventing, innovating, designing, implementing and improving. These activities are supported by techniques that guide the work of designers and developers in generating models and prototypes. Other techniques and procedures are used in working toward the optimization of the products, processes and systems that emerge from the efforts of developing.

Key Concepts and Terms

applied research
automation
consequences
consumables
continuous production process
cost
criteria
demand and supply
designing
developing
discovery
dynamic
effectiveness
efficiency
implications
innovation
invention
iterating
limited fuels

lodestone
magnet
maximizing the criterion
model
net profit
optimizing
patent
processes
products
prototype
quality control
reliability
renewable fuels
renewable materials
research and development
structures
tools and devices
trade-off process

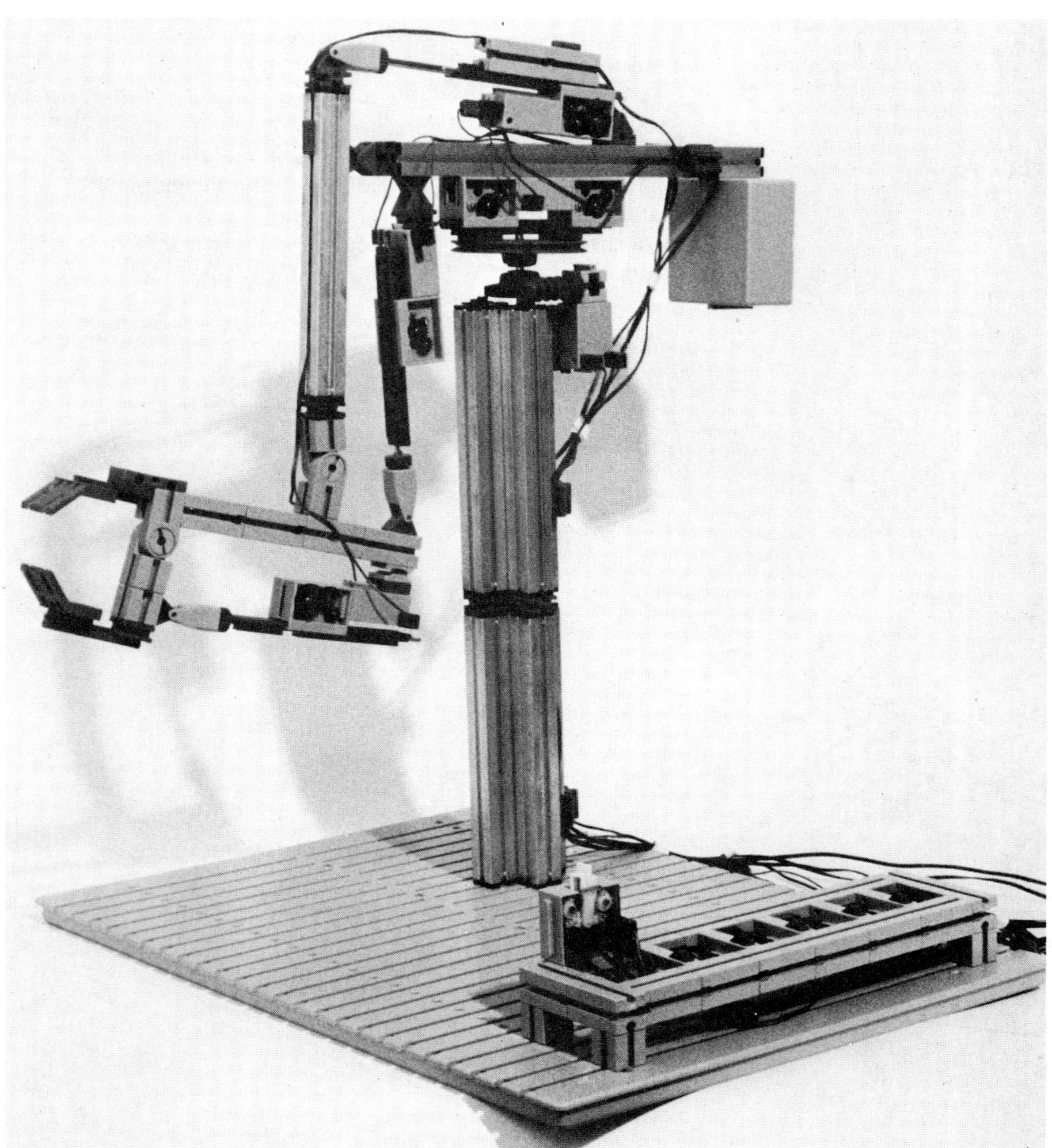

The robot is interesting not only because it can perform some tasks normally done by humans, many machines do that. But the robot integrates several technological advancements that enable it to make specific decisions and then control its own actions.

Chapter 14 Controlling

cybernation

Control is very important in technology. Devices and systems must provide for human regulation from outside the system. It is also useful for devices and systems to be *cybernetic.* A cybernetic system is self-regulated. For example, as the wind shifts, it exerts force on the tail of the windmill. This swings the windmill blades to face into the breeze. The windmill has taken full advantage of the wind velocity and the blades continue to rotate briskly.

A windmill operates as a cybernetic device. It adjusts and controls itself as conditions change.

A refrigerator is another example of cybernetics. Suppose the switch has been turned off so that the refrigerator is not running. Turn it on again and the refrigerator is now on "automatic pilot." It will adjust itself to maintain the proper degree of coldness.

gyroscopic action

A toy gyroscope spins on its tip without falling over. It also resists any change in direction. This gyroscopic action of any spinning wheel is what makes it possible to ride a bicycle without holding onto the handlebars, though this is not recommended. **Gyroscopic action** is important in the cybernetic control of airplanes.

In the early days of the airplane, an inventor named Lawrence Sperry entered an airplane in an international competition held by the Aero Club of France. A prize of 50,000 francs was to be awarded for the safest airplane in the competition. Sperry's entry had a gyroscope attached to the airplane's controls. Remember that the gyroscope is a rapidly spinning

The tail of a windmill is designed to point the propeller in the right direction.

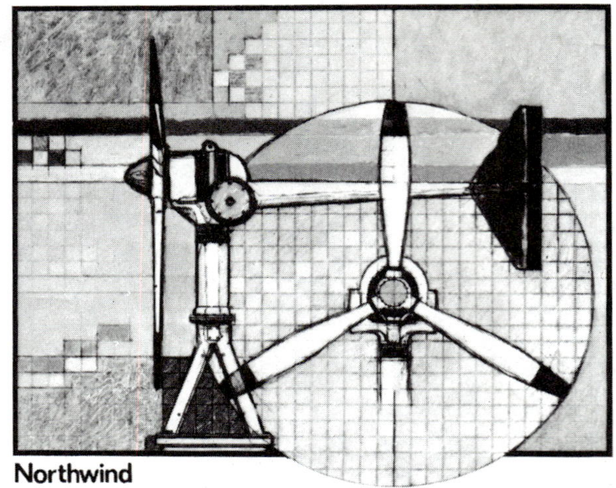

Northwind

device that does not like to have its direction changed. Sperry fastened wires from the gyroscope to the controls of his airplane. If the airplane tipped to the left, the gyroscope would pull the controls to the right and cause the craft to become level again. By connecting the gyroscope to the airplane controls, Sperry was able to make an airplane that was self-correcting.

To demonstrate his confidence in the new system, Sperry had his mechanic crawl out on one wing during the flight. As he and the mechanic sped by in the airplane, Sperry removed his hands from the controls and stood up in the cockpit. He waved with both hands at the judges. The gyroscope pre-

This toy gyroscope is very similar to those used to keep airplanes level in flight.

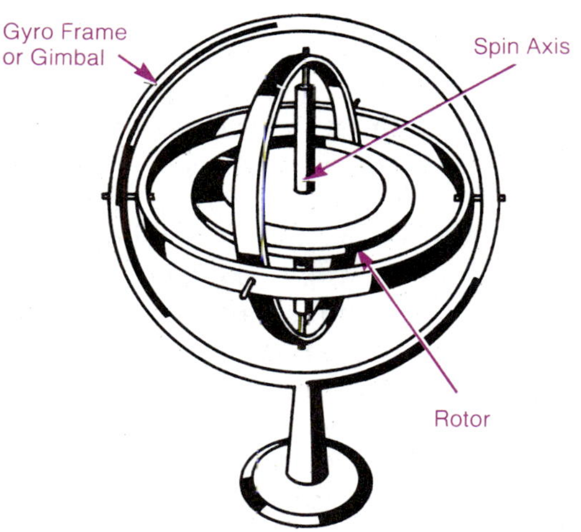

vented the plane from tipping under the weight of the mechanic. History did not record what Sperry's mechanic thought about the stunt, but Sperry won the prize.

Another definition of cybernetics is "the automatic control of a machine toward a desired goal." The tail orients the windmill to face the wind. The controls on the refrigerator cause it to work only enough to reach a certain temperature setting. Sperry's gyroscope keeps airplanes flying straight and level. When we can set a machine to accomplish a task on its own, without further human adjustment, we have a cybernetic control system. Cybernation, however, requires feedback.

Feedback

Some important questions must be answered before automatic machines are possible. How does the machine "know" when it has reached the desired goal? For example, how does the windmill know when it points directly into the wind? How does the refrigerator know when the correct temperature has been reached? How does the airplane know when it is flying straight and level? In all cases, the answer is in the concept of feedback.

feedback

In our three examples, adjustments continue until the machines reach the desired state of operation. The goal is sometimes called the **steady state.** In the example of the windmill, the tail continues to turn until the windmill faces directly into the wind. As the windmill swings around, the force of the wind against one side of the tail is reduced. Changes caused in the orientation of the windmill cause immediate changes in the tail. These tail changes are feedback from the original change. The windmill reaches a steady state when the force

steady state

The windmill turns because there is more pressure on one side of the tail than on the other. When the force becomes the same on both sides of the tail, the windmill does not turn.

The gyroscope in the airplane is able to control the plane and keep it level in flight. As the airplane tips to the right, the gyroscope continues to spin in its level position. The control stick connected to the gyroscope pulls the left aileron up and the right one down. This will cause the airplane to straighten up and fly with its wings level again.

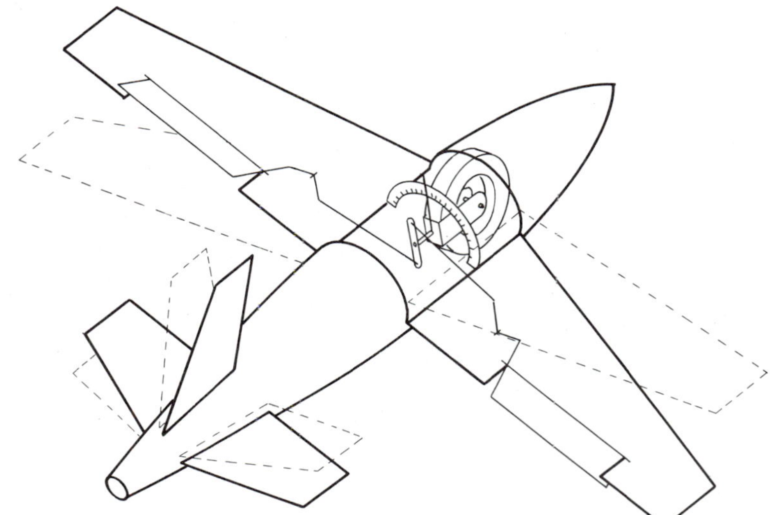

of the wind is the same on each side of the tail. This steady state represents the desired goal.

The refrigerator and the airplane also attempt to reach a steady state. The refrigerator has a sensing device that reacts as the temperature changes. When the temperature inside the refrigerator is warmer than the setting, the device turns on the cooling motor which drives the compressor and lowers the temperature. When the temperature is cold enough, the device turns off the motor. In Sperry's airplane, the steady state is established by the level position of the airplane when the gyroscope is first turned on. The gyroscope does not want to change its position, so each movement of the airplane is counteracted by the gyroscope so that it can move back to its original level position.

Logic and Decisions

Automatic control is not possible unless the machine is able to determine when the changes it makes are enough to reach the goal. Machines cannot "think" as people do. Cybernetic machines do, however, operate on the principle of **logic.**

logic

Machine logic is simply the unthinking response to controls. The tail of the windmill might unthinkingly turn the blades directly into the high winds of a storm. Although it was a logical move, the mill might be destroyed by the force of the wind. The feedback control system simply does what it has been designed to do. It is this lack of intelligence that can cause cybernetic machines to sometimes operate out of control.

Evolution of Cybernation

automation

In the simplest sense, cybernetic machines merely turn themselves off and on. This is the basic principle used in industries that operate by automation. **Automation** describes automatically controlled machines. "Auto" means "self." "Matic" means "control." Automatic means self-control.

Machines that operate automatically have been around for hundreds of years. Long before James Watt and the steam engine, automatic machines had been in use. Watt improved the steam engine by adding an automatic device to control the speed of the engine. His device, called the flyball governor, made the engine more useful. It was an early example of automation.

Automation has had a major impact on production. One of the first automated industries was set up by Oliver Evans, an American millwright. Evans used conveyors driven by water to move grain and other materials in a flour mill.

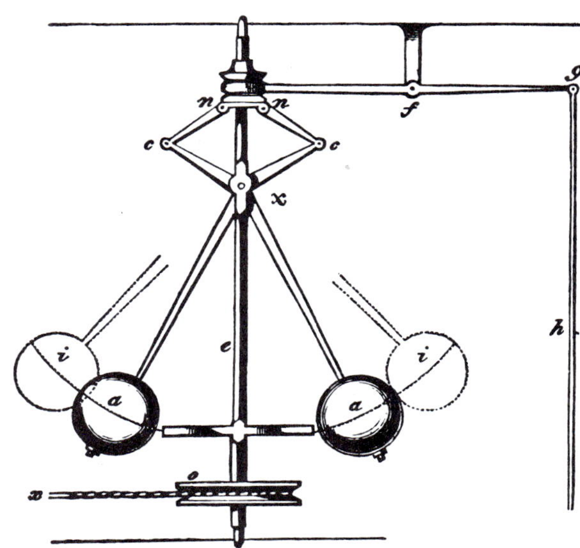

James Watt borrowed an idea from a grinding mill and used it as the governor for his steam engine. If the engine runs fast, the weights swing from position (a) to position (i). This causes (n) to be pulled down and (g) to go up. Rod (h) is connected to the throttle. The throttle cuts off part of the steam to the engine and slows it down. If the engine runs too slow, the opposite happens. The throttle opens and sends more steam to the engine.

One of the very earliest automated factories was Oliver Evans' flour mill. Evans used several different types of conveyors to control to feed grain into the grinders at the proper rates. Evans was able to make the process of grinding, the handling of materials and the "controls" of his conveyors and grinding mills work together in automated fashion.

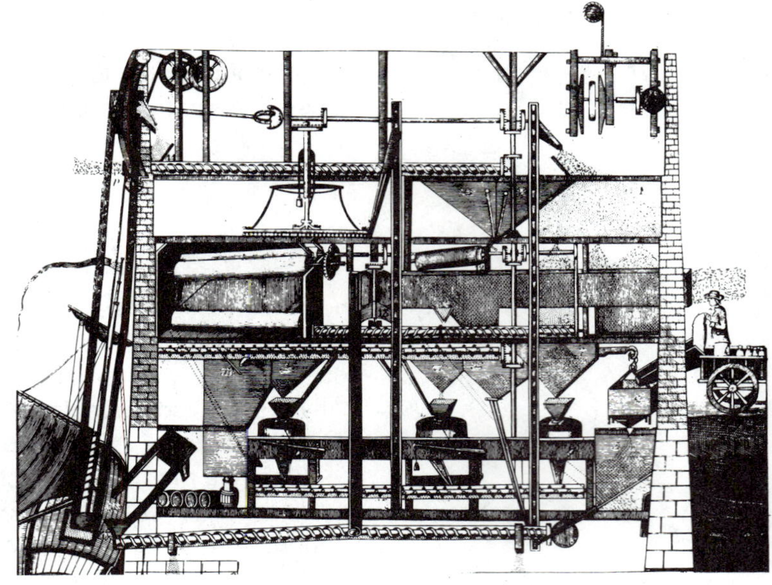

Evans' conveyors transported the materials to the machines that ground the grain and sorted the product. Simple automated devices such as those used by Evans have been available for hundreds of years. But, completely automated manufacturing has emerged only within this century. Within the past few decades, automation procedures have been applied to many different types of production systems. Automobile assembly plants and oil refineries are examples.

process
handling
control
manufacturing

Automation has three parts that work in close-knit harmony with each other; the **process** by which the product is being manufactured, the **handling** by which the materials are being transported to, through and from the machines performing the process and the **control** by which the process and handling are kept coordinated with each other. These three parts comprise a **manufacturing system.** The controls provide the feedback so that the system can sense changes in the process and can bring them back to a desired state of operation. Automation, therefore, is a very practical use of the principle of cybernetic control.

Mechanization and Automation

One of the earliest advances in production was the changing from muscle power to other forms of energy. Humans still had to sense the information, monitor the processes and provide the guidance and control of the machines and tools. As long as the production was the custom-making of individual items, it was alright for people to supply the guidance and control. But as demand grew for products, attention turned to developing more precise machines and more efficient energy production and distribution systems.

As people began to want more and more technological products, key mass production techniques such as interchangeable parts emerged. Mass production required high quality which in turn required information (feedback) on how accurately the items and parts were being made. As more and more complicated products were made, the manufacturing and quality control processes became more intricate. The most common answer to these problems was the continued division of labor and specialization on the assembly line. In this system, one person was employed to do one step of the process and to attend to feedback from only a few individuals at quality control stations on the line. This approach was often boring to the worker and expensive to the employer.

An alternative was to design devices and machines that could make quality control checks on the production processes. These machines would then send information as feedback to the production machines which in turn would make adjustments. This would improve the processes while the operation was going on. At first, the quality checks were made and feedback was provided through mechanical

Not all instances of automation or cybernation are necessarily complicated. The mechanism illustrated here is almost elegant in its simplicity. A device such as this fills the millions of boxes of cereal that are consumed at breakfast each day.

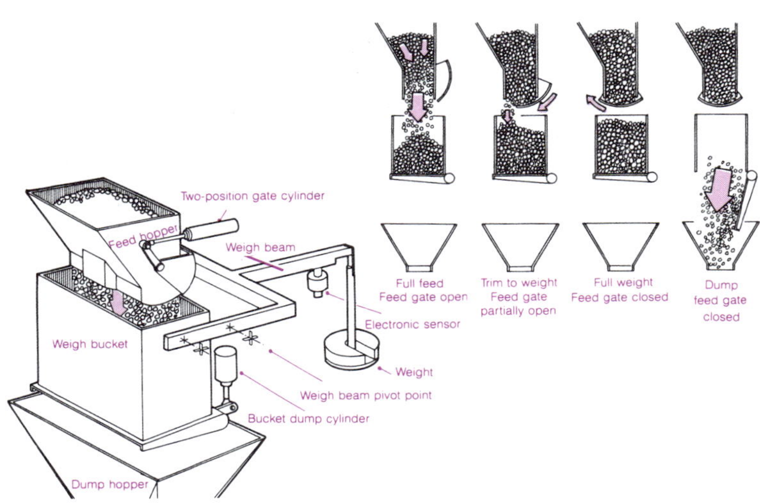

Robot manipulators generally fall into one of the four mechanical configurations pictured above. The hands or tooling at the end of the arms are developed to fit the special work to be done by the robot.

1. Rectilinear Coordinate Robot

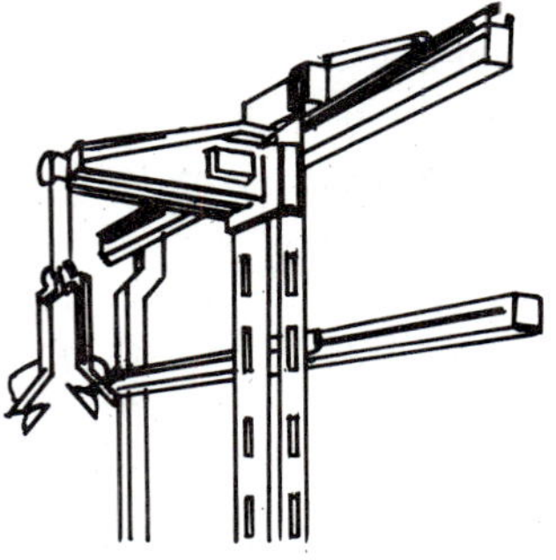

2. Cylindrical Coordinate Robot

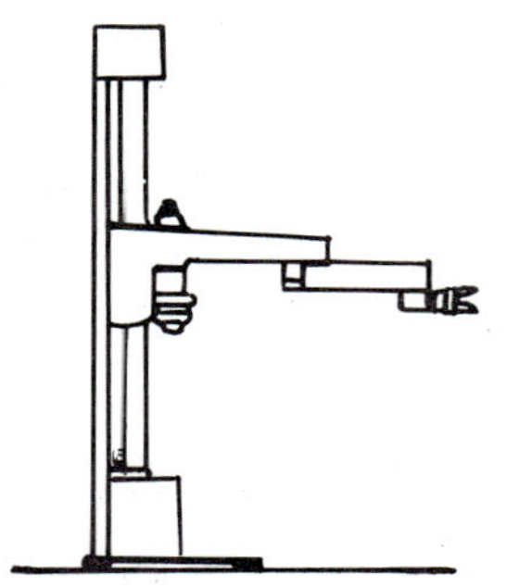

3. Spherical Coordinate Robot

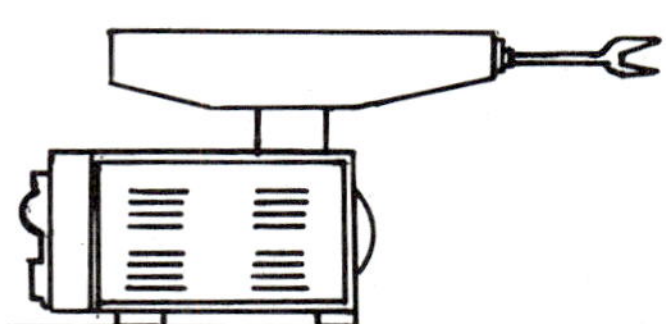

4. Jointed Arm Robot

memory

means. Improvements were made using electrical and then electronic devices. As electronic sensing, logic and memory devices became more sophisticated, the control of complicated processes was possible by machines. With the advent of robots, fully automated industries are now a reality.

Automation and Computers

computers

The first computers were analog in nature and were developed to measure processes. The speedometer in an automobile is an analog device that allows for the measuring of speed. Other examples of analog devices are the gas gauge in cars, a clock with hands and a pressure gauge in a boiler.

analog devices

Analog devices do continuous sensing and measuring of something that is happening.

Analog computers are often quite large. Many of the tasks for which analog computers were used have been taken over by digital computers. This is possible since the speed of the digital computer allows individual measurements to be made

Robot manipulators may be quite simple. In this case, the robot arm picks a part off the conveyor and places it in a press for punching.

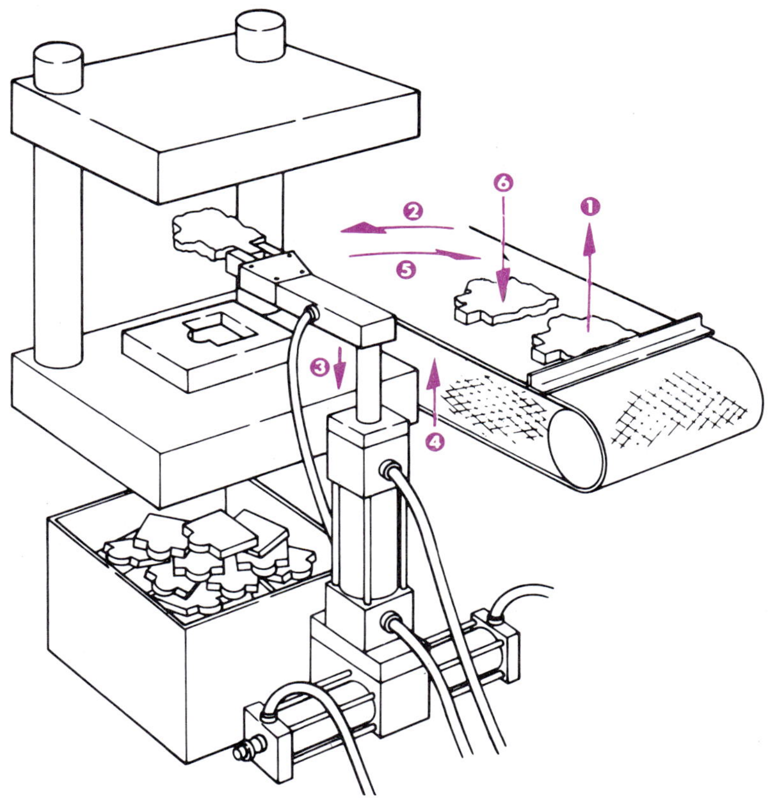

thousands of times per second. These discrete measurements merge into virtually a continuous measurement.

Digital measurements are discrete. In fact, digital means "unit." Our most common digital system works on a base of 10. Digital computers work on a digital system of base 2. The number is a base two system of "0" and "1." The common name for this system is **binary** because there are only two numbers used. The binary "0" and "1" are often represented by switches that are turned on (1) or off (0). Although the switches could be mechanical or electrical—as they were in the early stages of computers—they are now almost solely electronic. Electronic switches are extremely small, fast and cheap.

binary system

Automatic operations require more sophisticated devices with the capability of decision making. Some of the earliest examples of decision or logic devices were mechanical switches that were controlled by the level of water in a tank.

These devices worked just like the water valve in a modern toilet tank. The water valve is either "on" or "off." Many of the logic devices used today operate in a similar manner, but use electronics rather than water. More sophisticated decisions are possible by connecting simple switches in patterns that require combinations of "on" and "off" logic. The most common of these include *and, or* and *nor* circuits.

Logic circuits and devices allowed machines to make higher level decisions. These devices were very limited in what they could "remember." Memory devices such as the information cylinder in a music box, the punched paper roll in a player piano and the punched cards in automatic weaving looms served as models for new electrified memory. The memories at first were banks of switches, then small mag-
solid state
netic doughnuts. Now, even smaller solid state chips and magnetic bubble memories are used in computers. All of the above examples are digital devices that work on a binary system.

Memory and Programming

Early computers were huge in comparison to those in use today. Each memory bit was in the form of a relay and vacuum tube. Together, they were about the size of a small cereal box. Later, transistors provided a "bit" of memory in the size of a marble. They were hundreds of times smaller than the vacuum tubes. The transistor did the same thing as the relays—made on/off connections—but, with no moving parts. This was possible because the transistor was made of dynamic materials that could be made to act as a switch. Further reductions of memory size occurred as the dynamic materials continued to improve. Today thousands, or even hundreds of thousands, of individual bits of memory can be fitted onto a
chip
small "chip." A modern computer chip can easily pass through the eye of a needle!

Many small home computers have a memory capability of
byte
64k (64,000 bytes) or more. A byte is equal to 8 bits. 64,000
bit
bytes is roughly about a half million bits. If this small computer were built with the original vacuum tube and electrical relay system common 35 years ago, it would fill a moderate sized school building. Small computers that you can easily carry in your shirt pocket have hundreds, even thousands, of

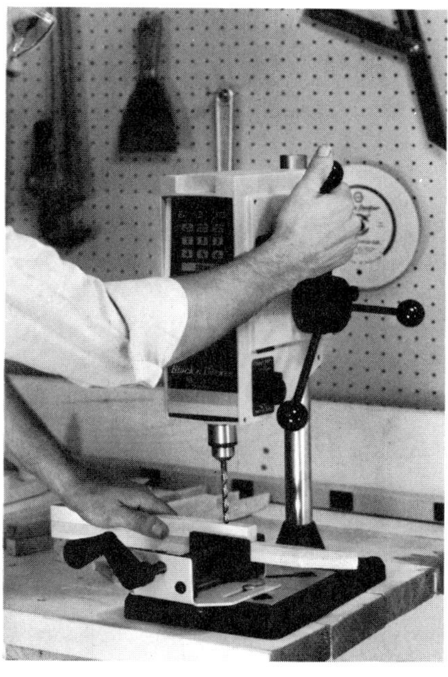

The invention of the transistor and its many improvements were important to the design and development of complex logic-control devices. These control devices help expand construction, transportation, communication and production systems. They are now used in hundreds of different products, such as a computer-controlled drill press, systems to help the handicapped become more independent, automatic controls for transport vehicles and control units in communication satellites.

With the increased use of electronics and computers, we hear a lot about memory devices. Modern memory devices are variations of memory devices that have been used over the centuries. One ancient memory device is the structure at Stonehenge, England which was used to remember and indicate the seasons of the year.

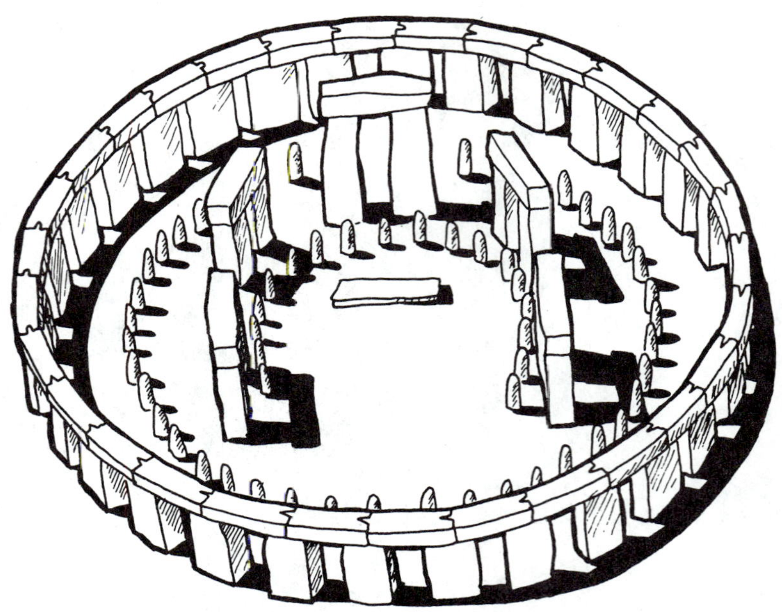

times more power than the first computers, and they cost less than one of the original "bits" of memory.

Memory is required to get any computer to perform a complex set of actions over and over again. It is also necessary to call up the proper sequence of actions at the proper time. The series of commands that tells a computer what to do is called a **program.** All programs may be thought of as having two parts, an internal program to control the operation of the computer and an external program to do the actual work. The internal program is determined when the computer is built. The specific procedures needed to control the internal workings of the computer are established by setting up the thousands and thousands of switches in specific patterns. This is accomplished by talking to the computer in a "machine language" of "0's" and "1's."

program

internal program

external program

Commands are sent to a computer through an external program. The commands can be as simple as touching the numbers on a microwave oven. The usual method of sending commands to a computer is by typing on a keyboard. A person can work at the keyboard and use the computer without knowing the machine language being used inside the computer. Gears in a standard car transmission also represent an internal program that is built into the machine. When you shift gears, you apply the external program by choosing the gear

By combining electronic memory and electronic controls, we are able to develop more sophisticated systems. In this system, we are able to give small droplets of ink an electrical charge. By controlling the input to the deflection plates, we can make the ink spell out any figure or symbol we want. A television set works the same way except that it uses scanning lines of light rather than drops of ink.

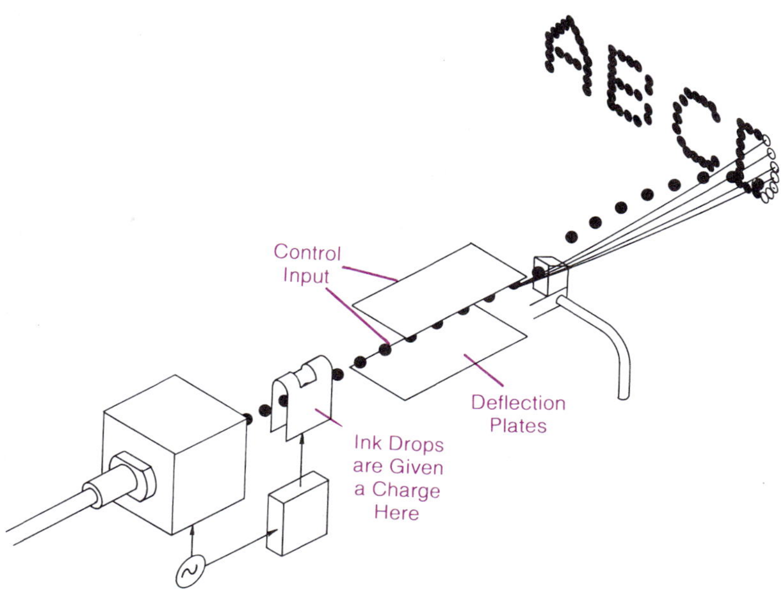

that you want. But, you can only choose the gears that were provided in the internal program. The same is true of a computer. You can only command at the keyboard what was provided for in the internal programming.

dedicated devices

Some microprocessors (small computers) are built so that their internal programs cannot be changed. These are **dedicated devices.** They only perform specific functions. Dedicated chips are used in microwave ovens, electronic ignitions for cars, computer watches, musical birthday cards, video games and pinball machines. You cannot program these chips to do any other tasks. The performance of these devices appears to be quite complicated because of their large memory. The machine must be programmed to understand that when "X" happens, they are to do "Y." When you play a game and make move "X," the dedicated chip takes predetermined action "Y."

One computer may also serve as a translator for another computer. The first computer takes the symbols typed in by the user and translates them into machine language that the second computer can understand and use.

Few people know the machine language well enough to do internal and direct programming. More people know how to program a computer by using "higher level" languages such as BASIC, FORTRAN, and COBOL. These are called higher level languages because they are more like normal English. Although almost anyone can learn to use a computer without knowing the internal programming, they are required to learn one of these new yet somewhat familiar languages.

More and more people have been affected by the increased availability of computers for everyday uses. In fact, you are being affected right now. The manuscript for this book was typed on a typewriter linked to a home computer. The computer was programmed as a word processor.

It is now possible to program the thermostat, the oven and other appliances to prepare your home before you arrive in the evening. The air conditioner may come on and cool your home to a preset temperature. Your dinner may be started and the lights turned on. You may even choose to have your favorite music playing when you arrive.

These possibilities have far less impact than the cybernetic operations of computers. Computers can control machines to do complicated procedures because of their fast memory and decision making capabilities. It was an onboard computer that controlled the landing of spacecraft on distant planets. Computers control sophisticated robots that perform amazing feats in manufacturing, including the assembly of parts for other robots. The robot exemplifies the merging of the three elements of automation—process, handling and control.

Systems

When all three concepts of processes, handling and control operate in one related system, automatic operation is achieved. The control parts of the system include the sensing devices that provide information. These are used to control and to change the processes. Other sensing devices provide information to direct the handling of the materials and parts. The materials are then converted through the intended materials conversion processes. Thus, process, handling and control can be united into one functioning system.

systems

Systems cannot operate automatically unless (1) there is information available in the form of feedback, (2) the parts of the system work together as a whole and (3) the system is self-

Technological systems exist at many levels. A simple system may be part (a sub-system) of a larger system that in turn may be part of a still larger system. The submersible vehicle consists of several sub-systems that when fitted together make up a larger system. The submarine does not exist alone either, but requires a larger ship as a tender. Together, they make up a system for undersea exploration.

1. Operator/observer

2. Pressure hull

3. Life support/controls

4. Exostructure

5. Ballast and trim

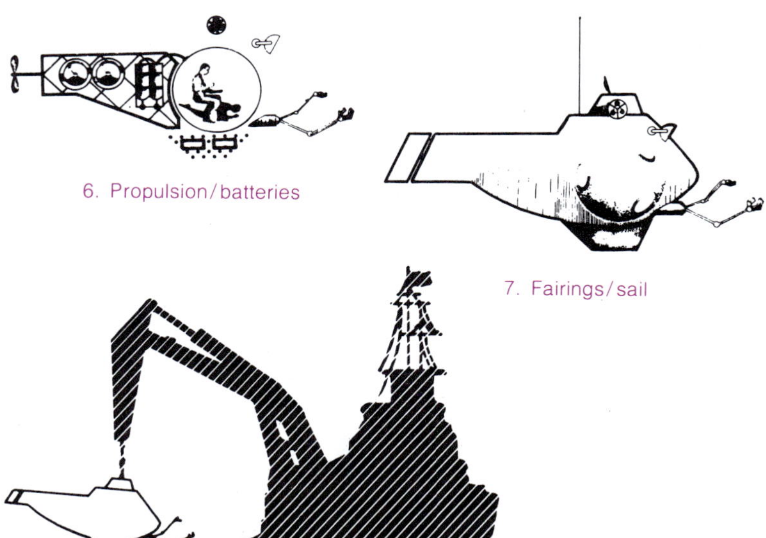

6. Propulsion/batteries

7. Fairings/sail

correcting. Without a self-correcting system, it is possible for a system to "self-destruct." In the refrigerator example described earlier, a relatively simple mistake of attaching wires incorrectly would result in a self-destructing system. If the control device incorrectly senses that the refrigerator is warm when in fact the refrigerator is quite cold, then the cooling system would never shut off. The cooling system would run until the compressor burned itself out or some other part of the machine failed.

Systems have many parts. For example, a system for producing ice cream requires an ice cream making machine and

Systems Model

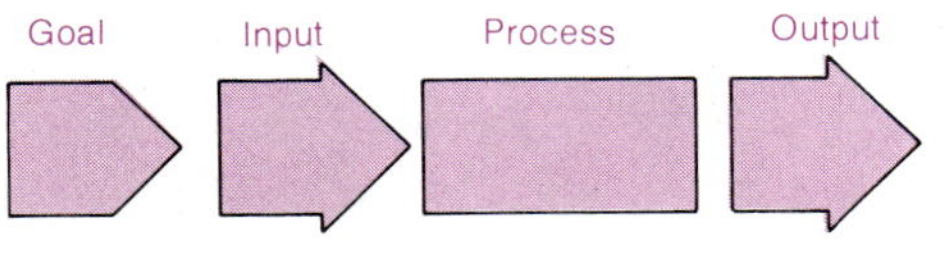

OPEN-LOOP MODEL

OPEN-LOOP

Here you see a Systems Model in its most simple form. It is an "open-loop" system and has no capacity for "feedback". Without feedback it is impossible to know how the system is operating. An open loop system is like riding a bicycle blindfolded. Worse than that the bike would have no handlebars, gear shift or other means of guidance. The only control available would be to start and stop the bike.

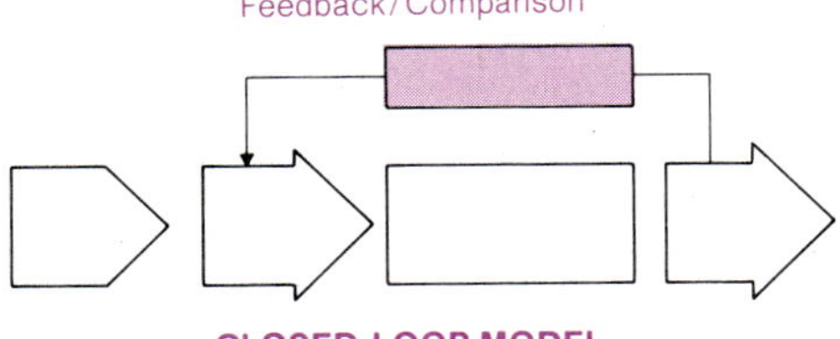

CLOSED-LOOP MODEL

FEEDBACK/COMPARISON

This level of the Systems Model adds "feedback/ comparison". Feedback takes a small sample of the system's output and changes it into information to help the operator understand what is happening. In the example to the right, feedback/comparison is provided by sensing a small amount of energy while the bike is in motion. That energy operates the speedometer. This allows the rider to know how fast she is going.

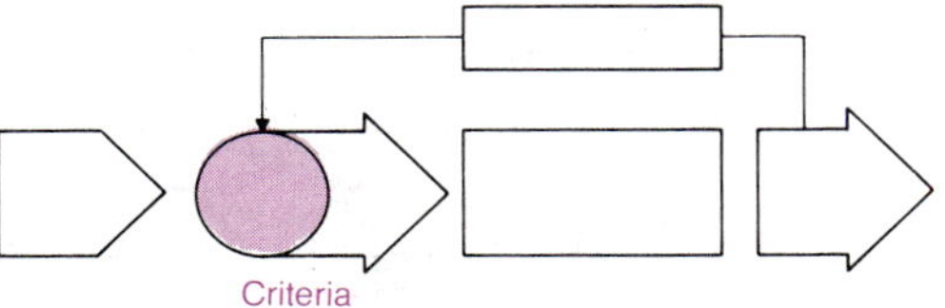

CLOSED-LOOP MODEL
(with Criteria)

CRITERIA

Knowing how fast she "is" riding is not useful to the bike rider if she does not know how fast she "should" be riding. Questions like: how fast or slow? how early or late? how near or far? how little or much? tell us how the system should operate. They provide "criteria" that the system should meet. In the example of the bikerider, only one standard is apparent—the speed limit. If she observes that criteria, she should go no faster than 25 miles per hour. Most systems will have several criteria to be observed.

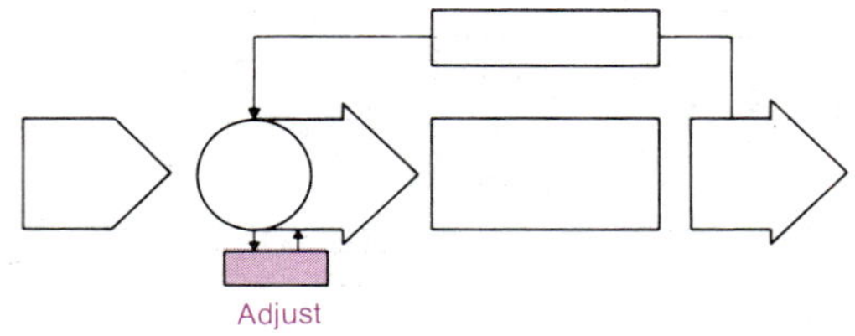

CLOSED-LOOP MODEL
(with Criteria and Adjustments)

CONTROL/ADJUST

Obviously if you are going to ride a bike you would like to know you can make it go where you want. The controls of a system allow you to make adjustments to its operation. The idea is the same as the "guidance and control" subsystems of machines introduced early in the book. As shown in the illustration, such additions to the bike allows the rider to "adjust" the direction and speed the bike will travel.

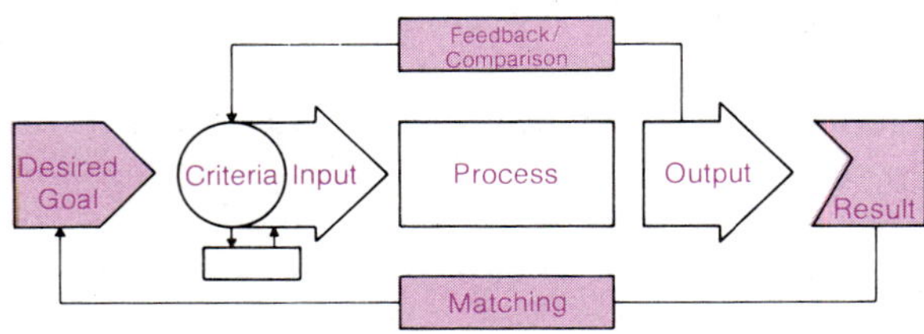

CLOSED-LOOP MODEL
(with Criteria, Adjustments and Goal Matching)

GOAL MATCHING

Did you get what you wanted? At this level the systems model provides you with some way of comparing what you accomplished with what you set out to do. The systems model must allow you to compare the "achieved" result with the original "desired" goal. If the result you achieved is different than you desired, you may be able to recycle and try to improve the match.

The Model in Use

GOAL (LEVEL I)
To arrive home by bike.

GOAL (LEVEL II)
To arrive home knowing how fast you have traveled.

GOAL (LEVEL III)
To arrive home safely knowing you have stayed within the speed limit.

GOAL (LEVEL IV)
To arrive home safely staying within the speed limits by 5:30 P.M. after picking up a gallon of milk and dropping off a book at the library and going by Rachel's house for a short visit.

GOAL (LEVEL V)
This goal is the same as Level IV. You decide to revise it, however, and not go by Rachel's house because you do not have enough time to do everything and still do it safely. You may not be pleased with the "actual" goal you achieved since it does not match your "desired" goal. You may decide on other actions to achieve or replace that missing aspect of your goal. You may call Rachel later in the evening or you may set a new goal for the next day that includes a trip to see Rachel.

its sub-systems. The materials required to make the ice cream are also needed along with the supportive production, transportation, communication, construction, exploring and developing systems. The production of ice cream, even by such simple methods, requires a complex support system.

integration

The work of humans in designing and controlling large systems is made easier by the computer in two ways. First, the computer is used to construct a simulation of the system during the design and improvement phases. Second, the computer is used to record the data during the operation of the system. Through simulation, the system can be modeled and experimented upon safely and economically. It would be unwise and sometimes impossible to try out a new approach without first conducting a simulated trial run. How would you like to be the first passenger on a new airplane that had never been tested? A new truck design may be simulated. So may a massive rerouting of trains, or a new system that combines air, rail and automobiles. The integration of these elements into compatible systems is possible only after initial modeling and experimentation through computer simulations.

The significance of the computer as a part of a cybernetic system lies in its ability to juggle many decisions without confusion. The larger a system becomes, the more information that must be handled to monitor and control that system. In reality, the handling of the data can be managed only with the aid of powerful computers. The recent development of new and smaller computers promises to make the task even easier and cheaper.

Summary

The concepts of control and cybernetics relate closely to concepts of communication. The feedback aspect, so important in communication systems, often takes the form of mechanical devices. Communication between two or more people does not necessarily require a machine, but the principles are the same. Cybernetics refers to communication and control as applied to machines and human systems.

Technological activities such as transportation, production and communication can also become systems. These are not usually automatic, but can occasionally be described as cybernetic in nature. Transportation, production and communication can fit together to form a cybernetic manufacturing system. Such a system must be capable of sensing changes in sales and preferences or of causing changes in preferences through advertising campaigns. A cybernetic system is possible only if there are channels through which feedback can flow to the decision points in the system so that the necessary changes can be made.

Key Concepts and Terms

analog devices
automation
binary system
bit
byte
chip
computers
control
cybernation
dedicated devices
external program
feedback
gyroscopic action
handling
integration
internal program
logic
manufacturing
memory
priorities
process
program
solid state
steady state
systems
variables

Summary—Unit Three

Changes in technology are made by exploring, developing and controlling. New ideas, approaches, objects and knowledge are important to the operation and improvement of technology activities. Exploring, developing and controlling close the circle of the larger technology systems. These activities provide knowledge and innovations for extending the larger systems.

Exploring and developing are especially important because they increase our knowledge of technology applications. Technology can be defined as the application of our knowledge, tools and skills to solve practical problems and extend human capabilities. New ideas and practical knowledge may emerge from exploring and developing. These ideas may be used within the activities of constructing, transporting, communicating and producing. These four activities are tied together by an elaborate web of connections. Information flowing through these connections establishes closed loop, complex systems.

As the processes become more and more complex, we need the help of the "artificial intelligence" provided by computers and robots. A worldwide communication system or an interstellar transportation system are possible. They depend upon the right parts to make the larger systems self-

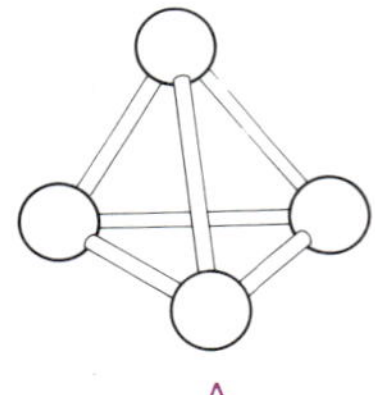

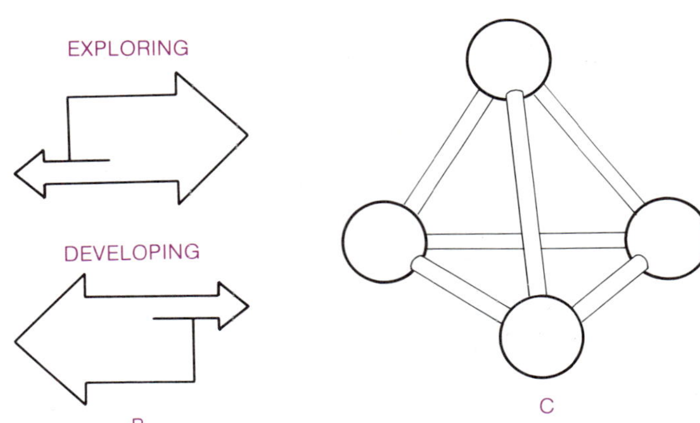

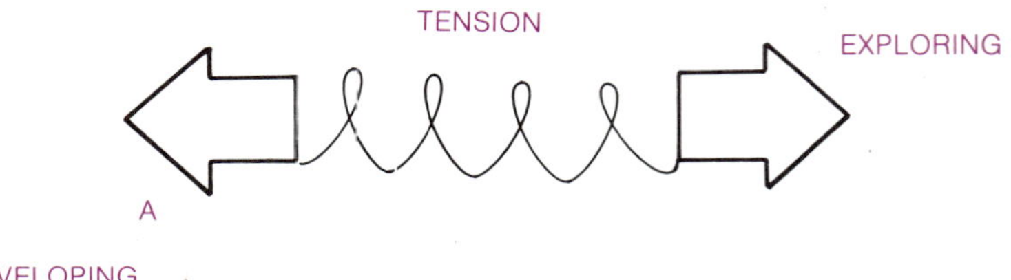

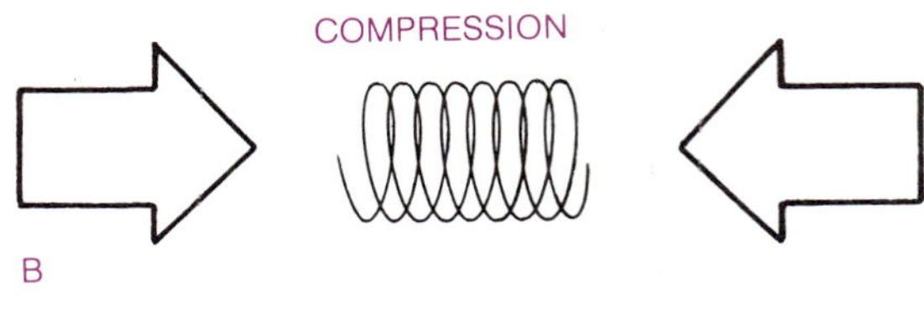

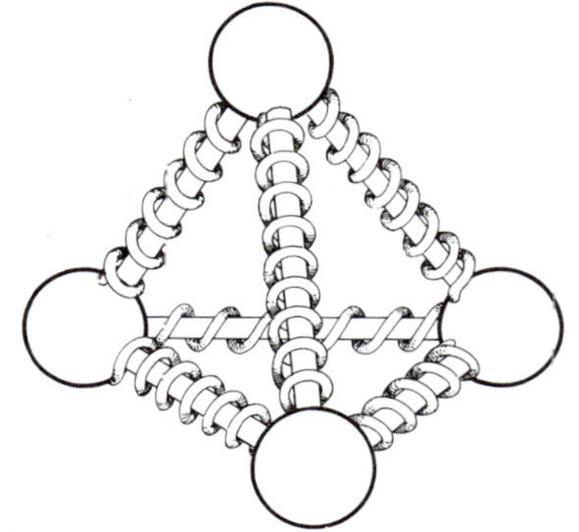

(A) The four systems of technology tend to expand as the exploring activity looks for and introduces new knowledge into the larger system. The activity of developing uses that knowledge and creates new innovations that tend to expand and enlarge the four systems.

(B) Using the compression and tension model introduced in earlier chapters, we can compare the expansion of the four systems to the stress of tension. This tension is created as efforts in exploring and developing find ways of improving and strengthening the systems.

(C) The tension of growth is held somewhat in check by a counteracting compression created by the activity of control. Controlling the system allows it to grow at a safe rate. In the final unit, we will look at the impacts of technology and try to determine if technology has been under control and, if not, how it might be.

regulating. The growth of more and more sophisticated systems appears to be characteristic of technological development. This growth is possible only with the information gained through exploring, developing and controlling.

UNIT IV The Impact of Technology

Introduction

The preceding chapters serve as a foundation for this unit. This unit explores the impact of future technology on people and nature. Consider the undesirable and the desirable effects of technology. There are both intended and accidental effects. These characteristics of technology are related to the processes and problems of making decisions about our lives.

Chapter Fifteen provides a set of future-oriented techniques and explores how they might be used. The chapter provides a glimpse into possible future events. This sets the stage for what might be the most important chapters of the book—*Consequences* and *Decisions.* In these chapters we ask you to question where technology is taking us, and what changes it is making in our world. We describe the difficult but extremely important process of making decisions related to technology. We challenge you to participate in that decision-making process.

Richard Hess

Chapter 15 Futures

Technology is changing our world. It will continue to do so in the future. With each generation of humans the rate of change is quickening. Technological advances that we enjoy today would have amazed our great-grandparents. What do you suppose life will be like fifty or one hundred years from now? Wouldn't it be interesting to know what the future will bring? Perhaps if we knew what possible changes were in store for us in the future, we could make more intelligent decisions today. In this chapter we will explore what it means to study the future.

Humans have always tried to predict the future. History shows that in each civilization there were a few people who had extraordinary ability to "see" into the future. Sometimes their predictions came true and sometimes they did not. In recent years, the art of forecasting the future has developed into a kind of science. A sizeable number of people now specialize in studying the future. They are called **futurists.**

predictions

futurists

Modern futurists do not try to predict what the future will be. Instead, they describe possibilities, or "alternative futures," that people should consider. The major purpose of future studies is to explore what possible changes lie ahead. Having done so, futurists hope that people will be more likely to choose the wisest course of action.

alternative futures

The major tool for understanding the future is the past. Humans have passed through many historical ages that lead through doors from one room to another. The next room has not yet been completed. We stand on the threshold of that new age.

In a democracy, citizens must participate in making decisions about technology. To make intelligent decisions, we must know how the technology works. It is helpful to be able to forecast what possible effects there might be in the future.

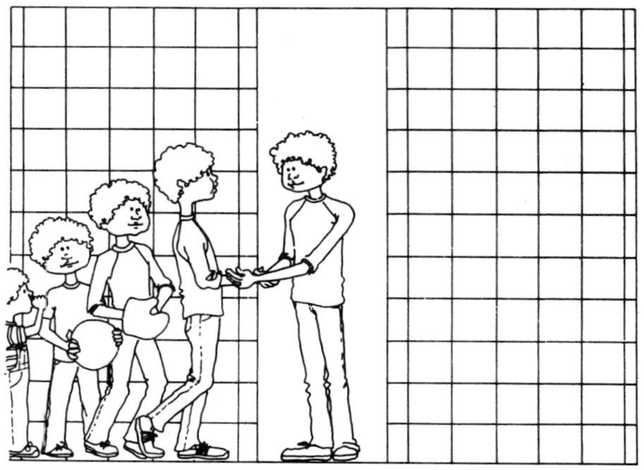

(a) We know there must be something called the "Past" and a "Future". We see neither of them since we exist between the "Past" and the "Future" in a slice of time we call the "Present."

(b) You gain knowledge from the experiences you had in the Past and use that knowledge to gain insights into the Future. Understanding the Future is very difficult, however.

(c) There would appear to be a prankster who lives in the Future whose job it is to make things happen that we do not expect. We call that prankster "Murphy". You probably have heard of Murphy's most famous law—"Whatever can go wrong will go wrong."

(d) You can study the future and attempt to identify possible events that could happen. You will find that short range events (1–5 years) are somewhat easier to project and have a better chance (a high probability) of being forcasted than do longer range events.

The future is something we never experience directly. It is possible to study the future but first we need to know some of the ground rules about that study.

Looking Toward the Future

projections

Because the future hasn't yet happened, we need special methods to study it. Most of the methods involve making **projections.** Projections into the future are educated guesses about what *might* happen.

Some of the projections are short-ranged. That is, they look at what might happen within the next few years. Mid-range projections are guesses about the next ten to twenty years. Long-range projections attempt to forecast what might happen as far as fifty years into the future. Obviously, short-range studies of the future are more reliable. That is, the events predicted are more likely to come true. Over the short run, not very many events are apt to be unexpected. Long-range projections are very uncertain, because there is much more time for events that could happen unexpectedly.

events

trends

One method of studying the future requires information about **trends.** By taking samples of similar events, such as the number of electronic inventions created each year over the past century, a trend can be determined. For example, futurists have identified a trend in which computers are being designed to be smaller and more powerful each year. They predict that this trend will continue for at least the next twenty years.

trend analysis

With your information about the elements of technology, you can conduct a "trend analysis" on one of the elements—say materials. You can start by studying material-related articles about research and development in science and industry. If your readings go back for several years, you may notice new developments in improved plastics, ceramics, aluminum, steel and other materials. In addition, you may note that materials can be made on a "custom" basis to fit the needs of the manufacturer and the products. Based on these trends, you project that more custom-designed materials will be used in manufacturing. This, in turn, may result in improved products.

analogies

A second method of studying the future involves use of **analogies.** Analogies are things and events that have some similarities between them. Because they are similar, whatever happens in one event may also happen in the other.

Some futurists, for example, believe that an inexpensive form of synthetic oil soon will be developed. If it is, then the United States and other Western countries will no longer be

TREND

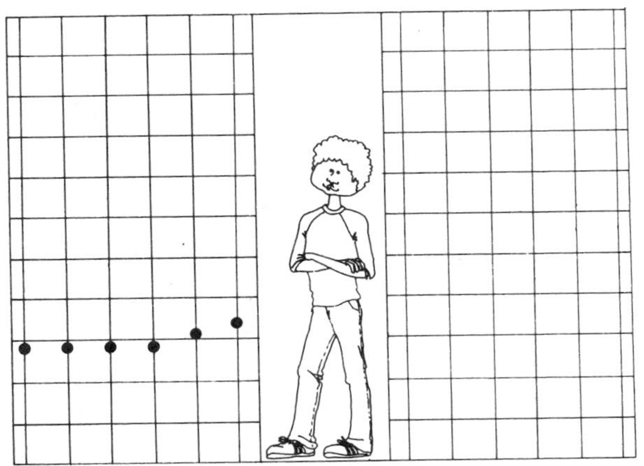

(a) First, you study related events from the past such as the growth and development of the telephone. In this case you try to identify the increase in the number and sophistication of telephones.

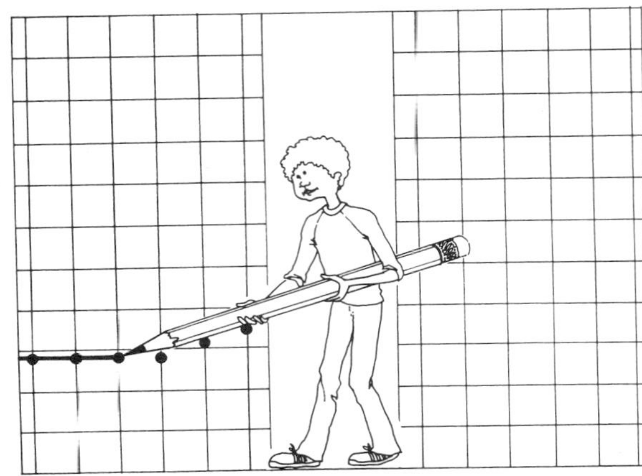

(b) Second, you attempt to identify if there was a pattern of growth. If so, you plot out that pattern of historical growth up to the present.

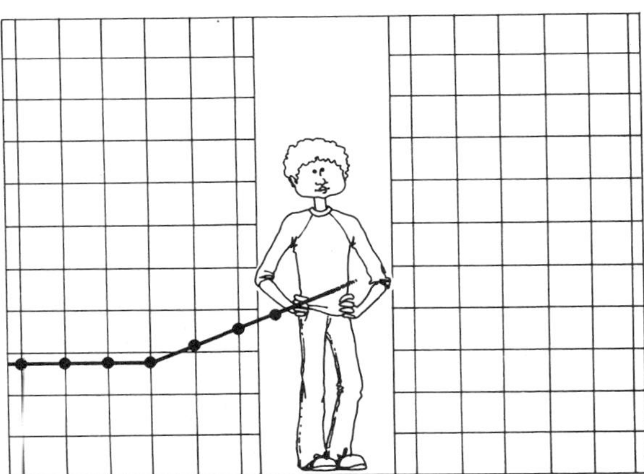

(c) Third, you attempt to identify if that pattern sets a "trend" or direction that may extend into the future.

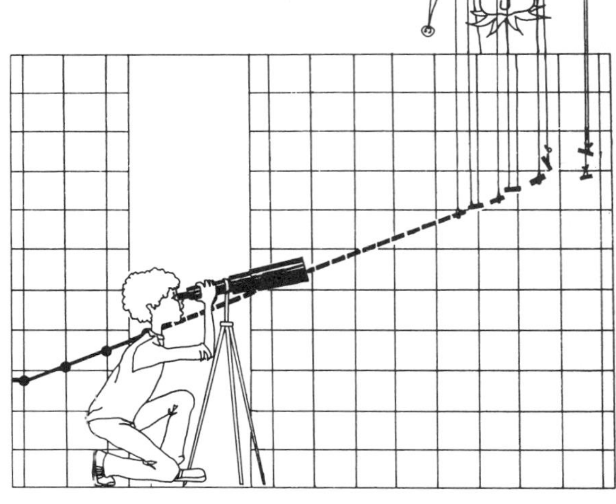

(d)

(d) Finally, if there is a trend, you attempt to project where it will lead in the future. Remember that long-range projections are much more difficult to make than those that are short-range.

One method for making future projections is by using "trends." A more sophisticated name for this type of projection is a trends analysis.

From your experience and study, you could probably project that the number of telephones will continue to grow, and that the phones will become more sophisticated, cost less and be more portable. It is possible that someday we could carry a telephone wherever we go.

First Commercial Telephone

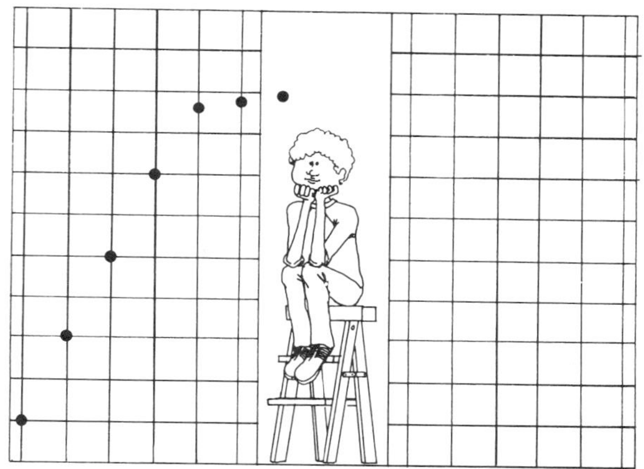

(a) It is necessary to identify a Trend that we can use as an Analogy to look at something of interest. In this example we will look at Trends in "Experimental" aircraft designs for the homebuilder as a potential Analogy for looking at Trends in commercial aircraft design.

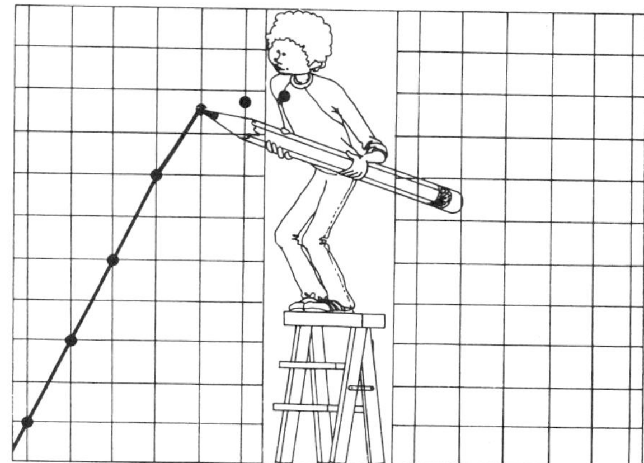

(b) After you have studied the importand developments in experimental aircraft design, you will need to identify the direction and pattern on those Trends. Next you must chart out the Trend(s) that is to serve as your Analogy.

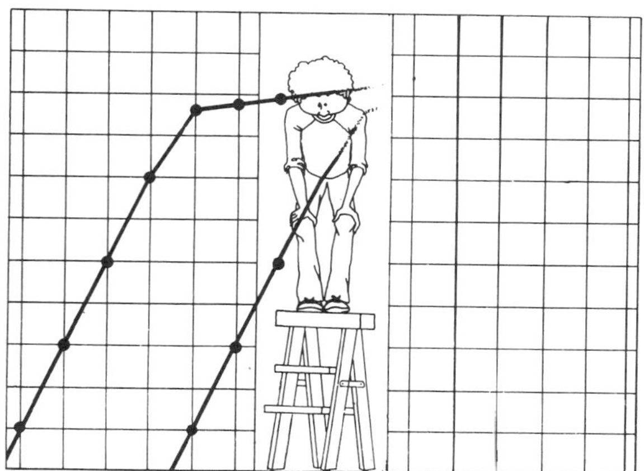

(c) This Trend of past events can serve as an indicator of a trend of later events. It is necessary that the second Trend of events follows a similar pattern of change.

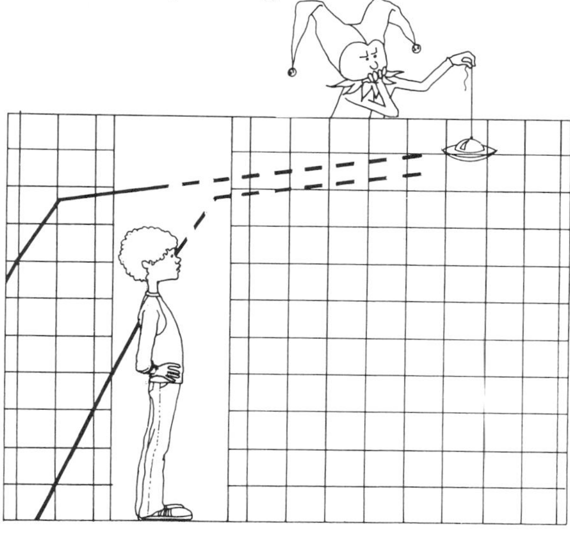

(d) If the analogy works the earlier trend will provide a pattern of Analogy for projecting the second trend and events yet to come. Remember that the Future, like Murphy, is a slight-of-hand expert and what you see is not necessarily what you get.

Another method of making future projections is the "analogy" approach. In this approach, a trend of one set of events is used to project the trend of another set of events.

From your study, you may be able to identify an upcoming change in the design of commercial aircraft. The designs of one person, Burt Rutan, along with the work of a small group of people, produced this aircraft design. That design is now revolutionizing the design of many commercial aircraft.

forced to pay high prices for foreign oil. This prediction is based upon similar events that happened during World War II. At that time, natural rubber made from rubber trees became very expensive and was in short supply. As a result, American scientists developed a synthetic rubber from available chemicals. The price of tires and other rubber products soon dropped.

Suppose you used an analogy to consider the future developments of processes. It would be possible to study the new developments in science that can be used to change materials, energy and information. These information developments of today may well be the basis for the innovative processes of tomorrow. For example, in your library you may find articles about the work of scientists who have isolated microbes that can live in liquid plastics and epoxies. The microbes appear to multiply slowly and attach themselves to the molecules of the plastics and other organic materials they may touch. Using this instance as an analogy, you might predict that a new biological glue will be developed that makes pieces of wood "grow together."

surveys

Delphi study

A third future study method relies on taking **surveys** or polls. These are not the kind of political opinion polls so often reported in the news. Instead, they are surveys of the educated guesses of scientists and inventors. The Delphi survey is one of these methods. If we wish to know when to expect a cure for cancer, for example, we might ask the top ten cancer researchers for their best estimates. The findings and ideas from each researcher are sent to all the members of the group with a request that they all reconsider their statements in light of what the other people have said. The process continues until a consensus (agreement) is reached.

You could use a survey to predict what new types of tools and machines might evolve in the next few years. You could talk to or write to hardware store owners, engineers and researchers to find out what new tools and machines they think will be on the market soon. As your experts see what others think, you could give them the chance to revise their ideas. By recycling through the process several times, you would have better information for your predictions. For example, one of your experts might propose that a laser-operated multipurpose cutter soon will be developed. If the idea appears sound, the other participants in your poll will support it. If they think the development is unlikely, they will reject the idea.

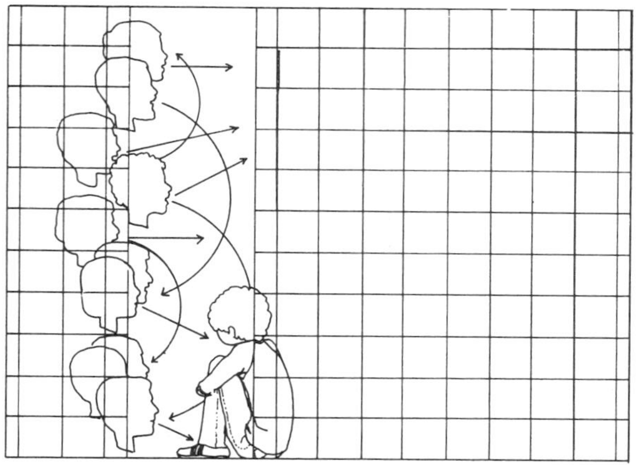

(a) Polling starts by conducting a survey or interview of quite a few people. If you wanted to know what the car-of-the-future might be like, you would survey people who were experts or somewhat knowledgeable about automobiles and aspects of technology that impact upon them.

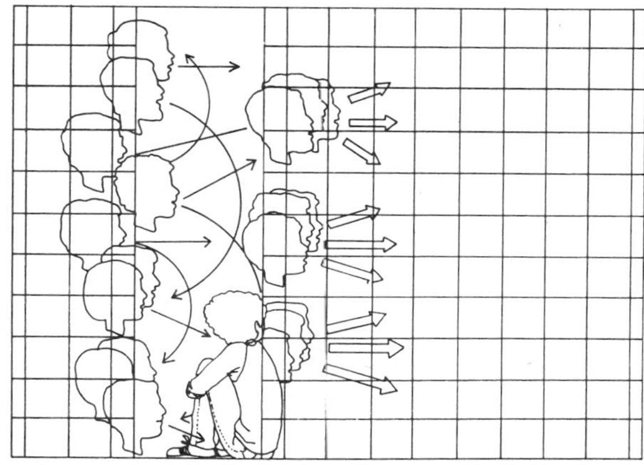

(b) After you conduct your survey or interviews you can share the statements made by the large group with all the individuals who are participating in the poll. Based on how others responded individuals may change their ideas.

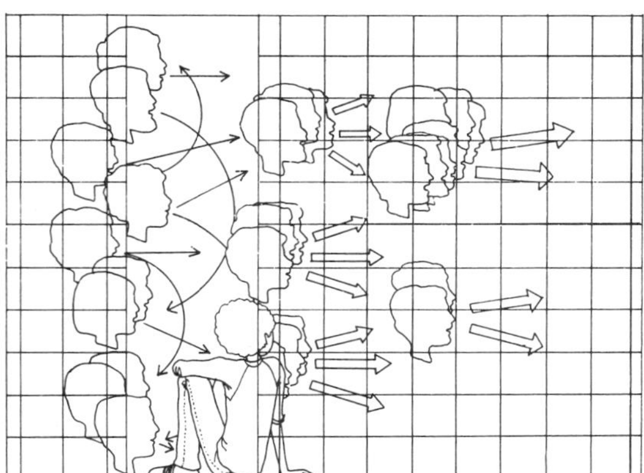

(c) The revised ideas of the total group can also be shared with everyone else. This in turn may cause individuals to consider what they have said and perhaps to change their ideas.

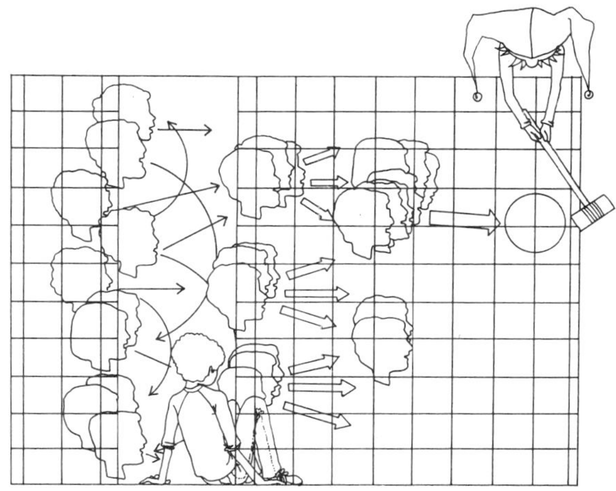

(d) This cycle can be repeated until there is no significant change in the positions taken or ideas presented by the participants.
A larger group of your participants may agree on what the car-of-the-future might be. A smaller group may disagree. Only time, and Murphy, will tell which group if any was correct.

Another approach in forecasting the future is the "polling" technique. In this approach you use the background and insights of a group of experts to reach a general agreement on what might happen in their field of experience.

If you used the polling approach, you might find that experts base many of their ideas on advancements that are identifiable now. New and available materials, controls and knowledge of how an automobile moves through our ocean of air represent a few of the emerging enfluences that will change the design of the car-of-the-future.

networking
decisions

The fourth method uses **networking** and **decisions.** This is the identification of events that must happen if a goal is to be reached. For example, if Americans are to decrease their use of oil, then a number of prior events must take place. For one thing, gasoline consumption must be reduced. For that to happen, people must use more economical cars or drive fewer miles by car. At the same time, substitutes for heating oil and other petroleum products must be found. The list could go on and on. Each event normally requires that many other events also happen.

decision tree

When all the necessary events are placed on a chart, the result is called a **decision tree.** Like a tree on its side, each branch (event) leads to other branches (events). When the total chart or picture of events is plotted, it becomes more clear what people will need to do to reach the goal.

You can use the networking approach. You can learn what is required to halve the amount of energy used in your home. Each energy saving idea and its cost must be listed. You must determine which of the steps you want to use. Then you will decide which of the steps must be completed first, which are second, third and so on. Your decisions can be laid out in graphic form for the network. At the last step you have reached or surpassed your goal. If the project takes a long time you may have to revise your decisions and the network.

future histories

Future histories are a fifth means of studying the future. The process begins by pretending to be alive at a time far in the future. By using their imaginations, futurists can then look back, so to speak, and write a "history." The future history describes the events that lead up to that imaginary time in the future. Many science fiction writers use this method to develop their stories. The chief value in this method is the directed use of imagination, somewhat freed from the predictions of "experts." An example of a futures history about humans and machines is presented here.

Future History: Humans and Machines

The year is 2050 A.D. Breakthroughs in research about the structure and function of the brain and body were made in the early years of the twenty-first century. This knowledge combined with what was learned in the 1990's made possible new and intricate relationships between humans and machines. Machines that assume full responsibility for the lesser functions of humans are now commonplace. Many of these first developments

DECISIONS

(a) You try to picture an event in the Future as the first step in seeing if it might happen. Perhaps you try to imagine what it would be like to live in a home or a country that was entirely energy independent from anyone else.

(b) After we have a sharp image of the future event we try to imagine the final decisions to reach that event. It can be helpful to picture the decisions as a tree with each branch a decision to be made.

(c) Each decision becomes an event and each event requires earlier decisions. Your decisions spread out like a tree. You are, in fact, developing a Decision Tree.

(d) Finally the limbs of our Decision Tree will reach the Present. This will allow us to see if specific key decision have been made that will help us reach the desired event in the Future.

You can also forecast the future using the "Decisions" technique. In this approach, you look at the many different choices that must be made to achieve a desired goal in the future. Then you determine the chance of these choices being made and implemented. If there are gaps in the decisions, it is unlikely that the desired event will take place in the future.

We could work to become energy independent in our homes and in this country, not requiring energy from outside sources. Many thousands of specific events would have to fall into place before that could happen. Pictured here is one such event. The invention by a teenager of a low-cost solar collector that allows the use of solar energy for cooking and heating water. This is only one leaf on the Decision Tree.

FUTURE HISTORY

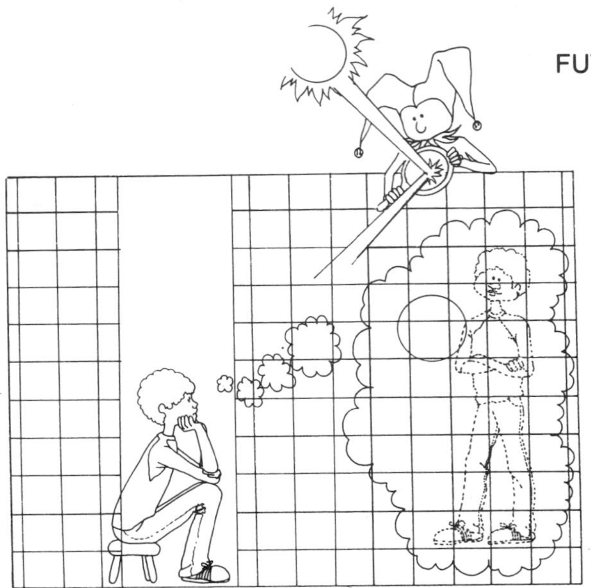

(a) A Future History is similar to a Decision Tree. This approach requires you to journey in your imagination to a future time and place and look back at the present. Consider what our communication system might be like by the year 2000.

(b) From our vantage point in the Future we write a history of the events leading up to that future time. If you imagine that all telephone-type communication will take place by satellites, LASERS and fiber optics, then you will need to describe how such a system could evolve.

(c) In our Future History we link the events together like links of a chain. When we finally reach the present we can decide if our chain of events make sense.

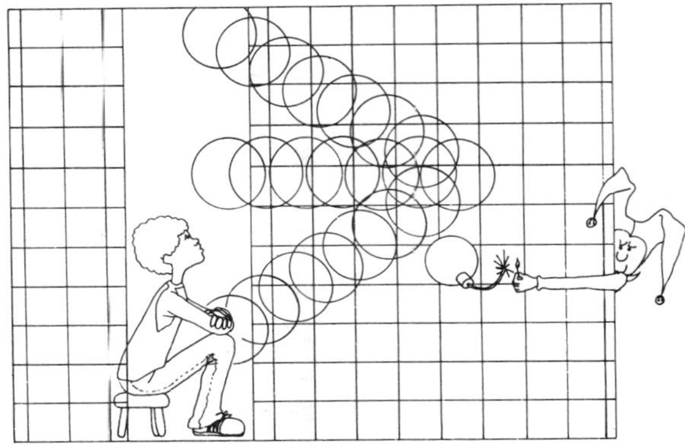

(d) Usually we write several Future Histories that could lead to the same final future event. As always our projections can be drawn off course or even blown up by events that we did not consider.

In the Future History approach, you climb aboard an imaginary time machine, journey into the future and look back at the present. Try to describe the string of events that link the two different times. This will allow you to weigh the possibility that the events might, in fact, take place.

Imagining and writing Future Histories allows us, in effect, to turn the clock foward and then back again. This can be a valuable tool in trying to imagine what could evolve in a technical system such as communication.

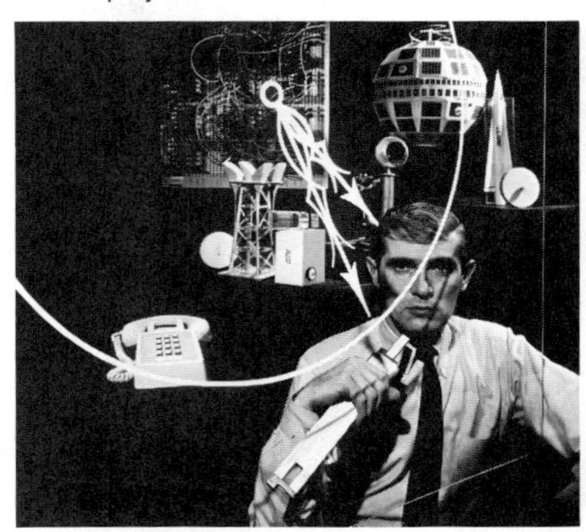

grew out of the use of computers, microprocessors, and robots in the 1980s.

The new "Cybuters" now supplement most of our rational decision-making. These machines are helpful in making ethical and moral decisions, although only at the simplest levels.

The responsibility for producing goods and products has fallen to a very small number of individuals who appear to thrive on the decision-making responsibilities required in large cybernetic "productories" (called "factories" back in the twentieth century). Many other people have become involved in individual and team efforts of another sort. They design and fabricate unique creations that do not lend themselves to mass consumption and production.

Most notable at this time is the sophisticated relationship between humans and machines. It is commonplace to find a human and a machine that have been designed as a one-of-a-kind system. Such "humachines" can amplify the human's physical and mental abilities. Together they can accomplish tasks that neither the human nor the machine could accomplish alone.

This approach to machine and system design came after many years of trying to build computers as "intelligent machines." The practice finally gave way to the development of Cybuters as extensions of intelligent people.

scenarios

Scenarios are also used by futurists. Scenarios are imaginative "pictures" of what could happen in the future. Usually, several scenarios are developed for a single event. Scenarios ordinarily identify a set of events and the consequences of those events. You might, for example, develop a scenario describing what would happen if a nearby river were dammed up to form a large lake. Think of the boating, fishing and other recreational possibilities that the lake would provide. You might describe the new houses that would be constructed along the lakeshore. There might be many new jobs created. Cheap electric power might be provided by the dam.

On the other hand, you might "picture" another scenario for the same event. In that scenario, good farmland is flooded over. People are forced to move from their family homesteads and their houses are torn down. Tourists move in. The quiet

FUTURE SCENARIO

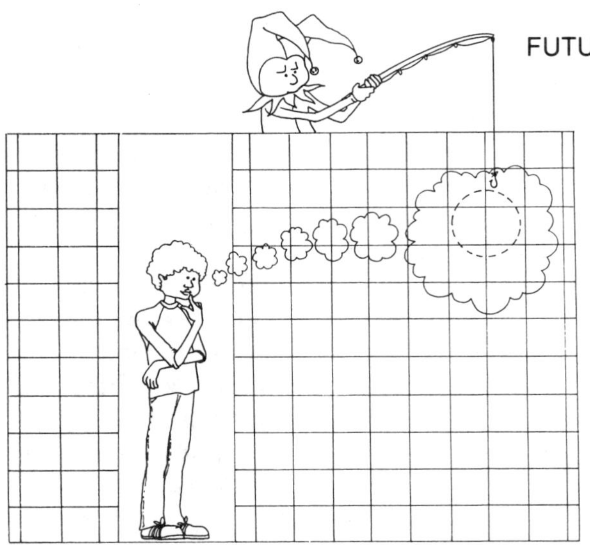

(a) A Future Scenario starts like a Future History by imaging a point sometime in the future. Such a scenario could also be built upon several related Future Histories. In all cases you try to envision a set of related events of a given time in the future.

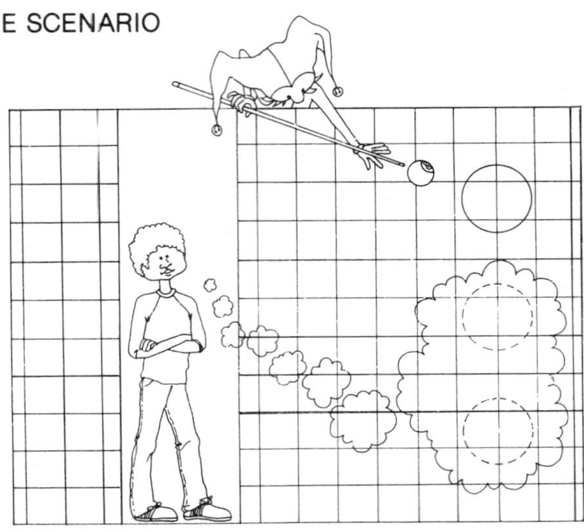

(b) You continue to build your Future Scenario by identifying more and more possible events connected with colonizing the Moon. We project these events in the same general time frame in the future.

(c) After you have drawn together the many related events, you try to describe how they are connected. You join them like links in a chain. Your description should begin to provide some insight on what it would be like to be on the Moon during its early stages of colonization.

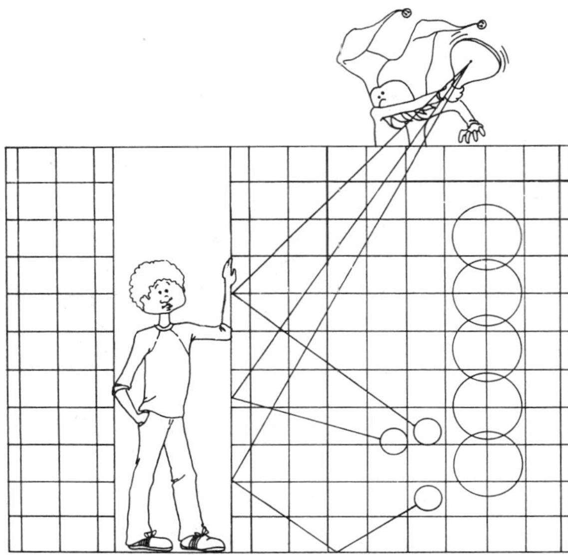

(d) As more and more connections are made between the various events, your description of the time slice will become more complete. That description is your Future Scenario. Remember that the Future can "play" havoc with any projected description. It is common practice, therefore, to develop several scenarios for a time slice that interests you.

The Future Scenario approach also allows you to imagine a time in the future. Through the Future Scenario you are able to draw a general picture of an identified slice of time in the future. Consider what such a scenario would be like that described how your children colonized the Moon.

village is changed by the influx of more people. This results in such inconveniences as traffic jams and the need for a costly new sewage treatment plant. This scenario is different from the first. It predicts many unwelcome consequences if the dam were built.

Many possible scenarios can be developed for a single event. If several are used, better and more reliable results are likely. What actually will occur is usually a combination of predicted effects. Awareness of possible bad results is very important if steps are to be taken to avoid them.

A scenario on information in the future is presented in very positive form below. What are some of the undesired consequences you think might be described in a related scenario?

Scenario: Instant Information

The year is 2001 A.D. Access to instant information has its roots in twentieth century developments such as the telegraph, telephone, television, computer and cybernetic systems. New advances in communication have increased the range and capability of satellites that orbit the earth and other planets in the solar system. Some are probing deep space. Data from these satellites is fed into one central computer. There it is analyzed. It is provided on-call to interested space scientists and technologists.

Everyone is equipped with a personalized information machine. This portable, high-yield device is given the same number as the identification code of its owner. The device is provided shortly after the human is born and is used in the early stages of life to monitor the intrusion of disease-carrying bacteria and viruses. Later, the device becomes one of the major teachers for the human. Eventually it acts as an aid to memory and value decisions. The information machine can scan past decisions made by the individual. The machine can indicate variations from the person's established patterns. This feedback allows the human to determine quickly if a decision is good or if the individual has unwittingly made a choice that conflicts with important basic values.

modeling

Modeling is the technique of charting or mapping the key elements of a complex series of events. A computer is often used for modeling because it can handle a large amount of data. As you might guess, any large event, such as the con-

MODELS

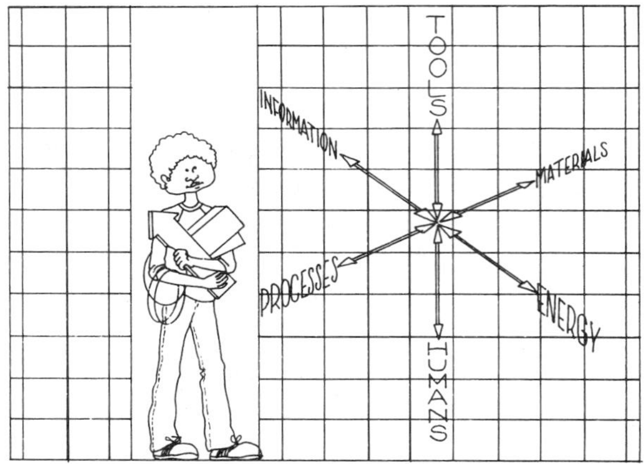

(a) Before we can Model some aspect of the future, we must identify what "parts" our model will have and how these parts "relate" or fit together. The Elements of Technology provides six specific parts and how they are related that can be useful in Modeling aspects of technology in the future.

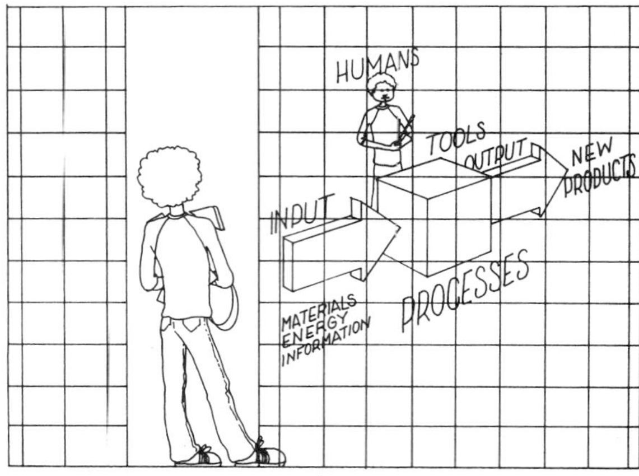

(b) Models must be capable of changing to meet some desired goals (criteria). The more goals that are met and criteria satisfied represents the best system. In this example we will try to get the most products (output) from the system at the least cost (input).

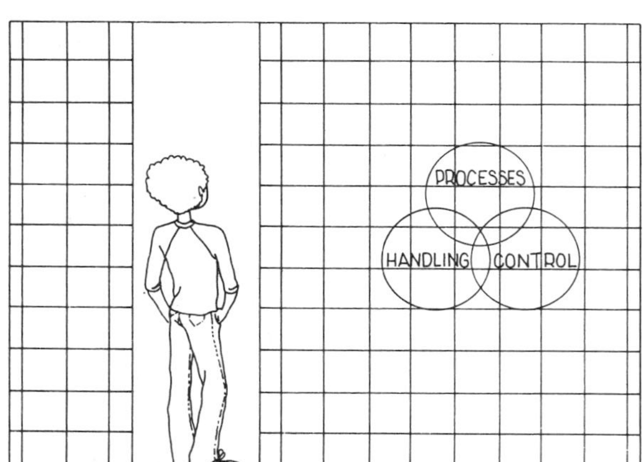

(c) All Models must deal with limits (contraints). The constraints restrict the range within which we can look for solution. If we are trying to establish an automatic production system three major constraints of Processes, Handling and Control must be considered. Our solutions will fall within these contraints.

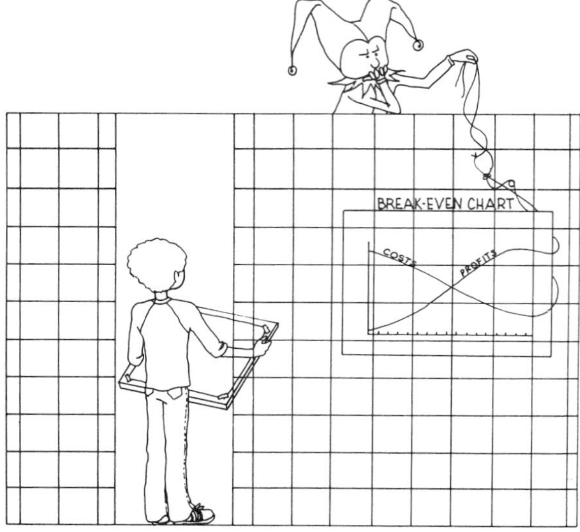

(d) To be successful our Model must help us find the best (optimum) solution to what we want (criteria) within the limits (constraints) of what is possible. In this case we are trying to have more profits (output) than costs (input).

Models allow you to experiment with what might happen in the future. By changing a part of the model you can observe how the other parts might change. Models allow a type of simulation of what might happen in the future.

Modeling is an important process of future planning, especially when there are large amounts of resources invested. If a group is considering establishing a new company or a new production line within an existing company, modeling can provide important insights into the possible success of the project.

(a) Determining the effects of decisions—made in the present to cause something to happen in the future—is very difficult. The process is like trying to knock over some dominoes that we can't see and can't be sure are even there to be knocked over.

(b) Determining the effects is difficult because we can not cause the desired effects directly. It is necessary to plan many decisions and events that we hope will cause the effect we want. We can plan the events by asking "what if this and this were to happen?" The dominoes are possible events.

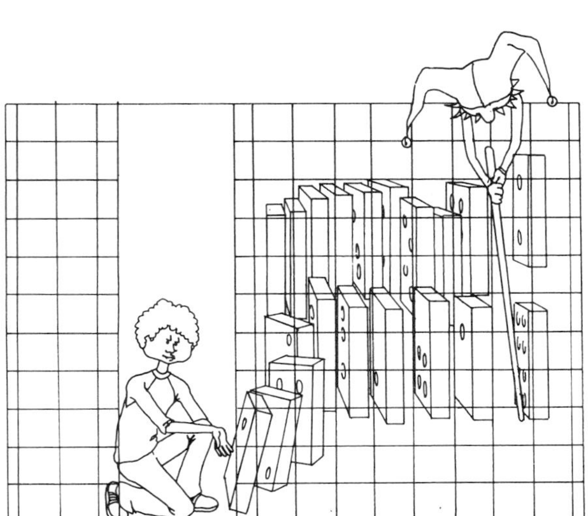

(c) After we have made our plan of decisions we can start the sequence by initiating the first event or knocking over the first domino. It is impossible to know exactly what will take place and what will interfere.

(d) It is only after the events have run their course that we will know their effect. Some events we wanted to happen may not occur. Some effects of other events may not become apparent until years later.

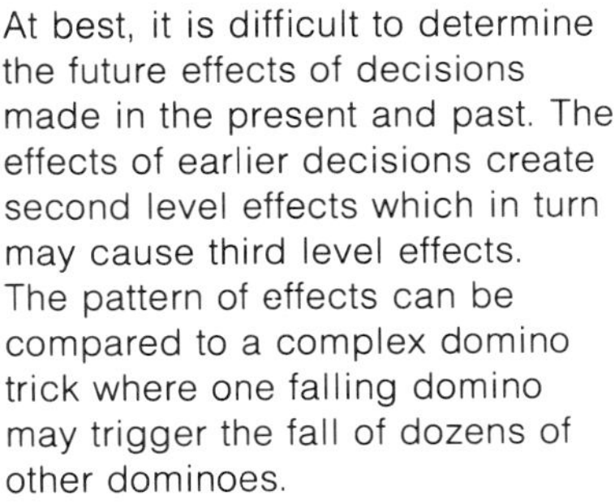

At best, it is difficult to determine the future effects of decisions made in the present and past. The effects of earlier decisions create second level effects which in turn may cause third level effects. The pattern of effects can be compared to a complex domino trick where one falling domino may trigger the fall of dozens of other dominoes.

struction of a dam, includes a number of smaller events. The smaller events—hiring workers, for example—have additional events related to them. Once a model is developed, the connections among the events become clearer. If one of the events should change, the model allows futurists to predict which other events might also change.

Projecting into the future by modeling is possible only if a model is available. The more complete the model, the better the projections will be. You could base many predictions on the model of the elements of technology presented in this book. The model has not been presented fully. It indicates only some of the relationships among the six elements. Because of this, the projections would be limited.

The more specific we are about the ways tools, materials, processes, energy, information and humans can work together, the more specific our predictions would be. For example, we know that a key connection in the model is between the "signals" of information and the "guidance and control" subsystem of machines. Predictions related to automation, cybernetics and intelligent machines will draw on that connection. The links between other parts of the model, such as types of energy and human subsystems, are less precise. Predictions relying on these links will be weak. If our model is converted into mathematical form, the projections will be more specific, but more limited.

Technology Assessment

New technologies are being developed each day. In some cases, new products and processes are being discovered faster than we can learn how to deal with them. Nuclear energy sources, for example, are being developed even though we are not yet sure of their long-term effects on human health. Because of this, it is becoming more important that the possible effects of new technologies be evaluated before they are put to use. The process of systematically evaluating

technology assessment

new technologies is called **technology assessment.** This is a study of the immediate and the delayed effects which result when an aspect of technology is introduced or changed. It is a study of technology's planned and unplanned effects on individuals, society and our ecosphere.

We can profit and learn from past experience. Predicting the future is a difficult task and success is not guaranteed. We have no alternative, however, since the future is too important to be left to chance.

The need for technology assessment is not new. The Roman aqueducts, built nearly two thousand years ago, were constructed almost entirely with muscle power. Hoists, winches and a variety of other mechanical devices were available but were seldom used. Apparently the Romans believed it was more important at the time to get the job done than to develop new machines. This kept large numbers of workers and slaves busy but did little to advance the level of technology. Water delivered to the city of Rome by the aqueduct system was distributed to most people through public fountains. The nobles had their water piped into their homes. The pipes were made of lead, which is poisonous. The water system slowly killed the leaders of Rome through lead poisoning. Some historians believe this was one important cause of the fall of the Roman Empire. If the Romans had invented and used technology assessment, they might have avoided this.

There was a time when all new technologies were widely accepted without question. Today people are beginning to be more cautious. The Congress of the United States now uses an Office of Technology Assessment. The people in that office investigate the possible effects of new technological development. The effects on people and the environment are carefully examined. Detailed reports are made to Congress. These help our legislators make decisions related to the new developments.

Special interest groups and consumer unions also help watch over new technologies that might be dangerous. Sometimes technology assessment slows down the process of implementing new technologies. For example, when a new fuel processing plant is proposed, it may require years of public study before construction can begin. During this time, numerous citizen groups and government agencies study and debate the good and the bad aspects of the plant.

Summary

Technology is changing our world. Like a snowball rolling downhill, technology is growing faster and faster. Each new invention or process makes other developments possible. The new developments in technology bring change. Change is sometimes difficult for people and societies to accept.

New technologies can be evaluated for their potential effects and consequences. This evaluation, called *technology assessment,* helps us avoid future dangers to humankind.

As the pace of change quickens, it becomes more important than ever to forecast and plan for the future. A new field called *future studies* has evolved. People who specialize in this field (futurists) have developed several methods of forecasting, including:

- Trend projection
- Analogies
- Surveys
- Networking
- Future histories
- Scenarios
- Modeling

Each of the methods can be useful for looking into the future. If the possible futures are identified, then people can make better choices in the present.

Key Concepts and Terms

alternative futures
analogies
decision
decision tree
Delphi study
events

future histories
futurists
impact analysis
modeling
networking
predictions
projections
scenarios
surveys
technology assessment
trend analysis
trends

Too often the consequences of our decisions about the use of technology result in the pollution of our world and detract from the quality of our lives.

Chapter 16 Consequences

Technology has had many different effects on people. Some of these effects or consequences have been desirable, some have not. Some of the effects have been intended, others have not. If technology is to be used to achieve our purposes, we must become better at predicting consequences *before* we act. Technology is a tool in itself. To use it effectively, *humans* must look ahead and make difficult choices about possible outcomes.

intended

predicting consequences

This chapter considers some examples of effects of technology. Special attention will be given to unintentional (accidental) consequences. The consequences of technology can be divided into three general categories. They include:

personal consequences

social consequences

ecological consequences

1. **Personal consequences**—the effects technology has on our bodies and minds.
2. **Social consequences**—the effects upon our homes, schools, workplaces, and all human organizations.
3. **Ecological consequences**—the effects of technology upon the physical world in which we live.

Personal Consequences

There is some evidence that early humans and some animals used tools. It appears that only humans learned to *make* tools for specific purposes. Scientists believe that success in making and using tools changed our ancestors in specific ways.

This may have been the beginning of the personal consequences of technology. Skilled hands are needed to make tools. Skilled hands cannot develop, however, if they are constantly used for walking. Skilled hands require a larger, more complex brain. The use of tools may have been a major cause of the growth of human's ability to think.

Early humans who could use tools were more likely to survive in a difficult environment, many scientists tell us. These humans were probably those who learned best how to use their hands and brains. These people were able to protect and feed their offspring better than people who couldn't use tools. As a consequence, tool-users probably had more offspring survive. Thus each generation was likely to have a higher percentage of children with hands and brains suited for tool use. The children could advance beyond their parents' development. Over many generations, humans have developed better tools. This, in turn causes us to continue our development.

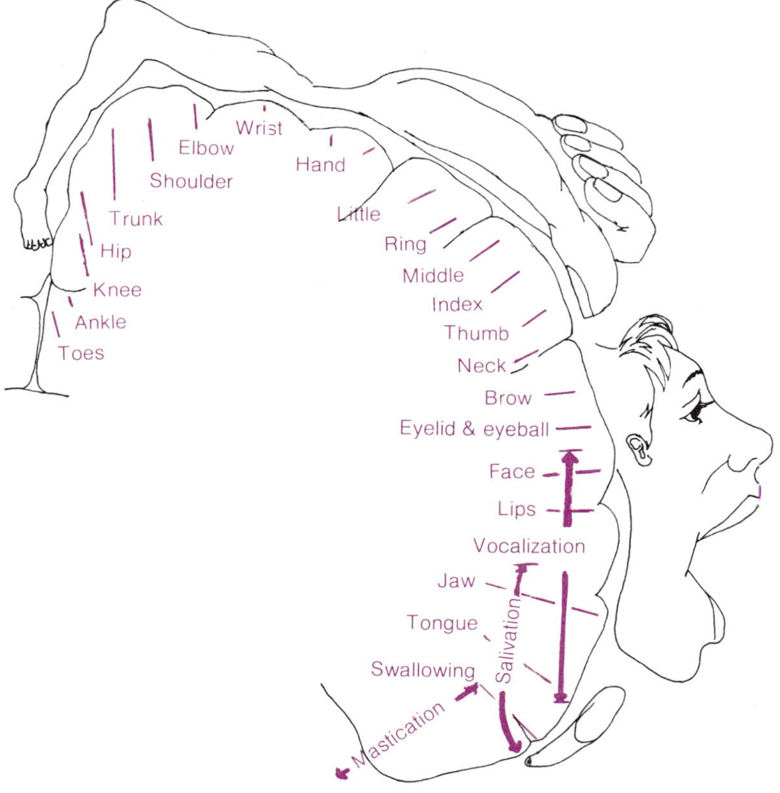

Research gave us the means to map the areas of the brain that control parts of the body. Some scientists propose that the larger area of the brain that is dedicated to the hand, especially the thumb, is a long term result of physically using tools.

Tools have played an important part in the ability of people to survive. Using tools, humans have learned to increase their power over natural problems. Use of tools helped early humans to accumulate and store more food. This left them more time to develop new ideas and communicate ideas to others. Some scientists tell us that as communication among humans became needed, those who could communicate may have been able to raise more offspring than those who couldn't. Their traits, the larger and more complex brains, became more and more typical in each new generation.

The brain has two major functions. It receives signals from some parts of the body, and sends signals to other parts. Each portion of the brain is responsible for a certain part of the body.

Relatively large portions of the brain control speech and hand operations. The development of oral communication (speech) was necessary so that humans could learn from one another. Tool skills required complex hand movements. The speech and hand skills probably developed more rapidly than any other skills. They remain important today, as humans rely on oral communication and hand manipulation for many day-to-day activities.

Physical Effects

physical effects

Tools continue to help us do our work more easily. We have become dependent upon them for even the simplest of tasks. Have you ever tried to open a can of soup without a can opener? If you have a home garden, you have probably used one of the first tools developed by humans—the hoe. Imagine how difficult it would be to do all the weeding by hand.

In almost every job we do, we reach for a tool to help us. Tools extend our use of physical strength. They eliminate some of the hard work experienced by our ancestors. Much of our heavy lifting and hauling is now done by machine. One of the consequences of our easier life today is that our bodies can get "soft."

In technologically-developed countries, physical fitness has become a significant problem. Students are bused to school, then required to participate in physical fitness classes. Adults drive home from work, then go jogging for exercise. Our great-grandparents would have thought that

The first automobiles sold to the public were produced in Germany by Karl Benz (1885). The impact of the automobile on the way we work and play has been awesome. It has reshaped our cities, our countryside, our economy, and, unfortunately our ecology.

was strange behavior. By making work easier for us, tools can also leave us weaker.

Technology affects our physical health through foods we eat. New varieties of wheat and rice have been developed. Farm machinery and techniques have been improved. More and more grain can be grown now to provide food for a growing population. Giant farms, thousands of acres in size, now produce what used to be grown in small farms and gardens. Huge tractors and other expensive pieces of equipment are needed to take care of the harvest. To compensate for the cost of machinery, plant disease and insects must be controlled by chemical sprays. Some of these pesticides remain in the food produced from the crops. Some of these pesticides have long-term effects on us.

nutrition

In a wild environment, nutrients cycle from the soil into plants and animals and then decay back into the soil. But nutrients taken from the soil by our food plants are not returned in our present food cycle. Current methods and demands on the food supply deplete our soil. To grow enough food from the available farm land, chemical fertilizers are used. These are costly and depend on a large energy supply. The foods are altered, too. Vitamins are added to flour to replace those lost in processing, for example.

vitamins

Without these chemicals, our present system of food production could not exist. The system is one of the most efficient in the world. But what are the consequences of mod-

food additives

Most inventions and innovations have important consequences. The availability of a low-cost, dependable sewing machine helped set in motion consequences that continue today. The sewing machine allowed families to produce more easily the clothes they needed. This led to an increase in demand for more woven cloth, which required more wool and cotton. The demand for cotton made slavery an economic necessity at the time. The quantity of cloth manufactured allowed a wider range of styles and colors. This led to people preferring the large selection and better quality available from mass-produced clothing.

ern food technology? We depend on a large energy supply to keep our soil replenished. Vitamins added to bread may add years to your life. Other chemicals added to keep the bread soft may be harmful to your body. Many fruits and vegetables are picked before they ripen. They are stored in coolers for shipment during the "off season." This method provides you with apples during the winter, but they are not as nutritious or tasty as those that have ripened on the plant. We have a lot of food, but are only now learning some of the consequences of our methods.

Emotional Effects

Technology changes people emotionally, as well as physically. Before the invention of the clock, time was not important. People worked during daylight and slept at night. Our concept of time was related to the changing seasons and their effects on plants and animals. When the sundial was developed, people began to divide each day into segments (hours). Arrangements could be made to meet someone at a given hour, not merely some time before sundown. Then people could begin to worry about being late for something!

After the mechanical clock was invented, people could organize not only their daylight hours but their nights as well. They could believe they were hungry because the clock said it was mealtime. A consistent schedule could be planned and shared with other people.

Think of how your life is regulated by time and timekeeping devices. What might happen if all the clocks and watches in the world mysteriously stopped running? Could our modern production system keep going?

efficiency

In high-technology nations, there has been a trend toward more efficient production. **Efficiency** may be defined as more goods produced with fewer resources. Ordinarily, efficiency depends upon specialization.

For example, rather than have one person make a complete refrigerator, each worker produces a small part. The parts then are put together on an assembly line. Each worker does a specific task.

job specialization
division of labor

This practice is called **division of labor.** It can be quite boring to do the same job over and over again. Bored people do not work very well and problems can develop. Sometimes,

Some historians believe that the clock played a dominant role in shaping our actions. Henry Ford introduced time into the assembly line by setting the "pace" of the work. The pace required that the workers meet the demands of the production system. The workers became like cogs on a giant machine.

on assembly lines, workers purposely attach a part incorrectly or leave it out altogether. Why do you think that might happen?

In earlier times, each worker built a complete product from start to finish. Often the worker knew the customer, personally. A barrel-maker (cooper), for example, shaved the wood staves, hammered out the hoops, then carefully fitted them together to form a finished product. Each barrel was a piece of craftsmanship, and there was satisfaction for the worker. But it was a slow process.

As the demand for barrels increased, production had to be improved. At first, some coopers specialized in stave-forming. Others made only the hoops. Still others did the assembly. Eventually, as many as six different workers were involved in the making of a single barrel. None of them ever

The division of labor and specialization, as seen on this assembly line, were integral parts of the American production system. The division of labor tended to change meaningful work into boring labor. This change in the worth of work resulted in people being very dissatisfied with their jobs. Recently, some manufacturing companies have reorganized their production systems. This plant has teams of workers build complete motorcycles. This approach helps workers recapture the meaning of seeing a complete job and the pleasure of knowing it was well done.

met the people who bought the products. As labor became more specialized, production of barrels increased. But, pride and satisfaction of the workers usually decreased. Quality also declined. Some industries have recognized this consequence of technology, and have tried new methods of production. The problem, called **worker alienation,** still persists in many industries.

worker alienation

The Consequences of Knowing

Books have had a profound effect on people's lives. In the centuries before printing technology was developed, important thoughts and events were recorded in song and story, if at all. As each generation passed, there were a few people (called **scribes**) who wrote down some of the things that happened. The handwritten books could be reproduced only if someone copied them by hand.

scribes

This was a very slow process. A few people isolated themselves in monasteries and spent their entire lives handcopying earlier religious writings. The few writings that existed were available only to leaders in the church and government. Most of the people were forced to learn from what little they were *told.* Most of the ideas and experiences of earlier generations were unknown to all but a few persons.

Johann Gutenberg developed a system of moveable metal

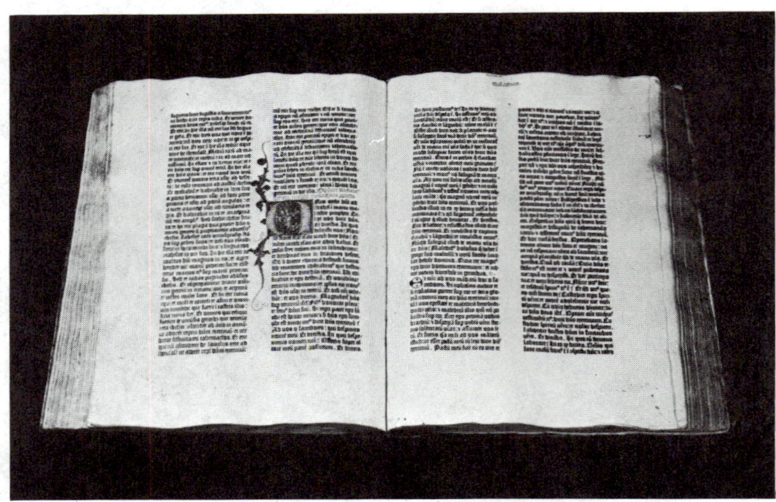

The availability of books allowed people to have the joy of learning new things.

type in about 1440 A.D. Books became available to many more people. The new printing process made it possible to reproduce cheaply thousands of copies of any book. The Bible was one of the first to be produced. This made it possible for everyone to have a copy of the Scriptures.

Reading became an important skill. Schools evolved so that people could learn to read and interpret the Bible. The consequences were tremendous. People began to question authority.

As more people learned to read and write, more books were produced. Scholars in science, mathematics, philosophy and religion were able to know what generations before them had learned. It became easier for them to share their new ideas with colleagues and with people in other fields of study. It became quite common to be "well-read" in all subjects, rather than in just one.

Today, modern printing processes turn out books, magazines and newspapers at a rate that would have amazed Gutenberg. Think of the billions of written materials that now fill libraries to overflowing. Many people cannot find time to read the few magazines to which they subscribe. And, can you imagine being assigned to read every book that was ever printed?

knowledge explosion

In the past seven years, the amount of information known to humans has doubled. It will double again in less time. This rate of increase has been called the **knowledge explosion.** Technology has made the knowledge explosion possible by providing faster and cheaper methods of communication.

emotional effects

For many people, the increase in collected knowledge and the increased speed of communication is causing an emotional overload. Even in relatively precise areas of study, specialists may not be able to keep up with advances. It is becoming more and more difficult to know what is happening in several areas of study. For all of us, now, there will always be more information available than we can ever learn.

computers

How can we cope with an ever-increasing collection of knowledge? Computers may hold the answer. Earlier in this book you learned about the potential use of small, home computers linked with television sets. The system was described as a new version of the library. Extraordinarily large amounts of information now can be transferred from information networks and stored in these microcomputers, to be called up on demand.

In the future, it may not be so important to remember large numbers of facts or data. It may become more important to know what kind of information you need and how to get it from the home or school computer. The availability of books in Gutenberg's time made reading very important. The development of modern microcomputers soon will make it necessary for everyone to be skilled at searching and analyzing information, as well as reading.

Books and computers are tools for thinking. The development of such tools has had an effect on the development of the human intellect. Historians estimate that the average scientist 200 years ago probably had a vocabulary equal to that of today's junior high school student. This doesn't necessarily mean that children of today are wiser. It does indicate some of the consequences of modern communication systems. Information is all around us in newspapers, radios, TV, magazines and movies. As needed, new words are made almost faster than they can be added to dictionaries.

Social Consequences

Beginning with the very first invention, technology has caused social changes. The technology of farming provided a

The explosion of knowledge is more manageable through the use of and the possible over-dependence on the computer.

The home computer can provide a range of services. There is some concern as to what impact these services will have on the family. What will happen when people have fewer and fewer reasons to leave the house for the things they want and need?

more reliable food supply than did hunting or foraging. Consequently, people began to form small villages instead of living as nomads. Animals were domesticated for food and to do work. For thousands of years the social structures of tribes and villages changed very little.

When the wheeled plow was invented in Europe during the sixth century A.D., things began to change. Before the wheeled plow, each family farmed its own small plot of land. The wheeled plow required large oxen for power. Families began to join with others to buy teams of oxen. They also combined their small plots into larger farms.

Especially in Europe, this practice led to the development

of a communal life style. People in the village began to specialize. Some farmed, others produced clothing, while still others made and repaired tools. All these social changes were related to technological advances.

co-ops

A similar pattern of change has occurred more recently on the plains of the United States. Owners of small farms have formed "co-ops." This helps them purchase the large machinery and storage bins required for agribusiness. Together they can store grain and sell it whenever the price is high.

unintended consequences

In both examples, a new set of social relationships was formed. People were required to give up some of their individual freedom to benefit from group cooperation. A measure of independence was lost as people became more dependent upon others. Technology, you see, can have both positive and negative consequences.

Inventions do not have to be big to be important. Some small devices can have powerful consequences. The stirrup is an example. Invented more than a thousand years ago, the stirrup changed the course of history. Before the stirrup, people rode bareback or with light saddles. When stirrups were attached to the saddle, the rider could more easily stay on the horse. The horse and rider could work together, thus using more of the horse's strength.

During the eighth century A.D. in Europe, kingdoms were warring among themselves for control of land and riches. A leader of the French, Charles Martel, observed that stirrups enabled his mounted troops to use long steel lances as battering rams. The stirrups gave them the power to overwhelm the soldiers on foot or the mounted warriors without stirrups.

To increase the number of mounted troops, Martel promised plots of land to those who would join his army. He soon had a vast force of mounted warriors. Martel and his son, Charlemagne (also called Charles the Great), built an empire that lasted for a century. When the army disbanded, descendants of the original mounted troops claimed the land they had been promised. Each landowner ruled a mini-kingdom, complete with knights, serfs and all the trappings of Camelot. And it all began with the stirrup.

Each of these examples—the clock, plow and stirrup—has influenced the development of humans and their social structures. The development of humans and technology is closely connected. Consider two ideas that are very important:

1. Technology is a product of humans.
2. Humans are influenced by technology. In a sense, we are "products" of technology.

It may be easy to accept the first notion, but what about the second one? It means that the shape of our hands, the volume of our brains, our social patterns and even the way we think are consequences of technology. Furthermore, technology will probably continue to change us in the future. We may not like some of the changes.

The stirrup was a very small technological innovation with major consequences. Would you believe that because of this device:

It was possible for warriors to fight more effectively from horseback. These warriors became the equivalent of the modern tank. Equipping these warriors with horse and arms was terrible expensive.

To cover these expenses, monarchs took large tracts of land from the church.

These tracts of land were given to those warriors who would fight for the monarch.

These mounted warriors became knights.

Stong castles were built for the safety of the people when other knights invaded their area.

The castles helped the knights become more and more powerful. Finally the knights rebelled from the monarch and set up their own small kingdoms.

The land, the social stucture and the government had gone through many important changes. The initial trigger for this change was the lowly stirrup.

Ecological Consequences

Environmentalists and ecologists are beginning to make us aware of what humans have been doing to our natural resources for centuries. They observe that for generations people have been using raw materials, processing them and

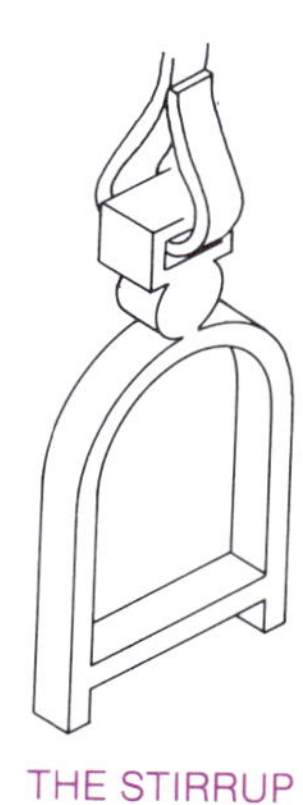

THE STIRRUP

LED TO THIS?

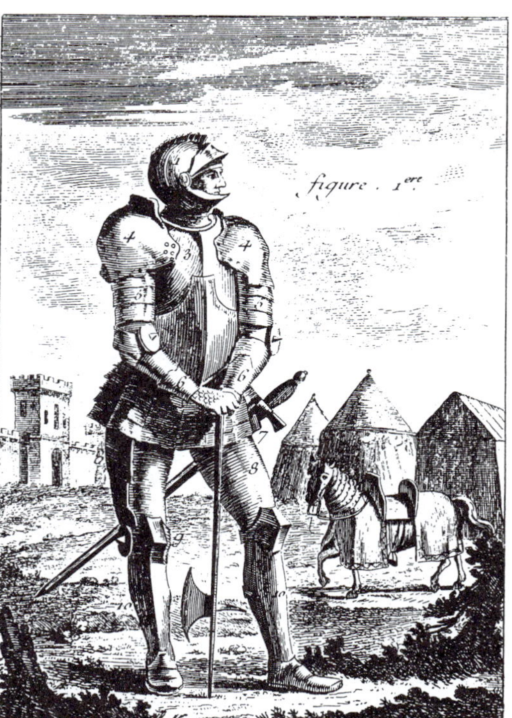

pouring pollutants in the rivers, oceans, air and soil. They point out that nature holds a delicate balance. Humans, they charge, are destroying plants, animals and even the air we breathe. Because it takes such a long time for ecological consequences to be discovered, it is difficult to know today how serious the situation will be in the future.

Some scientists point out that humans have always had an effect upon the environment. At the end of the last Ice Age, about 10,000 years ago, herds of wild horses on the North American continent disappeared. This happened in spite of an abundance of grass for food. The best guess is that the horses were slaughtered for food faster than they could reproduce. Finally, there were none left. More than nine thousand years later, Spaniards brought horses from Europe. The breeds of horses we now have in the United States are not native to North America, but are descendants of the Eastern Hemisphere.

Investigations by archeologists indicate that wherever humans have migrated, the balance of nature has been changed. About 12,000 years ago, humans moved across a land bridge that used to connect what is now Russia and Alaska. This seems to be the most plausible explanation why Alaskan Eskimos have physical and cultural characteristics akin to Oriental people. Prior to that time, the musk ox lived on the north and the south sides of glacial ice covering most of Canada. Soon after these Stone Age humans migrated into North America, the musk ox began to disappear. Because the northern side of the glacier was inaccessible to these early people, the animal survived and prospered there for centuries.

Some of the consequences of technology on plants and animals are easily observed. There are also important effects that we cannot see. Consider the following chain of relationships.

The earth is surrounded by a relatively thin layer of gases.

biosphere

This layer of air, the **biosphere,** sustains all living things. The air we breathe consists mainly of three important gases: oxygen (21%), nitrogen (78%) and carbon dioxide (less than 1%).

Bacteria and algae in the soil take nitrogen from the air and convert it to ammonia. Ammonia is poisonous to humans. Fortunately, other microorganisms convert the ammonia into nitrates. Nitrates are needed by green plants to produce pro-

During the Ice Age, humans migrated into the western hemisphere from what is now Russia. Prior to their arrival the musk ox were plentiful. Soon the musk ox dissappeared except in places where the glaciers made it impossible for humans to reach the animals. The patterns of human migration follow closely the patterns of extinction of large animals. This suggests that humans may well have been a major cause of that extinction.

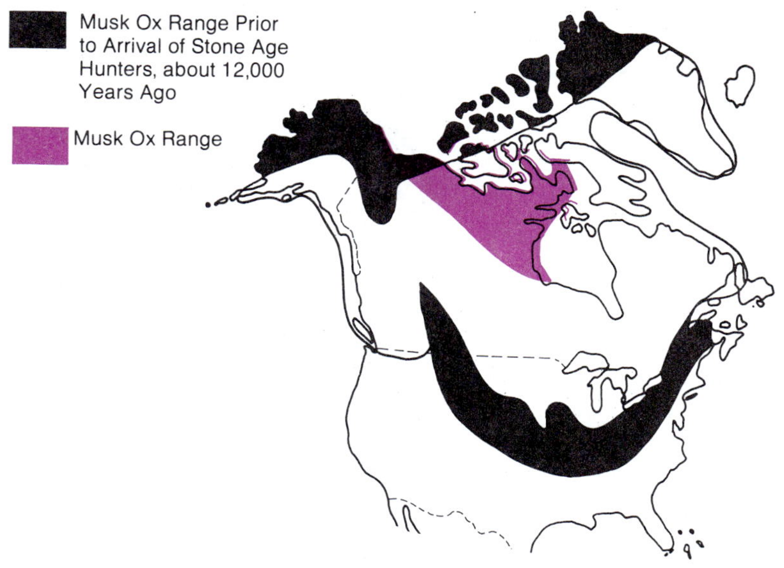

teins and oxygen. Animals in turn eat the proteins and breathe the oxygen. When animals and plants die, their protein is converted back to ammonia and the cycle continues. Life abounds! If ever there are not enough green plants left, humans will be in trouble.

The nitrogen cycle is a vital part of the ecosystem. The cycle provides for the circulation of protein-building nitrates among plants, animals and the soil. There is growing concern that humans and their technology are intruding seriously into this cycle.

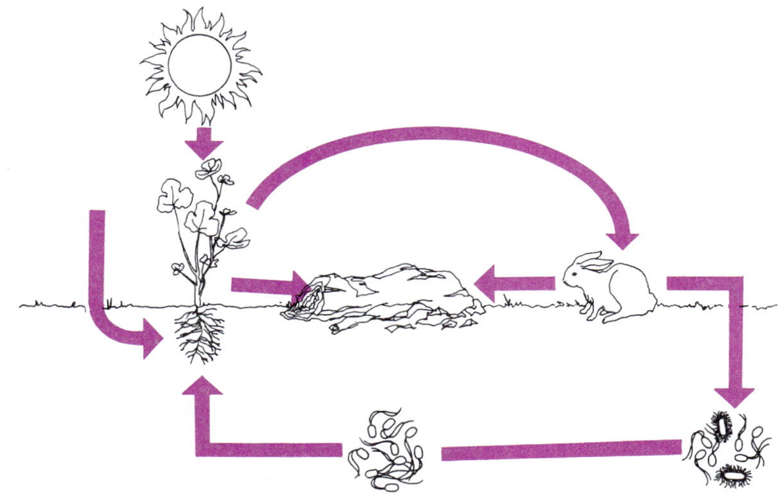

Through research and sometimes unfortunate experience, we are learning more and more about the consequences of using different chemicals. The use of pesticides on plants can result in traces of the chemicals showing up in our food. The great quantities of plastic products we use and throw away may decompose and contaminate our water. These are examples of important physical and ecological consequences of our technical systems.

Surely we recognize the importance of nature's life cycle in the biosphere. Or do we? For many years, vast amounts of a chemical called 2,4D ("Agent Orange") have been used in highly developed countries to kill weeds. During the Vietnam War, 2,4D was sprayed from planes to wipe out entire forests. Few questions were raised until years later, after the damage was done. Not only were green plants destroyed, but 2,4D turns out to be extremely dangerous for humans, as well.

Ecological consequences change the balance between the living and the nonliving things in the biosphere. Recently, scientists and concerned citizens have begun to investigate the possible consequences of many events such as:

- Excess burning of fossil fuels.
- Dumping of waste materials and garbage into rivers, lakes and soil.
- Nuclear testing and its resulting atomic fallout.
- Use of pesticides and weed killers.
- Oil leaks from ship wrecks or offshore drilling.
- Storage of nuclear waste.

There are difficulties in determining the possible consequences of technology. One difficulty is that the effects may become apparent *after* a harmful chain of events has been

Not all consequences of deliberate decisions are intended or desired. The steam engine shown here was built in England by a man named Stephenson. A decision was made to sell the engine to Germany. The name of the train was changed from the Rocket to the Adler (eagle). Germany eventually became one of the leaders in producing high-quality steam engines. This final effect was certainly an unintended, unplanned and undesired consequence of the decision for a technological transfer of hardware and knowledge from one country to another.

started. Once that sequence has begun, it may take a long time for it to be stopped, if it can be stopped at all.

For example, we are presently producing, using and discarding massive amounts of plastics each year. We do not know yet if those plastics will decompose and become harmless compounds. The decomposition of plastics is unique. There are some indications that plastics break down into new and dangerous compounds. Compounds never before found in nature are now emerging. We simply do not know what effects these may have. We will have to live with the consequences, whatever they are, for it is too late to stop the process.

Nuclear testing is another example of an activity which has long-term consequences. A large number of nuclear devices were exploded both underground and in the atmosphere between 1948 and 1964. Halfway through that period, some scientists indicated that we need not worry about the level of radioactivity in the atmosphere. By 1964, scientists found that radioactivity was poisoning our atmosphere, our soil, our food and even the milk we drink. They realized that we had reached a crisis. They stopped the cause of the crisis, nuclear

chain of events

testing. That could not stop the chain of events, however. For some years to come, we may be affected by the consequences. Two radioactive isotopes (strontium 90 and cesium 137) accumulate in our foods and in our bodies. Radioactivity can be harmful.

We are still waiting to see the effects of radioactive nuclear wastes. We already know that the materials in which nuclear wastes are now stored wear out and leak in less than a half a century. Nuclear wastes remain radioactive and potentially harmful for several centuries, however.

Consequences and Control

technology assessment

We do not have to be caught unprepared for the consequences of technological developments. Technology assessment (defined in Chapter 15) can help us plan for future consequences. We can learn to use technology assessment to try to foresee the immediate and the delayed effects of a proposed technological event.

Much care and study must be given to understanding how pollutants enter our food chain. From this understanding we may be able to develop methods of controlling and hopefully eliminating those pollutants. Pollutants may take many forms, such as herbicides, nuclear fallout, and discarded plastics.

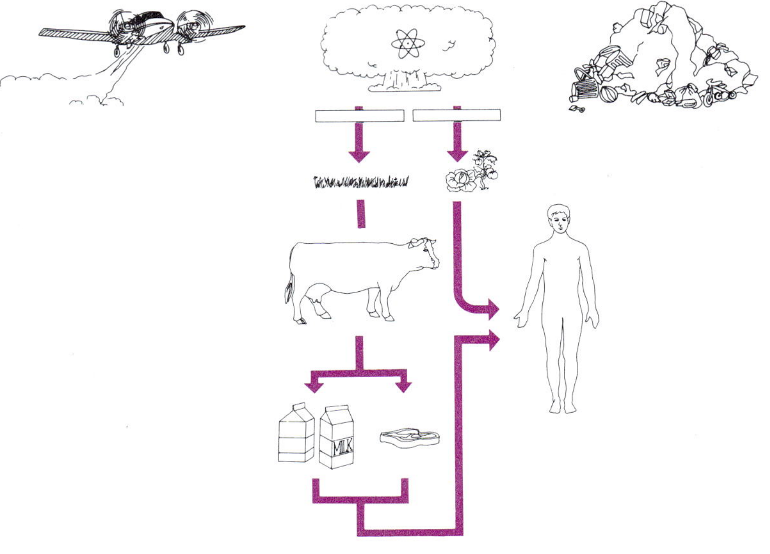

These delayed effects are called second order effects. They often occur in a sequence. They appear to interact like a row of dominos. The dominos sometimes fall very slowly. They may topple each other in strange and unusual patterns. A first order effect (the desired effect of the action) may cause **second order effects.** These, in turn, may cause others, and so on. Technology assessment is used to predict this string of possible effects. We would like to foresee positive *and* negative effects before the event happens.

second order effects

As with other future-oriented methods, technology assessment considers possibilities and probabilities. Even with its shortcomings, technology assessment is a potentially powerful tool for determining possible consequences of actions related to technology. It is now quite evident that our capabilities must be turned to making such predictions and assessments *before* technological decisions are made. The problems are not new. Our new asset is the availability of sophisticated information tools to help predict the consequences and decide if we want to live with them.

Summary

Technology can have a significant impact on each of us, on our social institutions and on our environment. Some of the consequences are helpful and some are harmful. It is necessary for us to plan ahead. We need to investigate all new technological developments for their potential effects.

Technology assessment procedures help us predict the consequences of new proposed technologies. When we understand the consequences, we can decide whether or not to use each new technological development. The benefits should outweigh the problems. If we do not learn all we can about new technologies then we are in danger of losing the choice about using them. We can find ourselves having to live with their effects for generations, without planning ahead.

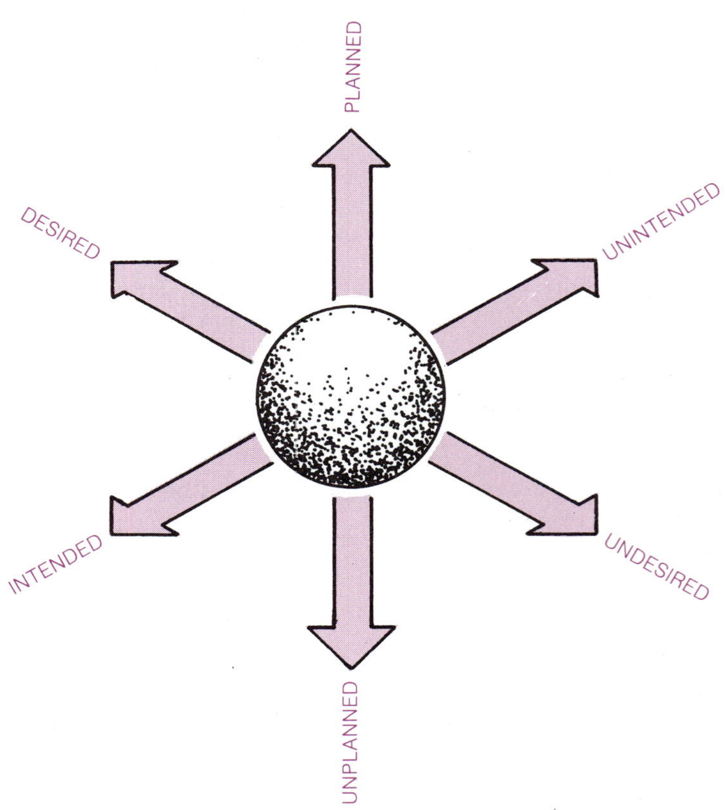

All technological decisions have consequences. The consequences may take six different dimensions. They may be planned or unplanned, intended or unintended, desired or undesired. The most dangerous consequences are those with unplanned, unintended and undesired effects.

Key Concepts and Terms

biosphere
chain of events
co-ops
computers
consequences
division of labor
ecological consequences
efficiency
emotional effects
food additives
intended
job specialization
knowledge explosion
nutrition
personal consequences
physical effects
predicting
scribes
second order effects
social consequences
technology assessment
unintended consequences
vitamins
worker alienation

Decisions

The preceding chapters present many concepts and ideas about technology. This chapter shows how your knowledge can be used to make decisions about technology as it affects your life and the lives of others.

hard technologies
soft technologies

You have learned that tools are an important part of technology. Many of the tools that have been discussed are considered **hard technologies,** such as machines and instruments. Decision-making skills are **soft technologies.** They are tools that can be used to change society. There are a number of decision-making tools. How well they are used may determine if the technology of the future will result in the consequences we desire.

Kinds of Decisions

decisions
alternatives

What is a decision? You make them every minute, every day. A decision is a choice among alternatives in the face of uncertainty. In making a decision, one course of action is chosen instead of others.

information

When you decide to buy a ten-speed bike rather than a three-speed, you make a decision based on the best information available. You probably consider how much long-distance riding you will do. You might consider how much adjustment and maintenance each bike will require. In the long run, your decision is an educated guess about how the bike will fit your future needs.

◀ You are faced with many decisions every day. Often you will not know if you are making the "right" decisions.

There is always an element of uncertainty in making decisions. Most people dislike being uncertain about things. Ordinarily, we try to find answers that are clear-cut. We want the "right answer" or the "truth" about technology. We want to choose either *this* or *that,* not one choice among many alternatives.

uncertainty

In reality, the world is not that simple. Most decisions in our technological world are selected from dozens of possible choices. Making decisions leads to accepting responsibility. When people make a decision they must accept the consequences. Dealing with uncertain consequences can be a challenge to the decision-maker. The only alternative is to let others make the decisions. If you do that, you may have to live with consequences you will not like and didn't choose.

choices

One type of decision is a choice among several alternatives, all of which seem to have desirable consequences. When this is the case, a person must decide which choice is the "most right." The process of making a choice is usually more complicated than flipping a coin. You will want to know as much as possible about each of the consequences and the likelihood (probability) that each will happen.

probabilities

options

Usually, the more you know about the options, the easier it is to make a choice. It still may be difficult to make a decision of this type. You may want to go roller skating *and* buy a new T-shirt, *and* put some money in your bank, even though you may have only enough money for one of these choices. To decide, you must think about the consequences (effects) of each possible choice.

A second type of decision is that made when one alternative is what you clearly want, and the other is something you

How many times have you had to make a choice between two things or activities, such as buying a new stero or spending your money at an electronic arcade?

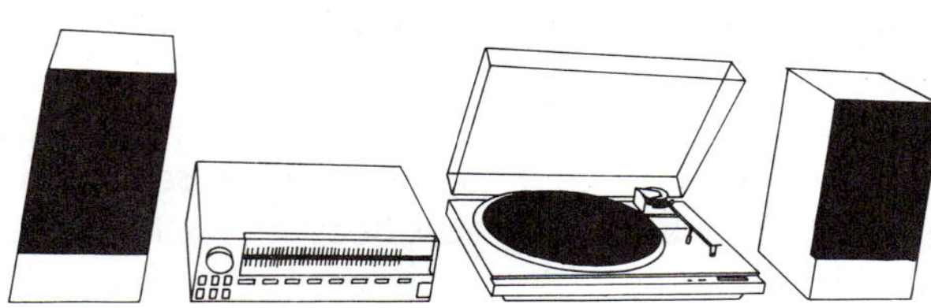

Decisions are easy when you have the choice between something you don't want and something you do. Stripping for coal may be undesirable, but reclaiming the land after the stripping may make the decision desirable.

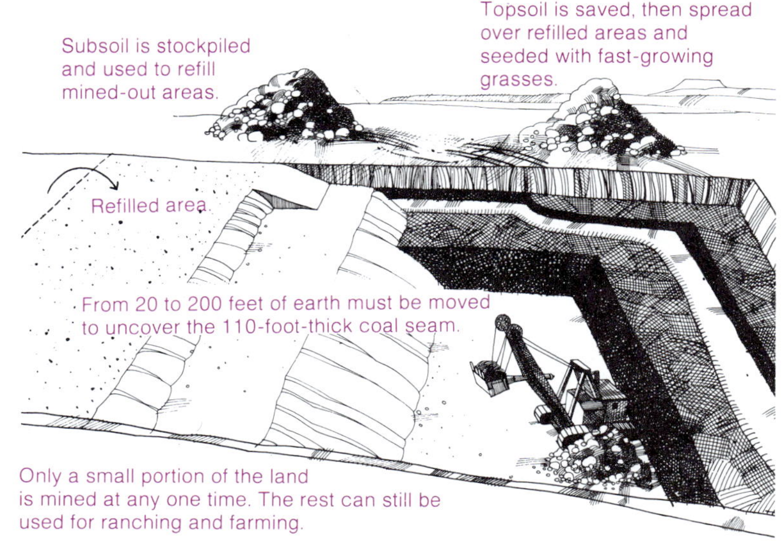

wish to avoid. This kind of decision is easiest to make, but only if you think about what you really want.

Instead of buying a new bike, for example, you might decide to continue riding your old one. The choice may be easier to make if you decide that to buy the new bike you must go without lunches, sell your coin collection and give up the money you were saving for a new stereo. You might also think about how careful you have to be to prevent your new bike from being damaged or stolen.

A third type of decision is the kind that people must make when all the choices seem to lead to unwanted consequences. This is sometimes described as "being between a rock and a hard place."

You may be faced with this kind of decision when you are told to make a choice between doing the dishes or your homework. Or, you might have to choose between going to the dentist or writing your term paper for English class. Anxiety and conflict may be evident, and you may try to postpone the decision as long as possible while searching for other alternatives. Usually, however, the decision must be made sooner or later.

When we make decisions, we must deal with probabilities. The more we learn about the probable consequences, the easier it is to make a decision. Because we are human, each

The hardest of decisions is when you must choose between two things or actions, neither of which you want. In this example, you will not want to see an increase in the amount of oil used by each person in this country nor will you want to see a decrease in the standard of living.

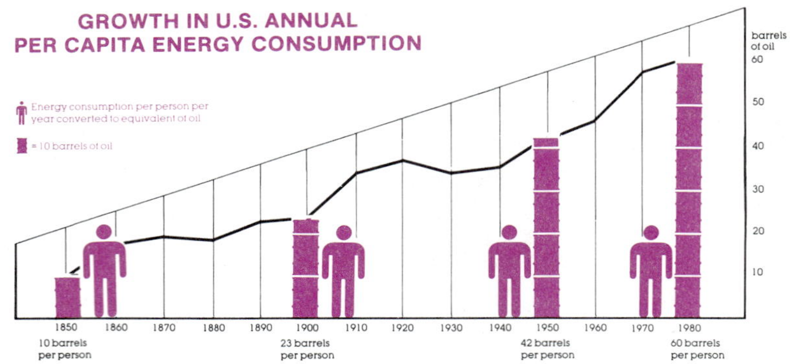

of us sees the consequences differently. That is why group decisions, such as those about building nuclear power plants, are so difficult to make. Different people "see" different effects and place different values on them.

Often the decision-making process is even more difficult when we realize that by *not* deciding, we have in effect made a decision. As a result of not deciding, we leave our future either to chance or to the decisions of others. If you choose not to act when someone needs your help, for example, your decision may have consequences. It seems that as difficult as it sometimes is, we cannot avoid making decisions. In technology, our decisions often have personal, social and ecological consequences.

In the face of uncertainty and change, we must make decisions based on the best information we can get. We will make some mistakes, but we can learn from past errors and avoid repeating them. Perhaps we can make better decisions if we develop our skills. One way to do so is to explore the different methods people use to make decisions.

How Do We Make Choices?

authority

advice

The four general bases for decision-making are **authority, feelings, logic** and **experience.** In the first approach, people make choices based on the **advice of others.** You may have decided to try a new brand of shampoo because a classmate recommended it. Manufacturers often pay famous people to endorse (recommend) their product, hoping that many customers will make a decision based on the star's "authority."

In making some decisions it might be helpful to ask an authority on the matter. Directions come in many forms and cover many subjects.

On more serious matters, we often rely on the advice of a physician, a religious leader, parents or a trusted friend. Before purchasing a product, some people consult a consumers' magazine to see how it and similar products have been rated. Decisions based on authority usually are made when we do not have the technical knowledge or experience to judge the probable consequences ourselves. As technology becomes more complex, you may want to become informed about how things work so that you can make intelligent decisions. Without such knowledge, you will be forced to rely more and more on the authority of others.

feelings

A second approach involves making a choice based on **feeling.** In this case, people choose not to rely on what others in authority recommend, but upon what pleases them. Such decisions are often the ones that "feel right."

The method involves no particular information. It is a kind of personal sense of what is best. Young children often make these kinds of decisions because they have not yet gained the knowledge required to make choices based on facts.

Sometimes, a feeling or sense of what is right is helpful. A decision to invite a new classmate to a party might be based on your feeling that it is the right thing to do. There may be no special information available nor authority to consult. The results can be positive. At other times, however, decisions

Sometimes we make the choice to invest time and effort in a project, such as the recycling of materials. Making these decisions involves feelings about the value of the activity.

based on feeling are made only according to what pleases the individual at that moment. Retail stores are aware of shoppers' tendency to buy products on impulse, and often construct flashy displays to catch their attention. Do you think decisions about technology should always be based on feelings?

logic

As a third way to decide, you can rationally consider the consequences of the decision. Using this method, people **logically** weigh the probable effects of a choice and judge the worth of each effect. The process usually involves **if-then statements.** *If* you buy the cheaper shoes, *then* will they last long enough? *If* you install a wood-burning stove in the family room, *then* who will cut the firewood? Rational consequence decisions require that people know how and why things work. If you know how a furnace thermostat works, you can better decide if it might save money to set it at a colder setting each night.

experience

A fourth method of making decisions is the use of **experience.** People sometimes make choices this way by remem-

A decision to build or buy the best product for your needs at the lowest cost requires choices based on a rational approach. Determining the best surfboard, for example, requires a careful study of the shape and design of the board. A final decision may be that you build your own surfboard.

bering an experience in similar circumstances. A well-driller, for example, often decides where to try for water by recalling earlier successes. Professional baseball pitchers keep a notebook to remind them of the pitches that have worked against opposing batters. In each case, the action and past consequences are used as a guide for new decisions.

Experience can also be used to indicate if a new choice is

In many cases, you will draw upon your background and past experiences, including failures, to make decisions.

necessary. You might choose a new tool to tighten a loose bolt on your bike. If it doesn't seem to be working, you probably will select a more familiar tool. This process is similar to the idea of feedback discussed in communication earlier.

Each of these methods of choosing may be used to make decisions about technology. In some instances, people make choices because a respected authority indicates which decision is best. In other cases, choices may be made on a feeling about the situation. The decision can be based on the probable consequences. It is possible to make choices based upon previous experience in similar situations.

Each method has certain advantages and disadvantages. Each is appropriate for some situations and not for others. If you develop some skill in using all four methods, you will always have the necessary "tools" to make a decision, no matter how difficult.

What Questions Must be Answered?

Most decisions require some consideration of **goals, resources, actions** and **results.** The decision-making process may start with any of the components, as long as all four are included.

goals

What results do we want? This question refers to the **goals** we wish to achieve. People often begin with this component, when faced with a decision.

short-term consequences
long-term consequences

Goals can include things we would like to have happen soon (short-term), and those that may take years to achieve (long-term). One of your short-term personal goals may be to bring home a good report card. A long-term goal might be to own your own successful business.

There are also short-term and long-term goals related to technology. Several years ago, people living near Lake Erie decided that the lake would "die" if steps were not taken to stop the pollution. Today, it appears that progress is being made. Would you consider the goal of cleaning up Lake Erie a short-term or long-term goal?

actions

What will need to be done? Decisions also require that necessary **actions** be considered. It would be foolish to set a goal to clean up the neighborhood baseball field without recognizing what work will be required. Technology decisions often require a lot of dedication and effort. The series of manned explorations of the moon, for example, required the cooperation of thousands of scientists, engineers and technicians.

Sometimes a decision between two choices must be made on the basis of what action is practical. If you wanted to have a bookshelf for your bedroom, you could build it yourself or buy it ready-made. If you don't have time (a resource) to build it (an action), you may decide (action) that buying (action) is your best choice.

resources

What **resources** will be needed? The general resources that support technological efforts include land, labor, capital, knowledge and time. These resources provide the needed support for the tools, materials, processes, energy, information and humans that are required for a technological act to happen.

The resources we choose will be directed by the desired outputs from the systems. These goals can then be matched to the resources we have available. In many cases our best choice is the one that uses the fewest resources.

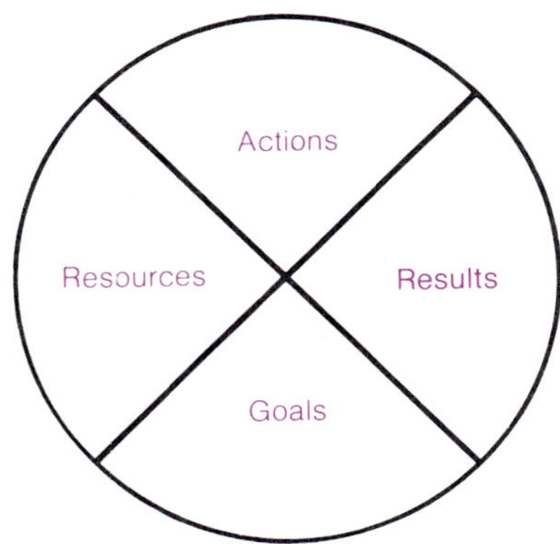

There are four related parts of a decision—goals, actions, resources and results. These form a cycle and interact during rational decision making. You may enter into the cycle at any point, providing you consider all four parts of the decisions to be made.

This often happens when a needed resource is in short supply. Car buyers, for example, are beginning to choose more fuel-efficient models because of possible gasoline shortages. The cost of the resource (gasoline) is also part of the consideration.

At other times, decisions may be made on the basis of an oversupply of resources. Once, when a large number of chestnut trees were killed by a blight, many homebuilders used the chestnut lumber for woodwork.

In this case, the choice was made to take advantage of the available resources. How would we look at garbage if we thought of it as a resource, not as a problem? Our choices would probably include more uses for garbage. We would avoid choices that included discarding it, a more costly solution.

Sites for construction of new industrial plants are specific forms of the land resource. The type and amount of available resources often decides which sites will be selected. Planners consider resources such as the availability of trained labor (knowledge) and energy sources. Also, the availability and quality of transportation, communication and technological systems are considered.

results

What might happen? As was discussed elsewhere in this book, every possible choice has some **results.** This may

PROBLEM SOLVING: Design and Decision-Making

The general systems model presented in the Controlling chapter is very useful in helping us reach our goals. It is helpful as a problem-solving model once the problems have been identified. But where do the problems come from? How do we know what a problem is? How do we know we have focused on the right problem? The following example should help clarify these points.

A Problem-Solving Case Study:

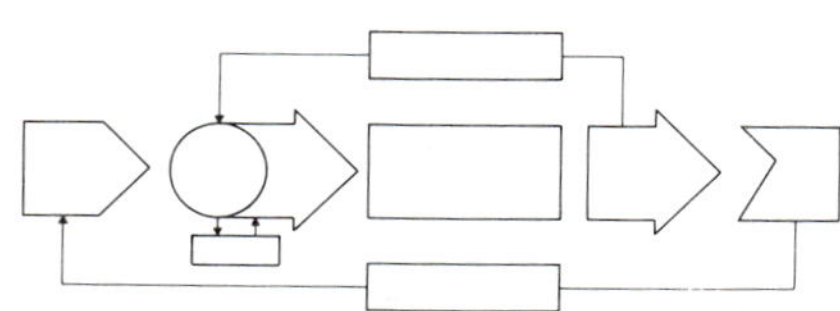

A teenager, who has recently completed a course in technology, has convinced the other family members that they should build a greenhouse. The greenhouse will allow them to reach their desired goal of having fresh vegetables and fruit year round that have been grown with few if any herbicides and insecticides. The family agrees to use some space that faces to the South to build a solar greenhouse. Knowing the value of planning in the solving of problems, the family has spent a good bit of time in developing a scenario for the building of the desired greenhouse. By using that scenario and the illustration of the greenhouse the family was able to determine (1) the cost of materials, (2) the time needed to build the real greenhouse, and (3) the projected costs for the operation of the greenhouse.

UNACCEPTABLE

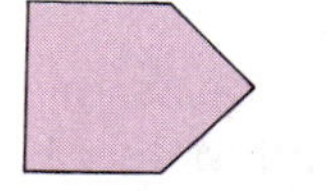

MISMATCH OF GOAL & RESULTS

At this point in the project all the family members agree that the cost is too high (and the amount of food would be more than the family could use). They are also concerned that one of their nicest trees would have to be cut down to provide the land required for the greenhouse. The actual results that the greenhouse project would achieve differs considerably from the desired goal. The goal and the results do not match in an acceptable manner. By agreement the family members decide that it is time to go through the complete problem solving process more carefully to look for possible alternatives and a more optimum solution.

MORE ACCEPTABLE

IMPROVED MATCH OF GOAL & RESULT

After recycling through the problem-solving process, the family found that there was a simpler yet more effective solution to their problem. A smaller hydroponic green house could be built in the opening provided by a large picture window. The hydroponic greenhouse does not require soil (it uses other growing media). The family will know what materials are used since they must mix the nutrients and chemicals required by the plants. Since each container of plants is quite light, they can be placed one above the other. This approach allows for a maximum output of fruits and vegetables in a minimum amount of space. By planting the family's favorite produce on a revolving schedule, it is possible to have the food they want when they want it.

The problem-solving process is called the I D E A T E procedure and includes six major steps.

1. I = Identify and define the problem.
2. D = Define the desired goal.
3. E = Explore possible, alternative solutions.
4. A = Assess the alternatives for an optimum solution.
5. T = Take action on best possible solution.
6. E = Evaluate the match of the desired goal and actual result.

(Recycle through the process as required to reach an acceptable match of the actual results and desired goal.)

The steps and decisions the family made in solving their problem included the following:

1. **I** = Identify and define the problem.
The problem—The cost of fresh vegetables and fruits, grown with a minimal use of herbicides and insecticides, is too high. The produce is not always available throughout the year. Securing the produce requires regular trips to the local grocery store or to the supermarket.

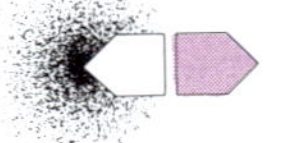

2. **D** = Define the desired goal.
The goal—To have fruits and vegetables at home that; are fresh, will be available on a continuing basis, were grown with a minimum of herbicides and insecticides, include the favorites of each family member, and cost less than the same produce in the supermarket.

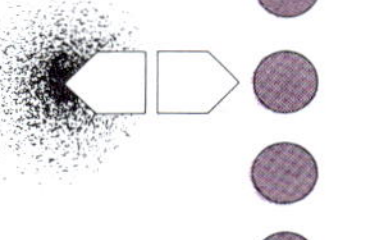

3. **E** = Explore possible solutions.
The possible solutions—Purchase produce from local greenhouse operator. Purchase produce at the local grocery store. Build a smaller version of the greenhouse originally planned. Build a larger version of the original greenhouse in a coopertive effort with several families. Grow vegetables in window boxes mounted inside the house. Consider new fresh-frozen fruits and vegetables. Set up a small hydroponic garden in the home.

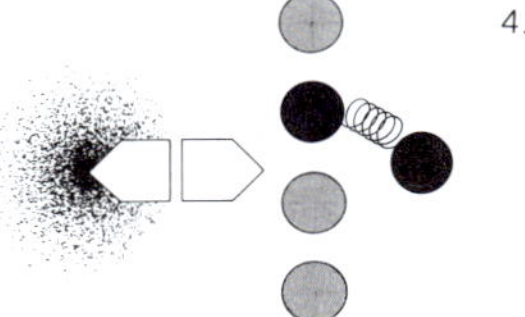

4. **A** = Assess the alternatives for an optimal solution.
The potential solution—All the alternative solutions could fulfill some of the requirements of the desired goal. Fewer of the alternatives could meet the requirement of having produce available in the home on a continuing basis. Fewer still could meet the requirements at a low cost. The most likely candidate for an optimal solution was the hydroponic garden.

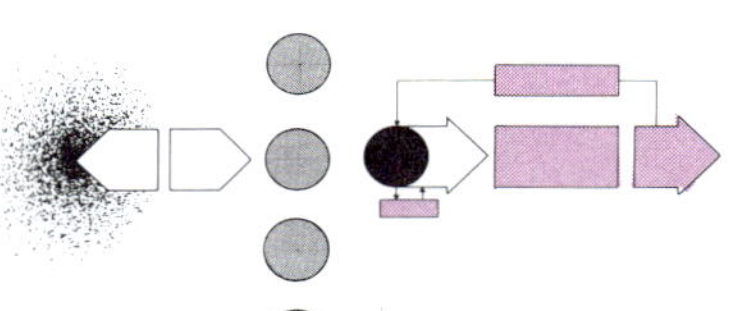

5. **T** = Take action on best possible solution.
The action—The systems model presented earlier provides the most efficient approach to solving the problem of setting up the hydroponic garden. The inputs for the hydroponic garden as a system would include tools and equipment: containers, tubing, pumps, fans, light fixtures, sensors, control devices; materials: support medium (replacement for soil), nutrients, water; energy: sunlight, artificial light, heat; information: signals from sensors to control the pumps, fans, lights; humans; designers, builders and operators of the system, consumers of the products. All of these inputs when brought together in an appropriate manner implement the growing process as well as all of the support processes of heating, cooling, pumping, lighting, metering, monitoring and controlling.

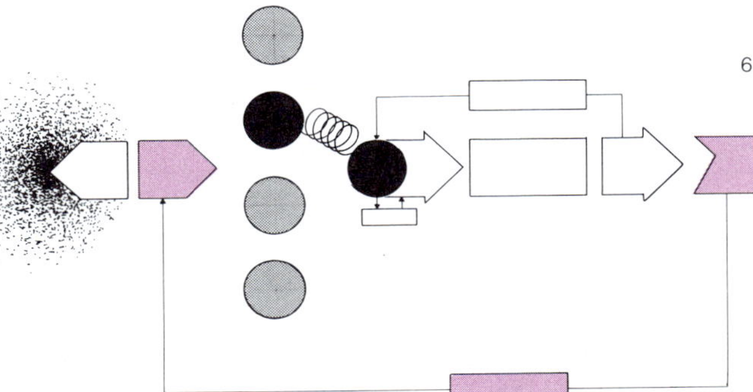

6. **E** = Evaluate the match of the desired goal and actual results.
The evaluation of the match.—The success of the problem-solving effort is determined by "how well" the output or results match the desired goal. If there is an acceptable match, the project is considered completed in a satisfactory manner. The results that would be achieved through the construction and operation of a hydroponic greenhouse appear to make a good match with the family's desired goal.
(In some instances when the final results fall short of the desired goal, it may be necessary to modify the desired goal and settle for a slightly different goal.)

predictions

effects

first-order effects

You may achieve a goal through any of several different actions. One action may achieve several goals. An action may use several resources and result in many consequences. The difficulty of decision-making should not be underestimated.

be the most difficult component in the decision-making process. It is not always possible to predict clearly the consequences of each option.

Who can be sure what will happen if your school decides to have your lunches prepared by a fast-food restaurant? Perhaps the consequences would be better than building a new cafeteria in the school, if that is the alternative. As with all decision components, however, it helps to have the necessary information. You can make the decisions about what to drink if you know that too much coffee or cola makes you jittery. The information and decision *can* help you improve your health.

It is often helpful to consider a sequence of effects of your actions. As we have said, these are called first order effects, second order and so on. Consider, for example, having a choice between using herbicides (plant-killing chemicals) on a vacant lot or cutting the weeds by hand. The immediate results may be the same (fewer weeds). There may be second order effects, however. You might save time by using the herbicide, but the chemical may leach into your water supply or into a local stream or river. Both the human and fish populations may be harmed. If you cut the weeds by hand, you may

GOALS	ACTIONS	RESOURCES	RESULTS
Produce Fresh Vegetables Get Exercise	Gardening Planting (Mulching or Hoeing)	Seeds, Tools, Fertilizer, Time, Growing Season	Fresh Vegetables, Sunburn, Sore Muscles, Blisters
Produce Fresh Vegetables Save Money and Time	Pest Control (Spraying or Bug-Eaters)	Tools Sprayers Chemicals Pesticides	Larger Yields Reduced Costs Killing Bees and Birds
Produce Fresh Vegetables Get Sun Tan	Cultivation (No Till or Weeding)	Tools Hand, Power Chemical Nitrogen Herbicides	Increased Yield Good Sun Tan Health Risk to Chemicals and Sun
Produce Fresh Vegetables Stay Out of Sun	Gardening (Greenhouse or Hydroponics)	Greenhouse Equipment Chemicals	Vegetables Year-Round Increased Costs Work, Knowledge Health Risks to Chemicals but Not Sun

second-order effects

avoid those unwanted second order effects. You may also work off some excess energy, achieve some satisfaction and get a suntan in the process. But you may have to give up the time you would prefer to spend on sports. Many decisions involve second order effects.

One way to make decisions is to chart the possible actions, events and their effects. This approach is similar to the design approaches discussed earlier. Decision components can be listed in order to show the interrelationships (connections) among actions and consequences. This may even cause you to rethink your original goals.

The decision chart may help you focus on the necessary information, but it will not make a decision for you. Decisions are rarely made in a straight-line sequence of goals–actions–resources–events. More often, decisions are reached by a

Decisions are seldom made or should not be made without some thought. In this case the thinking takes place as a cycle of (a) examining goals, actions, resources, and results, (b) deciding on a proposed action or goal, (c) searching for alternatives, (d) considering choices, and (e) after recycling through the process several or many times, making your decision.

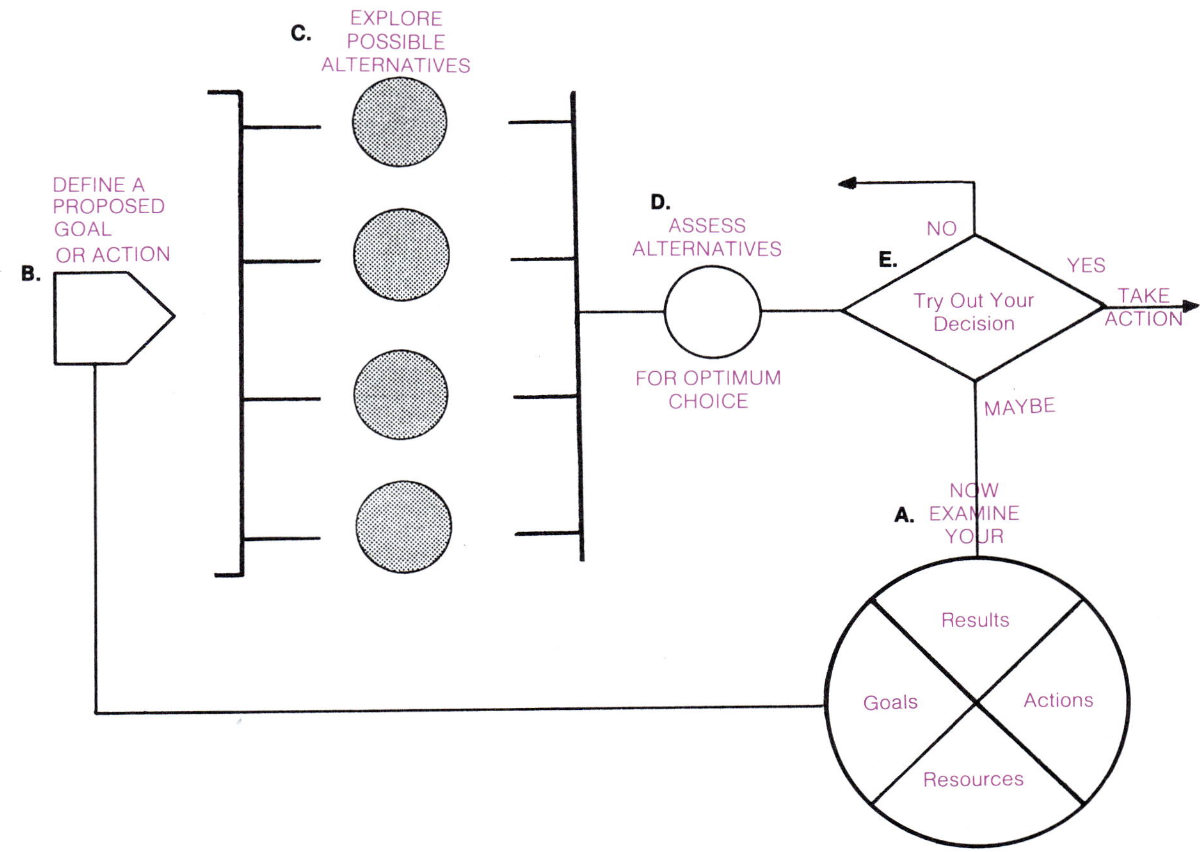

sequence of actions

continuous cycling of the process. After going through the process several times, the best choice may become evident.

Controls and Constraints

Decisions are not made in a vacuum. Normally there are personal, social and ecological factors involved. Each factor may bring certain limitations to the decision-making process. Often there are limits to our knowledge of the options, our resources and our ability to foresee consequences. We cannot dwell on the past but must consider present possibilities.

external controls

Sometimes our choices are limited by factors we cannot change. These are called **external controls** or *constraints.* For example, you had no choice over when and where you were born. To some extent, you may have little choice over where you now live and where you go to school. You may not always have control over the actions of other people who are involved in the decision. Many decisions must be made involving external factors such as these.

internal controls

Internal controls are those things which you are able to change, at least in theory. They may include personal attributes such as the skills and stamina needed to bring about the desired consequence. If you are able, for example, you might choose to ride your bike or walk to school rather than to be transported by bus or car. Others who have physical handicaps may not have that choice. They are limited by internal (personal) circumstances. You might consider buying soda in returnable bottles to save energy and materials. That

Controls and constraints rule out some of our possible decisions. The decisions that are open to us may not bring about the goals and effects we want.

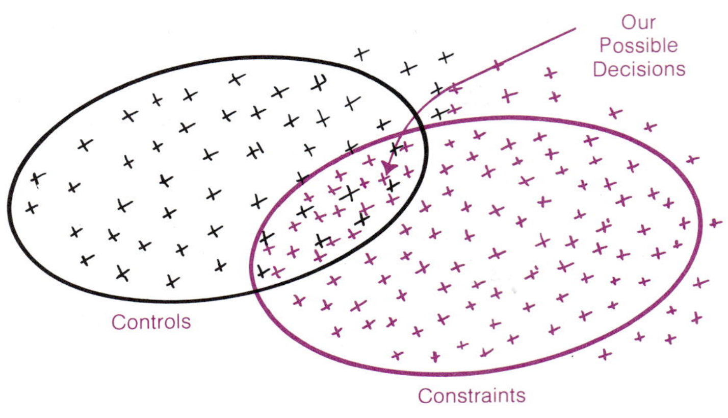

controls choice is possible if there are no external controls—if returnable bottles are available. The internal control in this case is your own willingness to make an extra effort.

values This leads us to consider the relationship between **values** and decisions. Throughout this book you have seen how people use technology to alter the environment. In the decision-making process, technology can be used to obtain information that will help make better choices. computer simulations Computer simulations, for example, determine the possible consequences of building a new superhighway around a town. The fact remains that *humans* decide which factors to consider. The ultimate decisions are human ones.

For example, years ago, the United States constructed a system of underground missiles. There were some people who thought that technical devices should be designed to fire the missiles automatically in the event of attack. Others believed that humans should remain in control. The issue was one of values—which judgment should be trusted, human or machine? The system was built to require a series of human decisions, including the President's, before the missiles could be fired.

The "Right" Decisions

In our society, we often speak of what is "right." The term can mean many things, including what is ethical and moral and what is technically correct (true). You have probably been taught many "shoulds" (how the world should be and how people should act). These "shoulds" reflect our values. Unfortunately, decisions that people make do not always fit what they *say* they value. Actions do not always correspond to words.

Centuries ago, the world population was small. Materials, food and energy resources seemed plentiful. People defined their society as those people who were like them or who belonged to their group. This feeling of community *did not* include people who spoke another language, who were from beliefs another racial group or who had different religious beliefs. Decisions were made in terms of the immediate effects upon the local group.

In modern societies, decisions about technology are not so easily made. We are learning more about the total human

population on the earth. We are becoming more aware of how far-reaching our decisions actually are. Personal decisions often have social consequences, and vice versa. Decisions made in the Middle East countries may have profound effects on other parts of the world. For many of us, a feeling of community involves all people now living and those yet to come.

We are beginning to see that many of the earth's resources are limited (finite). We realize that decisions must be faced about sharing these resources with billions of other people. Who should have access to the earth's resources? What about future generations? In the past, there was little concern for the long-term future. Recently, however, decision-makers have begun to examine possible effects centuries ahead. These "futurists" urge us to consider all of our decisions in terms of what effects there may be on our children's children.

All this may make our technological choices even more difficult. The only alternative, however, is *not* to decide. Would you like to see the world continue as it is now?

Summary

The "soft" technology of decision-making is serious business. Even our small, day-to-day choices may have far-reaching effects. You may base some of your decisions on advice from authorities, such as parents, teachers or religious leaders. Other decisions may be based upon your own feelings and beliefs. If you know the potential consequences of your decisions, logic may help you make choices. Finally, you may fall back on your experience from similar situations to help you decide. In all cases you need to acquire as much information as possible. That way, your decisions will be the best possible under the circumstances.

constraints

Decisions require that we consider at least four factors: goals, actions, resources and consequences. Often there are constraints or limitations on each of these factors. Choices must be made accordingly. A decision may require that we cycle several times through the decision-making process.

Ultimately, the "right" or "best" decision is the result of human judgment. The process is becoming more difficult all the time as we become aware of the interconnectedness of our decisions. The decisions we make about technology may have serious consequences for the rest of the world and for future generations.

Many of our technological decisions should consider what is fair to us, what is fair to others and what is fair to those who will live here in the future. Some of our resources are limited. If we deplete these, future generations will be at a real disadvantage.

Key Concepts and Terms

actions
advice
alternatives
authority
beliefs
choices
computer simulations
consequences
consideration of others
constraints
controls
decisions
effects
experience
external controls
feelings
first-order effects
goals
hard technologies
information
internal controls
logic
long-term consequences
options
predictions
probabilities
rationality
resources
responsibility
second-order effects
sequence of actions
short-term consequences
soft technologies
uncertainty
values

Epilogue

Many people these days are enjoying the fruits of technology. And yet, there is increasing concern about where technology is taking us. Some say we are controlled by technology. Some are overwhelmed by the difficulty of making decisions about technology. They throw up their hands and say an individual cannot do anything about it. Many people buy and use what they want and do not consider long-term consequences. And some people think about technology and take part in planning and deciding.

We hope, now that you have read this book, that you are ready to use your head rather than throw up your hands. We hope that by reading this book you have increased your knowledge about technology. We would like to have you understand that technology is a very human endeavor. Technology is a tool for human use. As with any other tool, it can cause damage if the user is unaware, unskilled or unwilling to exercise caution.

Technology can benefit people greatly, if used with understanding and concern for long-term consequences. It takes willingness to make difficult decisions. We are learning more and more about people, the principles by which things work and the processes of applying our knowledge. We are learning to be more precise in reaching our goals.

Making decisions about technology is a complicated process. It is easy to sit back and let others make the choices. It is easy to be overwhelmed and wonder what one person can do. We would like to suggest several things you can do to help.

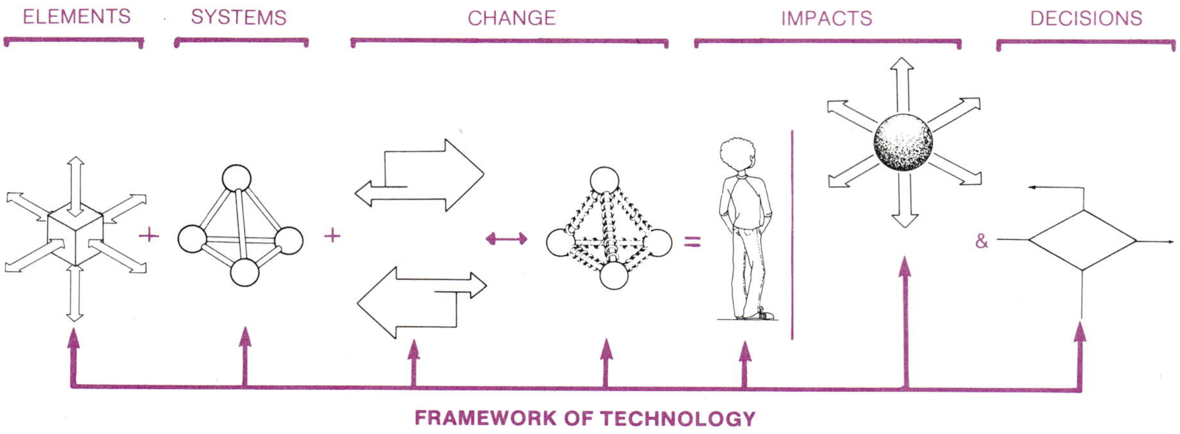

ELEMENTS SYSTEMS CHANGE IMPACTS DECISIONS

FRAMEWORK OF TECHNOLOGY

A simplified graphic picture of the framework of technology. The elements are found in the systems of technology. The systems grow but are controlled by influences both inside and outside technology.

(1) Develop your technological skills and understanding to the highest degree possible, given the resources you have. You are among the most privileged group in the world. You have better access to the information, tools, materials and processes of technology than do most people. You live in the most highly developed technological society in the world. You have the chance to become a leader by being a smart producer and user of technology.

(2) Develop your survival skills and understanding. It is often easier to make difficult decisions if you know that you can live with the choices. In the introduction to this book you were asked to pretend that all electricity was turned off and all fossil fuels were unavailable for two weeks. Suppose that really happened. Suppose there was a power failure and the energy networks closed down. If you are technologically prepared, you will be able to cook your food, stay warm, get where you want to go and get along until the power again becomes available.

Suppose there were a strike and the food supply system were cut off. Could you go into the country and find edible plants and catch animals for food? Do you know how to take care of yourself if the transportation system were to break down? Transportation moves food and fuels, as well as people. Hopefully, you will not have to test these skills, but you have begun to develop them by reading this book.

Survival skills can help people to live in their environment with limited technological resources. This kind of training can help you avoid the "technology trap." That is the trap of being helpless if the technology systems break down for a time. We hope that you are now better equipped than you were.

You may want to learn more about technology. There are many opportunities to do so by reading, watching television specials, taking more technology courses in school and participating in clubs and educational organizations.

(3) Interpret technology to others. We have indicated how rapidly information is growing and how technology changes people's lives. The pace of change may make young people the experts on technology. Older people may be required to *unlearn* old ideas before they can grasp the new. Be patient as you help others learn. You may be in a better position to ask pertinent questions about technology than are your parents, grandparents and teachers.

(4) Consider personal, social and ecological consequences when you go about your life every day. You might ask yourself several questions:

(a) Is this made of non-renewable materials? If so, can I conserve or recycle it?
(b) What is the energy used for making and operating this tool or machine? If it is from a nonrenewable source, can I find another way of reaching my goal? If not, how can I conserve that energy?

In your choice of food you may decide to eat fewer processed foods and less meat that has been artificially "fattened up." You may wear clothes designed to keep you warmer in winter so you can be comfortable in buildings that are minimally heated. You may help insulate your home and take advantage of solar heating techniques. You may choose to use mass transportation, walk, bike or form a car pool. Many technological decisions have ecological consequences. As you continually remind yourself of these consequences, you may change some of your behavior and urge others to change theirs also.

Industries are often reluctant to spend the money to clean up their wastes. The money spent means reduced profits or increased prices. Yet many industries have responded to documented evidence of problems caused by industry. Groups of concerned students and adults can make the difference.

Vacant lots might be turned into playgrounds or community gardens. You might plant a small garden just to find out how it is done. Grass clippings, leaves and hair from a barbershop or beauty shop can be recycled to nourish the soil and to control weeds. You can learn to use organic bug and pest controls.

Locally, you can ask questions about your environment and help improve it. With your friends, perhaps you can hold a car wash or bake sale to earn money. You can spend it on paint to cover graffiti or on other improvements in your area. You might urge local business owners to contribute the materials while you help with the labor. You can organize a litter cleanup day or ask your elected officials to promote research about groundwater contamination.

You are best able to decide where to put your individual energy and resources. You will want to look at the possibilities and probabilities before you decide what action you wish to take.

(5) Lend your support to larger groups that are organized to have impact on industry and government. Many citizens' groups have shaped local, state, national and worldwide decisions about our uses of technology. Some examples of positive actions include stopping the construction of a dangerous intersection, lobbying for a better sewage treatment center and boycotting a company that produces unsafe products. You may fail to reach your goal the first time, but the process of applying pressure and asking questions often changes similar decisions in the future.

Consumer groups bound together on a statewide or nationwide basis have been effective in influencing the decisions of industries, businesses and government. Consumer groups can be powerful, partly because they represent a large number of people. These large groups are customers and voters, so decision-makers tend to listen. Political decision-making is very important. It can set directions and decide actions about technology. It is important to elect representatives who are knowledgeable about technology and its potential consequences.

It is not enough simply to elect individuals to make technology work for us. We must also observe continually. You can keep track of the decisions of your political representatives. This way people can know whether their wishes are being observed.

Groups have organized to serve this purpose. The League of Women Voters and Common Cause are examples of such groups. Even if you are too young to vote, politicians and groups that watch them are often glad to have your help. You could volunteer to distribute literature or do other kinds of work for a candidate or organization you support. Your questions at home and in your community can be valuable to alert politicians that people are concerned about the consequences of technological decision-making.

(6) Remember that even if you cannot cast a political vote, you vote every time you spend money. As you decide what to purchase, you tell companies what you do and do not want. Whether you join or form groups, or focus on developing your decision skills, you do have the potential to make a difference.

You can be action-oriented and be involved in the decisions of industry, government or other groups. You can also be an advocate and work to promote what you believe in. You can spread information. Your choice of what effort to join requires a working knowledge of technology, its parts, its activities and its impact. We hope that this book has laid the foundation upon which you can build.

The first decision that must be made, however, is whether or not you will get involved. If you decide not to engage in the scuffle, you automatically lose control of an important aspect of your life. A decision *not* to act has many consequences, too. The choice is yours.

ACKNOWLEDGEMENTS

If we were to acknowledge all the people who contributed to this effort, through their direct assistance or indirectly through their ideas, suggestions and support, this section might well be larger than the book.

Two people must be singled out, however, for their role in disclosing the potential of technology as a vibrant area of study and work. They are Delmar W. Olson and James C. Durkin, valued colleagues and friends.

The illustrations, developed with painstaking care to portray a maddening array of concepts, are a major feature of this book. Our sincere appreciation to Marc Dabe and Tsun Tam for their illustrations and insights. Particular thanks go to Patricia Hutchinson for her graphic design, her illustrations, her patience and her humor.

Thanks to the many students, past and present, whose questions and questioning helped develop and strengthen the conceptual framework for this book. We would like also to thank those who struggled through the many rough drafts and suggested many worthwhile improvements: Caleb Crowell, Marshall Hahn, Kaye McCrory, Jim Prunier, Cecily Selby, Ray Shackelford and Roni Todd. Thanks to Barbara Lach who was our word processer before we got our word processor.

And, finally, a special thanks to all the agencies, corporations and individuals who so willingly provided reference materials and illustrations.

Index

A

B

C

D

E

F

G

J

K

L

M

N

O

P

Q

R

S

T

U

V

W

X

Z